Biodiversity and Ecology

Biodiversity and Ecology

Edited by Neil Griffin

SYRAWOOD
PUBLISHING HOUSE

New York

Published by Syrawood Publishing House,
750 Third Avenue, 9th Floor,
New York, NY 10017, USA
www.syrawoodpublishinghouse.com

Biodiversity and Ecology
Edited by Neil Griffin

© 2019 Syrawood Publishing House

International Standard Book Number: 978-1-68286-701-3 (Hardback)

Cataloging-in-Publication Data

Biodiversity and ecology / edited by Neil Griffin.
    p. cm.
Includes bibliographical references and index.
ISBN 978-1-68286-701-3
1. Biodiversity. 2. Ecology. 3. Biodiversity conservation. I. Griffin, Neil.
QH541.15.B56 B56 2019
333.95--dc23

TABLE OF CONTENTS

PREFACE

In my initial years as a student, I used to run to the library at every possible instance to grab a book and learn something new. Books were my primary source of knowledge and I would not have come such a long way without all that I learnt from them. Thus, when I was approached to edit this book; I became understandably nostalgic. It was an absolute honor to be considered worthy of guiding the current generation as well as those to come. I put all my knowledge and hard work into making this book most beneficial for its readers.

Biodiversity refers to the variation of life forms on earth at the genetic, species and ecosystem level. It varies widely across the globe and from region to region. It is dependent on abiotic characteristics of temperature, precipitation, altitude, etc. Biological diversity exists at various levels. These include taxonomic diversity, ecological diversity, morphological diversity and functional diversity. Ecology is a sub-field of biology that studies the interactions between organisms and the environment. The study of biodiversity, its distribution and population of individual organism groups as well as the interaction between systems of organisms are areas of interest in ecology. It has practical applications in conservation biology, wetland management and natural resource management, among many others. The objective of this book is to give a general view of the different areas of biodiversity and ecology, and their applications. It brings forth some of the most innovative concepts and elucidates the unexplored aspects of these fields. For someone with an interest and eye for detail, this book covers the most significant topics relevant to these fields.

I wish to thank my publisher for supporting me at every step. I would also like to thank all the authors who have contributed their researches in this book. I hope this book will be a valuable contribution to the progress of the field.

Editor

Vertebrate bacterial gut diversity: size also matters

Jean-Jacques Godon[1*], Pugazhendi Arulazhagan[1,2], Jean-Philippe Steyer[1] and Jérôme Hamelin[1]

Abstract

Background: One of the central issues in microbial ecology is to understand the parameters that drive diversity. Among these parameters, size has often been considered to be the main driver in many different ecosystems. Surprisingly, the influence of size on gut microbial diversity has not yet been investigated, and so far in studies reported in the literature only the influences of age, diet, phylogeny and digestive tract structures have been considered. This study explicitly challenges the underexplored relationship connecting gut volume and bacterial diversity.

Results: The bacterial diversity of 189 faeces produced by 71 vertebrate species covering a body mass range of 5.6 log. The animals comprised mammals, birds and reptiles. The diversity was evaluated based on the Simpson Diversity Index extracted from 16S rDNA gene fingerprinting patterns. Diversity presented an increase along with animal body mass following a power law with a slope z of 0.338 ± 0.027, whatever the age, phylogeny, diet or digestive tract structure.

Conclusions: The results presented here suggest that gut volume cannot be neglected as a major driver of gut microbial diversity. The characteristics of the gut microbiota follow general principles of biogeography that arise in many ecological systems.

Keywords: Biodiversity, Biogeography, Gut, Fingerprint, Species-area relationship

Background

Among a number of parameters, the 'size' of an ecosystem is often assumed to have a key impact on the management of diversity. In fact, the species-area relationship is central to the ecological theory [1] and was first described for macro-organisms [2]. For bacteria, the species-area relationship is generally expressed in terms of habitat volume (i.e., volume-area relationship) and has been illustrated in liquid sump tanks of metal-cutting machines [3], membrane bioreactors [4] and tree holes (i.e., rainwater accumulated in holes at the base of large trees) [5]. However, until present, the microbial species-volume relationship has never yet been studied for gut or body size, even though vertebrate gut size covers a wide range of magnitudes. There is a 10^6 body mass factor between a tiny bird or a shrew and an elephant.

The vertebrate gut hosts a microbial community that fulfils many vital functions for the host: it enhances resistance to infection, stimulates mucosal immune defences, synthesizes essential vitamins and promotes caloric uptake by hydrolysing complex carbohydrates. The bacterial populations inhabiting the gut are complex, varying considerably from individual to individual and from species to species. However, gut microbial ecosystems are not a random association of microbes but are shaped by the host. A transfer occurs vertically from mothers to offspring or horizontally between individuals within a specific group. Such transfers have given rise to the long-standing co-evolution of microbiota and their hosts [6].

The benefit of bacterial diversity in the human gut has often been highlighted [7] and driving factors such as age [8], diverse lifestyles [9] and diet variations [10] have already been explored. Despite such an interest, the relationship between body mass and gut microbiota has never been explored whereas, in contrast, the positive

*Correspondence: jean-jacques.godon@supagro.inra.fr
[1] UR0050, Laboratoire de Biotechnologie de l'Environnement, INRA, 102 avenue des étangs, 11100 Narbonne, France
Full list of author information is available at the end of the article

links between the abundance of parasitic organisms or protozoal faunas and animal body size have been thoroughly referenced [11] [12]. The aim of the present study is to analyse a large bacterial dataset, comprising faeces collected from 71 different vertebrate species, in order to examine the effect of the volume-microbial diversity relationship in animal digestive tracts.

Methods

Sampling
All the animal samples were obtained from domesticated or captive populations in France (zoo, farm, aquarium, recreative farm or individual keeper). There is non-experimental research dedicated for this study, faeces samples were collected on ground with the animal keeper or animal owner without stresses for the animals. We obtained permissions from Lunaret zoo, Montpellier; Océanopolis, Brest; Réserve Africaine, Sigean; Mini Ferme Zoo, Cessenon sur Orb and consent from the animal owners (Jean-Philippe Steyer, Anais Bonnafous, Jean-Jacques Godon). Animal were living alone or in small groups (1 to 5). Furthermore, their food (meat, seeds, fruits or hay) were more standardized in comparison to wild diets.

Human stool specimens used in the present study were from infant and adult subjects included in international multicentric studies. Samples were collected between 2001 to 2005 and used on previous published studies. Infants samples were collected in the frame of the European project INFABIO (http://www.gla.ac.uk/departments/infabio/), ethical permission was obtained from Yorkhill Research Ethics Committee P16/03 and parents gave written informed consent [13]. Adults samples were collected in the frame of the European project Crownalife, the studies were approved by the Ethics Committee of Versailles Hospital Centre and written informed consent was obtained from all participants [14]. Approval for Institut National de la Recherche Agronomique to manage human-derived biological samples in accordance with Articles L.1243-3, R.1243-49 of "Code de la Santé Publique" was granted by the Ministry of Research and Education under number DC-2012-1728.

Faeces from 189 individuals belonging to 71 vertebrate species (31 mammals, 37 birds and 3 reptiles) were collected (Table 1). They were sub-divided into 80 categories according to species or to body mass (i.e., age (young–adult), sex (female–male), breed size (small–big–domesticated–wild), see Table 1). Body masses were provided by the breeder for large animals or obtained from literature for small animals. Body masses, along with diversity, were displayed with a logarithmic scale in order to highlight the linear shape of the power-law relationship. Except for the distinct dimorphism of male and female turkey samples, an average value of male and female

body mass values was used. Dwarf or young individuals from the same species were also classified in specific body mass categories. For example, human samples were divided into two body mass categories: babies between 1 and 10 months old (mean of 5.8 kg) and adults between 29 and 61 years old (set at 70 kg). Composite faeces samples were avoided except for those that could not provide enough material for DNA extraction (less than 0.5 g).

DNA extraction, PCR amplification and Capillary Electrophoresis Single Strand Conformation Polymorphism (CE-SSCP) fingerprinting
Genomic DNAs were extracted from 0.5 g of raw material using the procedure described by Godon et al. [15]. The V3 region of the 16S rRNA gene was amplified with *Bacteria*-specific primers and PCR products were analysed by CE-SSCP analysis using an ABI3130 Genetic Analyzer (Applied Biosystems, Foster City, CA, USA) in accordance with a previously described method [16]. All raw CE-SSCP data are available on Additional file 4.

Calculation of diversity and statistical computing
Diversity was estimated by the Simpson Diversity Index from CE-SSCP fingerprinting patterns. The Simpson Diversity Index was expressed as $D = 1 / \sum_{i=1}^{p} a_i^2$ where a_i is the relative abundance of each CE-SSCP peak p. This index was directly calculated from each CE-SSCP fingerprint [17] using the R StatFingerprints library [18].

Preference was given to the Simpson Diversity Index from CE-SSCP fingerprinting rather than the Richness estimation because: (1) neither fingerprinting nor sequencing data can provide a robust estimation of richness [19]; (2) the Simpson Diversity Index can be estimated accurately with CE-SSCP fingerprinting [17, 20].

A generalized linear model was applied to fit the relationship between body mass and diversity. ANOVA followed by Tukey post hoc tests were used for determining the statistical difference between (sub-) categories and body mass or diversity, both expressed in a logarithmic scale. All statistics were performed under R software (version 3.1.2) [21]. The calculation of the slope z was based on the exponent of the power-law relationship as follows: diversity = c weight^z.

Results and discussion
The bacterial diversity of faeces from 189 vertebrates belonging to 71 species (31 mammals, 37 birds and 3 reptiles) was analysed (Table 1; Fig. 1). Analysis was only focused on diversity (Simpson Diversity Index), which can be accurately measured according to CE-SSCP fingerprinting patterns [15] (see the "Methods" section and Additional file 1). Apart from their phylogenetic position, animals can also be classified according to: (1) their

Table 1 Animal data ranked by body mass

Name (common name)	Phylogeny	Body mass (kg)	Feeding type	Type of digestive tract	Size of animal husbandry group	Diversity	SD	Number of samples
Taeniopygia guttata (zebra finch)	Aves, Passeriformes	0.012	Granivorous	Hindgut colon	Large	1.2	0.1	3
Serinus canaria (canary)	Aves, Passeriformes	0.024	Granivorous	Hindgut colon	Large	1.6	0.5	2
Ramphocelus bresilius (brazilian tanager)	Aves, Passeriformes	0.035	Frugivorous	Hindgut colon	Small	3.4	0.4	4
Melopsittacus undulatus (budgerigar)	Aves, Psittaciformes	0.04	Granivorous	Hindgut colon	Large	1.9	3.0	3
Ploceus cucullatus (village weaver)	Aves, Passeriformes	0.04	Granivorous	Hindgut colon	Small	2.3	0.0	2
Agapornis fischeri (Fischer's lovebird)	Aves, Psittaciformes	0.05	Granivorous	Hindgut colon	Large	1.5		1
Agapornis roseicollis (rosy-faced lovebird)	Aves, Psittaciformes	0.05	Granivorous	Hindgut colon	Large	2.1	0.0	2
Amblyramphus holosericeus (scarlet-headed blackbird)	Aves, Passeriformes	0.08	Carnivorous	Hindgut colon	Small	3.4	0.2	2
Nymphicus hollandicus (cockatiel)	Aves, Psittaciformes	0.08	Granivorous	Hindgut colon	Small	1.4	0.3	2
Guira guira (guira cuckoo)	Aves, Cuculiformes	0.14	Carnivorous	Hindgut colon	Small	2.9	0.5	4
Poicephalus senegalus (senegal parrot)	Aves, Psittaciformes	0.14	Granivorous	Hindgut colon	Small	2.6		1
Streptopelia decaocto (eurasian collard dove)	Aves, Columbidae	0.19	Granivorous	Hindgut colon	Large	2.4	0.5	3
Corvus monedula (eurasian jackdaw)	Aves, Passeriformes	0.22	Omnivorous	Hindgut colon	Small	1.8		1
Psarocolius decumanus (crested oropendola)	Aves, Passeriformes	0.3	Omnivorous	Hindgut colon	Small	3.8	0.3	3
Columba livia (pigeon)	Aves, Columbidae	0.3	Granivorous	Hindgut colon	Large	1.9		1
Gallus gallus (dwarf chicken)[a]	Aves, Galliformes	0.3	Granivorous	Hindgut caecum	Large	2.1	0.5	2
Tauraco erythrolophus (red-crested turaco)	Aves, Cuculiformes	0.35	Frugivorous	Hindgut colon	Small	3.8		1
Agamia agami (agami heron)	Aves, Ciconiiformes	0.46	Piscivorous	Hindgut colon	Small	3.6		1
Coracopsis vasa (vasa parrot)	Aves, Psittaciformes	0.5	Frugivorous	Hindgut colon	Small	2.4		1
Chinchilla laniger xChinchilla brevicaudata (chinchilla)	Mammalia, Rodentia	0.6	Herbivorous	Hindgut caecum	Small	4.2	0.1	2
Ramphastos tucanus (white-throated toucan)	Aves, Piciformes	0.675	Frugivorous	Hindgut colon	Small	3.6	0.4	3
Chrysolophus pictus (golden pheasant)	Aves, Galliformes	0.700	Granivorous	Hindgut caecum	Small	3.4		1
Cavia porcellus (domestic guinea pig)	Mammalia, Rodentia	0.8	Herbivorous	Hindgut caecum	Large	5.0	0.2	3
Anas acuta (northern pintail)	Aves, Anatidae	0.9	Granivorous	Hindgut caecum	Small	3.6		1
Elaphe guttata (corn snake)	Sauropsida, Serpentes	0.9	Carnivorous	Hindgut colon	Small	4.1		1
Lampropeltis getula (common kingsnake)	Sauropsida, Serpentes	1	Carnivorous	Hindgut colon	Small	3.3		1
Ara ararauna (blue-and-yellow macaw)	Aves, Psittaciformes	1	Granivorous	Hindgut colon	Small	2.8	0.1	2
Anas platyrhynchos (wild type duck)[a]	Aves, Anatidae	1.1	Granivorous	Hindgut caecum	Large	3.1		1
Neochen jubata (orinoco goose)	Aves, Anatidae	1.25	Granivorous	Hindgut caecum	Small	3.5		1

Table 1 continued

Name (common name)	Phylogeny	Body mass (kg)	Feeding type	Type of digestive tract	Size of animal husbandry group	Diversity	SD	Number of samples
Gallus gallus (chicken)[a]	Aves, Galliformes	1.5	Granivorous	Hindgut caecum	Large	1.7	0.5	6
Numida meleagris (guinea-fowl)	Aves, Galliformes	2	Granivorous	Hindgut caecum	Large	3.0	0.7	3
Branta sandvicensis (nene)	Aves, Anatidae	2	Granivorous	Hindgut caecum	Small	2.6		1
Oryctolagus cuniculus (domestic rabbit)	Mammalia, Lago-morpha	2.2	Herbivorous	Hindgut caecum	Large	4.5	0.7	3
Anas platyrhynchos (domestic duck)[a]	Aves, Anatidae	2.3	Granivorous	Hindgut colon	Small	2.3	1.4	3
Eudyptes chrysocome (western rockhopper penguin)	Aves, Sphenisci-formes	2.6	Piscivorous	Hindgut colon	Large	1.6	0.6	3
Testudo hermanni boettgeri (Hermann's tortoise)	Sauropsida, Testudines	3	Herbivorous	Hindgut colon	Large	5.6		1
Meleagris gallopavo (turkey female)[a]	Aves, Galliformes	3	Granivorous	Hindgut caecum	Small	2.2		1
Thylogale sp. (pademelon)	Mammalia, Marsu-pials	3.5	Herbivorous	Hindgut colon	Small	4.3		1
Cairina moschata (muscovy duck)	Aves, Anatidae	4	Granivorous	Hindgut colon	Large	3.3		2
Chauna torquata (southern screamer)	Aves, Anseri-formes	4	Herbivorous	Hindgut caecum	Small	3.4		1
Canis lupus familiaris (puppy)[a]	Mammalia, Car-nivora	4	Carnivorous	Hindgut colon	Small	2.6		1
Pavo cristatus (blue peafowl)	Aves, Galliformes	5	Granivorous	Hindgut caecum	Small	3.7	0.1	2
Anser anser domesticus (domestic goose)	Aves, Anatidae	5	Granivorous	Hindgut caecum	Small	3.5	0.1	2
Homo sapiens (baby human caucasian)[a]	Mammalia, Primates	6	Omnivorous	Hindgut colon	Small	3.2	0.7	15
Meleagris gallopavo (turkey male)[a]	Aves, Galliformes	8	Granivorous	Hindgut caecum	Small	3.8		1
Wallabia bicolor (black wallaby)	Mammalia, Marsu-pials	9	Herbivorous	Hindgut colon	Small	4.8		1
Hylobates lar (gibbon)	Mammalia, Primates	10	Frugivorous	Hindgut colon	Small	5.5		1
Aptenodytes patagonicus (king penguin)	Aves, Sphenisci-formes	13	Piscivorous	Hindgut colon	Small	2.8	1.1	4
Capra hircus (dwarf goat)	Mammalia, Rumi-nantia	15	Herbivorous	Ruminants foregut	Small	5.2	1.1	2
Canis lupus familiaris (medium size dog)[a]	Mammalia, Car-nivora	20	Carnivorous	Hindgut colon	Small	3.2	1.1	2
Ovis aries (dwarf sheep)[a]	Mammalia, Rumi-nantia	20	Herbivorous	Ruminants foregut	Small	5.7	0.7	2
Hippotragus equinus (roan antelope)	Mammalia, Rumi-nantia	20	Herbivorous	Ruminants foregut	Small	5.7		1
Tragelaphus streps (greater kudu)	Mammalia, Rumi-nantia	20	Herbivorous	Ruminants foregut	Small	5.5		1
Hystrix cristata (crested porcupine)	Mammalia, Rodentia	25	Herbivorous	Hindgut caecum	Small	5.7		1
Rhea americana (greater rhea)	Aves, Rheiformes	31	Granivorous	Hindgut caecum	Large	4.0	0.5	4
Ovis aries (sheep)[a]	Mammalia, Rumi-nantia	40	Herbivorous	Ruminants foregut	Small	5.0		1
Canis lupus familiaris (big size dog)[a]	Mammalia, Car-nivora	40	Carnivorous	Hindgut colon	Small	3.1		1
Pan troglodytes (chimpanzee)	Mammalia, Primates	40	Omnivorous	Hindgut colon	Small	5.3		1

Table 1 continued

Name (common name)	Phylogeny	Body mass (kg)	Feeding type	Type of digestive tract	Size of animal husbandry group	Diversity	SD	Number of samples
Dromaius novaehollandiae (emu)	Aves, Casuariiformes	40	Granivorous	Hindgut colon	Small	3.9		1
Capra hircus (goat)	Mammalia, Ruminantia	50	Herbivorous	Ruminants foregut	Small	7.0		1
Sus scrofa (dwarf pig)[a]	Mammalia, Suina	55	Omnivorous	Hindgut colon	Small	5.4	0.4	2
Lama glama (llama)	Mammalia, Tylopoda	55	Herbivorous	Ruminants foregut	Small	5.4		1
Homo sapiens (adult human caucasian)[a]	Mammalia, Primates	70	Omnivorous	Hindgut colon	Large	4.4	0.8	34
Sus scrofa (pig)[a]	Mammalia, Suina	100	Omnivorous	Hindgut colon	Small	5.8	1.1	4
Tragelaphus spekei (sitatunga)	Mammalia, Ruminantia	100	Herbivorous	Ruminants foregut	Small	7.5		1
Struthio camelus (ostrich)	Aves, Struthioniformes	120	Herbivorous	Hindgut colon	Small	4.4	0.2	3
Equus asinus (donkey)	Mammalia, Equidae	150	Herbivorous	Hindgut caecum	Small	5.3	0.4	2
Ammotragus lervia (aoudad)	Mammalia, Ruminantia	150	Herbivorous	Ruminants foregut	Small	6.1		1
Equus caballus (pony)	Mammalia, Equidae	160	Herbivorous	Hindgut caecum	Small	5.6	0.1	2
Panthera leo (african lion)	Mammalia, Carnivora	160	Carnivorous	Hindgut colon	Small	4.4		1
Equus zebra hartmannae (mountain zebra)	Mammalia, Equidae	350	Herbivorous	Hindgut caecum	Small	5.4		1
Syncerus caffer nanus (forest buffalo)	Mammalia, Ruminantia	450	Herbivorous	Ruminants foregut	Small	2.9		1
Camelus dromedarius (arabian Camel)	Mammalia, Tylopoda	500	Herbivorous	Ruminants foregut	Small	3.2		1
Bos grunniens (yak)	Mammalia, Ruminantia	600	Herbivorous	Ruminants foregut	Small	5.3		1
Tragelaphus oryx (eland antelope)	Mammalia, Ruminantia	600	Herbivorous	Ruminants foregut	Small	6.2		1
Bos taurus (cow)	Mammalia, Ruminantia	750	Herbivorous	Ruminants foregut	Large	6.2	0.7	4
Giraffa camelopardalis reticulata (somali giraffe)	Mammalia, Ruminantia	1100	Herbivorous	Ruminants foregut	Small	6.4		1
Giraffa camelopardalis peralta (nigerian giraffe)	Mammalia, Ruminantia	1100	Herbivorous	Ruminants foregut	Small	6.6		1
Ceratotherium simum (white rhinoceros)	Mammalia, Rhinocerotidae	2500	Herbivorous	Hindgut colon	Small	5.6		1
Elephas maximus (asian elephant)	Mammalia, Proboscidea	3500	Herbivorous	Hindgut colon	Small	4.9		1

SD standard deviation

[a] species with different sizes (young-adult, female-male, small-big or domesticated-wild)

diet (herbivorous, granivorous, omnivorous, carnivorous, piscivorous and frugivorous); (2) their metabolic body mass (from 12 g (zebra finch) to 3500 kg (Asian elephant)); (3) the structure of their digestive tracts; (4) and the size of the animal husbandry group (small and large). The present study focused on bacterial diversity, although changes within the structure of the bacterial communities were not taken into account. This study is also based on two assumptions: (1) the gut size should be proportional to the animal body mass, as has been demonstrated for herbivores [22] and birds [23]; and (2) the microbial diversity of faeces should be similar to that in the gut [24].

Results point to a correlation between animal body mass and microbial diversity (linear regression with a slope z of 0.338 ± 0.027; p value $<2.2 \times 10^{-16}$),

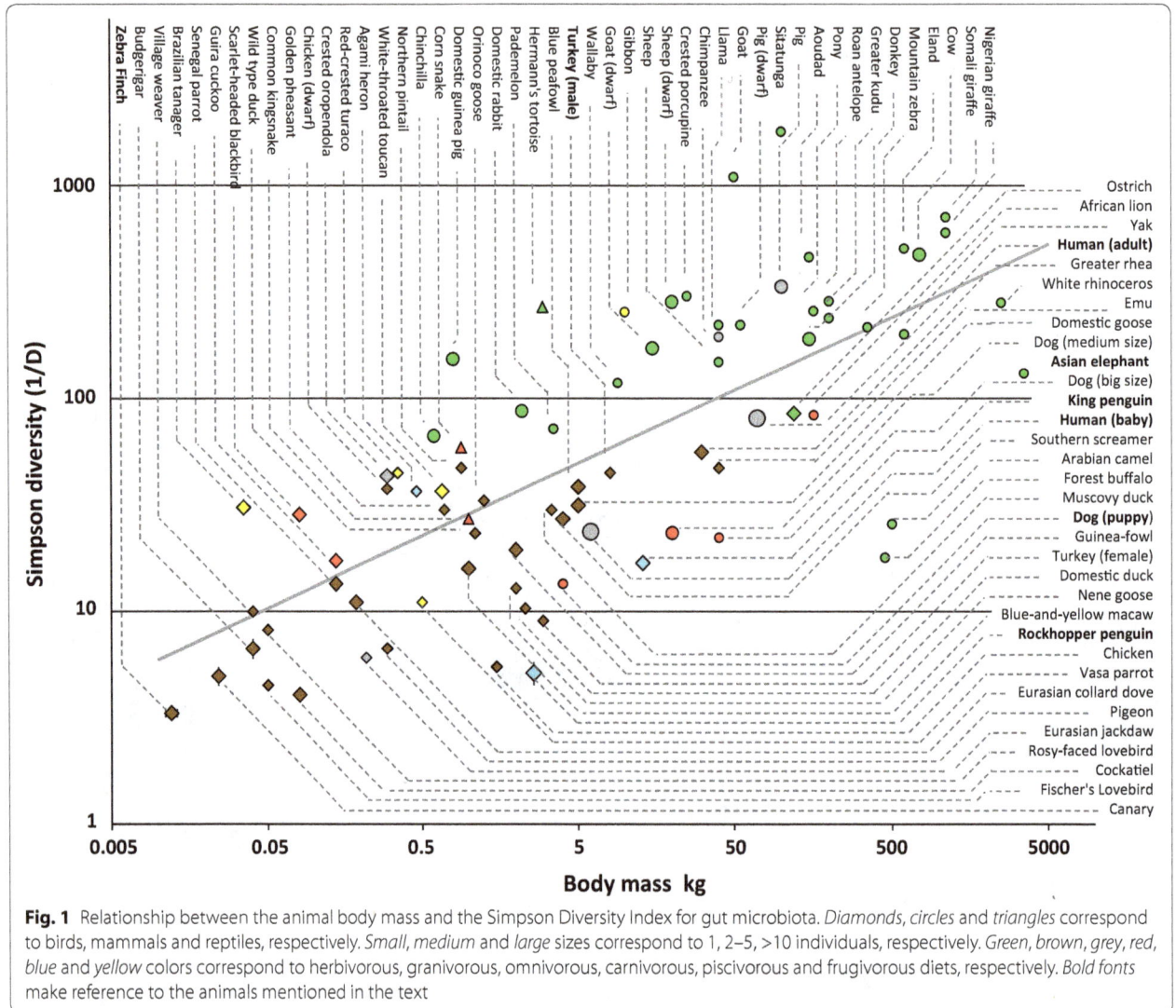

Fig. 1 Relationship between the animal body mass and the Simpson Diversity Index for gut microbiota. *Diamonds*, *circles* and *triangles* correspond to birds, mammals and reptiles, respectively. *Small*, *medium* and *large* sizes correspond to 1, 2–5, >10 individuals, respectively. *Green*, *brown*, *grey*, *red*, *blue* and *yellow* colors correspond to herbivorous, granivorous, omnivorous, carnivorous, piscivorous and frugivorous diets, respectively. *Bold fonts* make reference to the animals mentioned in the text

irrespective of the diet, phylogeny or structure of the digestive tracts (Fig. 1). Consequently, the use of a greater amount of samples over a wider size range confirms previous works on unrelated bacterial communities that have suggested the existence of a link between volume and diversity in tree holes [5], membrane bioreactors [4] and metal-cutting fluid sump tanks [3]. In the present results, the Simpson Diversity Index ranges between 3.3 and 1789.5, thus corresponding to a 5.6 log body mass range (Fig. 1).

A wide variability in the diversity between individuals for a given species was observed. However the average diversity value for species that were represented by several individuals was close to the regression line (Fig. 1). For example, the average diversity value for adult human microbiota (34 samples) was 80.8 with a standard

deviation of 294.2, and 23.7 ± 20.3 for the 15 baby human microbiota. As a matter of comparison, Trosvik et al. [25] observed a similar range of diversity (over 2 log-units of Shannon index) when analysing a time-series of 332 sequencings over 443 days, on a single male adult individual.

Animal gut microbiota covered a broad range of diversity ranging from 2.2 to 1808.0. This was comparable to the values found in various types of environment, like drinking water, raw milk, plant roots, activated sludge in wastewater treatment plants, compost or soil (Additional file 2). On one hand, the lowest diversity in gut microbiota varied around 2, similarly to those found in drinking water. On the other hand, the highest diversity in gut microbiota reaching about 1808 resembled the values found in soils (Additional file 2).

Table 2 Bacterial diversity and animal weight within sub-categories, correlation between diversity and weight, and slope of the relationship of the diversity versus log-weight

Category	Number of samples	Simpson diversity mean (SD)	Weight in kg mean (SD)	Pearson correlation between diversity and weight		Power law relationship diversity = c weight^z	
				cor	p value	Slope z	Confidence interval
Sub-categories							
Diet							
Carnivorous	13	32.0 (23.4)	19.0 (44.1)	0.277	0.359 (NS)	0.075 (NS)	–
Frugivorous	10	55.2 (68.5)	1.3 (3.1)	0.533	0.113 (NS)	0.234 (NS)	–
Granivorous	54	20.4 (21.7)	4.4 (9.4)	0.667	3.7e–08 (***)	0.298 (***)	0.205–0.391
Herbivorous	44	301.6 (342.3)	354.5 (668.1)	0.338	0.025 (*)	0.137 (*)	0.018–0.256
Omnivorous	60	111.3 (149.0)	50.5 (32.4)	0.542	7.7e–06 (***)	0.361 (***)	0.214–0.508
Piscivorous	8	20.4 (20.8)	7.5 (5.9)	0.030	0.944 (NS)	0.029 (NS)	–
Phylogeny							
Bird	85	25.1 (23.6)	7.8 (23.0)	0.456	1.1e–05 (***)	0.202 (***)	0.116–0.288
Mammal	101	194.9 (268.8)	183.3 (464.0)	0.415	1.6e–05 (***)	0.272 (***)	0.153–0.391
Reptile	3	119.3 (131.9)	1.6 (1.2)	0.964	0.172 (NS)	1.686 (NS)	–
Gut structure							
Caecum	46	70.4 (80.6)	26.1 (65.3)	0.528	1.7e–04 (***)	0.397 (***)	0.203–0.591
Colon	124	78.5 (122.5)	79.0 (382.7)	0.678	<2.2e–16 (***)	0.293 (***)	0.236–0.350
Rumen	19	484.9 (449.7)	411.8 (384.5)	–0.036	0.883 (NS)	–0.031 (NS)	–
Group size							
Large	85	93.1 (151.0)	65.8 (156.4)	0.734	1.3e–15 (***)	0.380 (***)	0.303–0.457
Small	104	137.2 (253.7)	130.6 (449.1)	0.632	6.6e–13 (***)	0.300 (***)	0.227–0.372
Age							
Adult	173	125.6 (222.2)	110.3 (364.6)	0.687	<2.2e–16 (***)	0.337 (***)	0.283–0.391
Baby	16	28.9 (20.1)	6.3 (2.0)	0.205	0.446 (NS)	0.427 (NS)	–
All	189	117.4 (214.3)	101.5 (350.0)	0.675	<2.2e–16 (***)	0.338 (***)	0.284–0.391

NS not significant

* low significance, *** high significance

This vast range of variations in gut diversity is often associated with factors that are different to the body mass: diet [10], phylogeny [26], digestive tract structure [27], age [8] [28], way of life [29], ethnic origin [30], state of health (immune system, pregnancy, obesity) [31] [32], or genetic background [32]. Among these parameters, age has been well documented as the major one to explain these variations and the diversity or richness between human baby microbiota and those of adults [33] [34]. However the size of the gut also varies during infant growth. In this case, a difference in the microbial diversity between infant (29.9 ± 20.3) and adults (106.6 ± 76.0) was observed, concomitantly with changes in body mass when comparing human babies (6.5 ± 1.9) and adults (70 kg). The same observation was made for young and adult dog samples (Table 1). Furthermore, when comparing two penguin species that only differ in their body mass (only adult specimens, with the same diet and living in the same location), the relationship between microbial diversity and body mass still remain valid.

The correlation between body mass and diversity has been assessed for homogenous sub-categories (Table 2 and Additional file 3), thus excluding the potential effects of the different parameters. Indeed, the 189 samples could also be analysed according to phylogeny (reptile, bird, and mammal), diet (carnivorous, herbivorous, granivorous, omnivorous and piscivorous), gut structure (hindgut caecum, hindgut colon and foregut ruminant), age (baby and adult), and size of the animal husbandry group (small and large). Except for the latter category, all of them depended on the body mass (e.g. body mass was related to phylogeny, related to age or to ruminants). Significantly positive body mass/diversity correlations were observed for each sub-category, provided that a sufficient amount of data was available (over 50 samples minimum per sub-category) (Table 2; Additional file 3). The significant slopes z of the mass-diversity relationships generally ranged from 0.202 ± 0.043 to 0.380 ± 0.039. As the herbivorous group only contained 44 samples, the interestingly weak body mass diversity correlation with a z value of 0.137 could not be correctly interpreted.

The observed slope z was similar to that reported for 'island' patterns of bacterial diversity such as metal-cutting fluid sump tanks ($z = 0.245$–0.295) [3] and tree holes ($z = 0.26$) [5] and varied within a similar range to that reported for plants and animals from discrete islands ($z = 0.25$–0.35). The slope z-values reported for continuous patterns (such as marsh sediment [35] with z-values between 0.02 and 0.04) are generally much lower than those reported for discrete habitats.

According to these results, which confirm the assumption that species and volume are related, guts can be compared to an archipelago, where microbes originating from feed tend to colonise the available niches provided by the gut. This is also in line with the MacArthur and Wilson biogeography theory [1]. Size, similarly to island environments appears to reflect the heterogeneity of the environment. Hence, a large gut size should provide more space, enabling a large microbial diversity to settle in [36].

Conclusions

The aim of this study was not to explain the genesis of bacterial diversity in vertebrate guts but was rather focused on producing evidence on the role of gut size in the maintenance of a level of microbial diversity. This work highlights the hitherto unexplored relationship between volume and diversity in the case of gut microbiota. Gut volume should henceforth be taken into account along with other parameters to explain the level of diversity. Finally, this work confirms the relevance of the microbial world when addressing ecological issues such as the relationship between species diversity and the size of the habitat [37].

Additional files

Additional file 1. Examples of representative CE-SSCP fingerprinting patterns from low-diverse to high-diverse samples.

Additional file 2. Simpson Diversity Index calculated from 671 CE-SSCP fingerprinting patterns. Samples were grouped according to their ecosystem of origin.

Additional file 3. Simpson Diversity Index as a function of animal weight for (A) all the samples and for the sub-categories; (B) diet; (C) phylogeny; (D) gut structure; (E) age and (F) group size. The slope of the linear regression is indicated.

Additional file 4. Raw 671 CE-SSCP profiles. Columns correspond to the samples. The 7359 lines correspond to the fluorescence intensity according to time.

Abbreviations

CE-SSCP: capillary electrophoresis single strand conformation polymorphism; PCR: polymerase chain reaction; 16S rRNA: 16S ribosomal RNA; ANOVA: analysis of variance.

Authors' contributions

JJG designed the experiment, JJG and JH collected the data, PA performed the experiments and contributed to the analysis of the data. Analysis and interpretation were carried out by JH and JJG. JH and JJG contributed to the first draft, which was completed by JPS. All authors read and approved the final manuscript.

Author details

[1] UR0050, Laboratoire de Biotechnologie de l'Environnement, INRA, 102 avenue des étangs, 11100 Narbonne, France. [2] Present Address: Centre of Excellence in Environmental Studies, King Abdulaziz University, Jeddah, Saudi Arabia.

Acknowledgements
We thank the following people and organizations who generously provided samples: Joël Doré; Thierry Gidenne; 'Reserve africaine', Sigean, France; 'Mini-ferme zoo', Cessenon/Orb, France; 'Oceanopolis', Brest, France; and Lunaret Zoo, Montpellier, France. We thank Biswarup Sen, Anais Bonnafous and Valérie Bru-Adan for technical assistance. INRA funded this research.

Competing interests
The authors declare that they have no competing interests.

References
1. MacArthur RH, Wilson EO. The theory of island biogeography. Princet Univ Press Monogr Popul Biol. 1967;1:202.
2. Arrhenius O. Species and area. J Ecol. 1921;9:95–9.
3. van der Gast CJ, Lilley AK, Ager D, Thompson IP. Island size and bacterial diversity in an archipelago of engineering machines. Environ Microbiol. 2005;7:1220–6.
4. Van der Gast CJ, Jefferson B, Reid E, Robinson T, Bailey MJ, Judd SJ, Thompson IP. Bacterial diversity is determined by volume in membrane bioreactors. Environ Microbiol. 2006;8:1048–55.
5. Bell T, Ager D, Song JI, Newman JA, Thompson IP, Lilley AK, van der Gast CJ. Larger islands house more bacterial taxa. Science. 2005;308:1884.
6. Ley RE, Peterson DA, Gordon JI. Ecological and evolutionary forces shaping microbial diversity in the human intestine. Cell. 2006;124:837–48.
7. Le Chatelier E, Nielsen T, Qin J, Prifti E, Hildebrand F, Falony G, Almeida M, Arumugam M, Batto J-M, Kennedy S, Leonard P, Li J, Burgdorf K, Grarup N, Jorgensen T, Brandslund I, Nielsen HB, Juncker AS, Bertalan M, Levenez F, Pons N, Rasmussen S, Sunagawa S, Tap J, Tims S, Zoetendal EG, Brunak S, Clement K, Dore J, Kleerebezem M, et al. Richness of human gut microbiome correlates with metabolic markers. Nature. 2013;500:541.
8. Yatsunenko T, Rey FE, Manary MJ, Trehan I, Dominguez-Bello MG, Contreras M, Magris M, Hidalgo G, Baldassano RN, Anokhin AP, Heath AC, Warner B, Reeder J, Kuczynski J, Caporaso JG, Lozupone CA, Lauber C, Clemente JC, Knights D, Knight R, Gordon JI. Human gut microbiome viewed across age and geography. Nature. 2012;486:222.
9. Schnorr SL, Candela M, Rampelli S, Centanni M, Consolandi C, Basaglia G, Turroni S, Biagi E, Peano C, Severgnini M, Fiori J, Gotti R, De Bellis G, Luiselli D, Brigidi P, Mabulla A, Marlowe F, Henry AG, Crittenden AN. Gut microbiome of the Hadza hunter-gatherers. Nat Commun. 2014;5:3654.
10. Vital M, Gao J, Rizzo M, Harrison T, Tiedje JM. Diet is a major factor governing the fecal butyrate-producing community structure across mammalia aves and reptilia. ISME J. 2015;9:832–43.
11. Kamiya T, O'Dwyer K, Nakagawa S, Poulin R. What determines species richness of parasitic organisms? A meta-analysis across animal, plant and fungal hosts. Biol Rev. 2014;89:123–34.
12. Clauss M, Mueller K, Fickel J, Streich WJ, Hatt J-M, Suedekum K-H. Macroecology of the host determines microecology of endobionts: protozoal faunas vary with wild ruminant feeding type and body mass. J Zool. 2011;283:169–85.
13. Fallani M, Amarri S, Uusijarvi A, Adam R, Khanna S, Aguilera M, Gil A, Vieites JM, Norin E, Young D, Scott JA, Doré J, Edwards CA, Team I. Determinants of the human infant intestinal microbiota after the introduction of first complementary foods in infant samples from five European centres. Microbiol SGM. 2011;157:1385–92.
14. Mueller S, Saunier K, Hanisch C, Norin E, Alm L, Midtvedt T, Cresci A, Silvi S, Orpianesi C, Verdenelli MC, Clavel T, Koebnick C, Zunft HJF, Dore J, Blaut M. Differences in fecal microbiota in different European study populations in relation to age, gender, and country: a cross-sectional study. Appl Environ Microbiol. 2006;72:1027–33.
15. Godon JJ, Zumstein E, Dabert P, Habouzit F, Moletta R. Molecular microbial diversity of an anaerobic digestor as determined by small-subunit rDNA sequence analysis. Appl Environ Microbiol. 1997;63:2802–13.
16. Wéry N, Bru-Adan V, Minervini C, Delgénes JP, Garrelly L, Godon JJ. Dynamics of Legionella spp. and bacterial populations during the proliferation of L. pneumophila in a cooling tower facility. Appl Environ Microbiol. 2008;74:3030–7.
17. Haegeman B, Sen B, Godon JJ, Hamelin J. Only simpson diversity can be estimated accurately from microbial community fingerprints. Microb Ecol. 2014;68:169–72.
18. Michelland RJ, Dejean S, Combes S, Fortun-Lamothe L, Cauquil L. Stat-Fingerprints: a friendly graphical interface program for processing and analysis of microbial fingerprint profiles. Mol Ecol Resour. 2009;9:1359–63.
19. Haegeman B, Hamelin J, Moriarty J, Neal P, Dushoff J, Weitz JS. Robust estimation of microbial diversity in theory and in practice. ISME J. 2013;7:1092–101.
20. Lalande J, Villemur R, Deschenes L. A new framework to accurately quantify soil bacterial community diversity from DGGE. Microb Ecol. 2013;66:647–58.
21. R-Core-Team. R: a language and environment for statistical computing. Vienna: R Foundation for Statistical Computing; 2014.
22. Clauss M, Schwarm A, Ortmann S, Streich WJ, Hummel J. A case of non-scaling in mammalian physiology? Body size, digestive capacity, food intake, and ingesta passage in mammalian herbivores. Comp Biochem Physiol A: Mol Integr Physiol. 2007;148:249–65.
23. Lavin SR, Karasov WH, Ives AR, Middleton KM, Garland T Jr. Morphometrics of the avian small intestine compared with that of nonflying mammals: a phylogenetic approach. Physiol Biochem Zool. 2008;81:526–50.
24. Michelland RJ, Monteils V, Zened A, Combes S, Cauquil L, Gidenne T, Hamelin J, Fortun-Lamothe L. Spatial and temporal variations of the bacterial community in the bovine digestive tract. J Appl Microbiol. 2009;107:1642–50.
25. Trosvik P, de Muinck EJ, Stenseth NC. Biotic interactions and temporal dynamics of the human gastrointestinal microbiota. ISME J. 2015;9:533–41.
26. Ley RE, Lozupone CA, Hamady M, Knight R, Gordon JI. Worlds within worlds: evolution of the vertebrate gut microbiota. Nat Rev Microbiol. 2008;6:776–88.
27. Stevens CE, Hume ID. Comparative physiology of the vertebrate digestive system. Cambridge: University Press; 2004.
28. Waite DW, Eason DK, Taylor MW. Influence of hand rearing and bird age on the fecal microbiota of the critically endangered kakapo. Appl Environ Microbiol. 2014;80:4650–8.
29. Dicksved J, Floistrup H, Bergstrom A, Rosenquist M, Pershagen G, Scheynius A, Roos S, Alm JS, Engstrand L, Braun-Fahrlander C, von Mutius E, Jansson JK. Molecular fingerprinting of the fecal microbiota of children raised according to different lifestyles. Appl Environ Microbiol. 2007;73:2284–9.
30. Kwok L, Zhang J, Guo Z, Gesudu Q, Zheng Y, Qiao J, Huo D, Zhang H. Characterization of fecal microbiota across seven chinese ethnic groups by quantitative polymerase chain reaction. PLoS One. 2014;9(4):93631.
31. Rescigno M. Intestinal microbiota and its effects on the immune system. Cell Microbiol. 2014;16:1004–13.
32. Turnbaugh PJ, Hamady M, Yatsunenko T, Cantarel BL, Duncan A, Ley RE, Sogin ML, Jones WJ, Roe BA, Affourtit JP, Egholm M, Henrissat B, Heath AC, Knight R, Gordon JI. A core gut microbiome in obese and lean twins. Nature. 2009;457:480–7.
33. Koenig JE, Spor A, Scalfone N, Fricker AD, Stombaugh J, Knight R, Angenent LT, Ley RE. Succession of microbial consortia in the developing infant gut microbiome. Proc Natl Acad Sci. 2011;108:4578–85.
34. Henderson G, Cox F, Ganesh S, Jonker A, Young W, Collaborators GRC, Janssen PH. Rumen microbial community composition varies with diet and host, but a core microbiome is found across a wide geographical range. Sci Rep. 2015;5:14567.
35. Horner-Devine MC, Lage M, Hughes JB, Bohannan BJM. A taxa–area relationship for bacteria. Nature. 2004;432:750–3.
36. Kassen R. The experimental evolution of specialists, generalists, and the maintenance of diversity. J Evol Biol. 2002;15:173–90.
37. Jessup CM, Kassen R, Forde SE, Kerr B, Buckling A, Rainey PB, Bohannan BJM. Big questions, small worlds: microbial model systems in ecology. Trends Ecol Evol. 2004;19:189–97.

2

Transient recovery dynamics of a predator–prey system under press and pulse disturbances

Canan Karakoç[1], Alexander Singer[2,4], Karin Johst[2], Hauke Harms[1,3] and Antonis Chatzinotas[1,3]* ⓘ

Abstract

Background: Species recovery after disturbances depends on the strength and duration of disturbance, on the species traits and on the biotic interactions with other species. In order to understand these complex relationships, it is essential to understand mechanistically the transient dynamics of interacting species during and after disturbances. We combined microcosm experiments with simulation modelling and studied the transient recovery dynamics of a simple microbial food web under pulse and press disturbances and under different predator couplings to an alternative resource.

Results: Our results reveal that although the disturbances affected predator and prey populations by the same mortality, predator populations suffered for a longer time. The resulting diminished predation stress caused a temporary phase of high prey population sizes (i.e. prey release) during and even after disturbances. Increasing duration and strength of disturbances significantly slowed down the recovery time of the predator prolonging the phase of prey release. However, the additional coupling of the predator to an alternative resource allowed the predator to recover faster after the disturbances thus shortening the phase of prey release.

Conclusions: Our findings are not limited to the studied system and can be used to understand the dynamic response and recovery potential of many natural predator–prey or host–pathogen systems. They can be applied, for instance, in epidemiological and conservational contexts to regulate prey release or to avoid extinction risk of the top trophic levels under different types of disturbances.

Keywords: Pulse disturbance, Press disturbance, Transient dynamics, Recovery, Trophic interactions, Protist, Bacteria, Predation, Prey release, Food web

Background

Disturbance is one of the key drivers of the dynamics and diversity of communities [1–3] and is defined as a discrete event in time killing or damaging individuals [4]. Disturbances occur in many natural systems with different strengths and durations. They are often classified as pulse disturbances (short-term events) or press disturbances (long-term events) depending on their duration in relation to the generation times of species [5, 6]. These different temporal patterns of disturbances are important for understanding the structural and functional community responses [7]. Press disturbances, for instance, can cause increasing variability in the relative abundances of species, whereas pulse disturbances can cause dramatic structural and functional shifts [8].

Besides the characteristics of the disturbance, the traits of the species and their biotic interactions are important determinants of community responses [9, 10]. However, the indirect impacts of disturbances caused by the biotic interactions are not well understood and are often overlooked. In particular, the trophic status in food webs plays a major role for the species response to disturbances [11–13]. Traits such as large body size, slow growth rate and

*Correspondence: antonis.chatzinotas@ufz.de
[1] Department of Environmental Microbiology, Helmholtz Centre for Environmental Research-UFZ, Permoserstraße 15, 04318 Leipzig, Germany
Full list of author information is available at the end of the article

low population size make top predators more vulnerable than other trophic levels. Studies have been shown that a reduced top-down control allowed prey outbreaks with cascading changes in ecosystem structure and function [14–16]. Similarly, in a microcosm study, increasing temperature led to increasing invasion success of a bacterial prey species due to the increased prey release from protozoan predation stress [17].

It is well known that long transient phases of population dynamics may occur in response to disturbances [18] and particularly strong or long-term disturbances may prolong these transient phases [19]. Among the ecological attributes known to affect transient recovery dynamics, the presence and availability of resources are particularly important [8]. It was previously hypothesized that the availability of alternative resources for the predator may increase the persistence of predator–prey systems [20]. Moreover, foraging behavior may be flexible and may change in disturbed environments [21]. Surprisingly, little is known about how the coupling of the predator to an alternative resource affects the recovery dynamics.

In this study, we combined microcosm experiments and modelling to investigate transient recovery dynamics of a simple microbial food web (consisting of predator, prey and a common resource). We exposed this system to disturbances, which we applied as increasing dilution rates. We contrasted two different disturbance regimes (i) a discrete and severe disturbance (pulse), and (ii) a long term and mild disturbance (press). We monitored the abundances of predator and prey before, during and after the disturbance.

In a second step we investigated using an ecological model the transient dynamics of both trophic levels under different disturbance strengths and durations beyond those applied in the experiments. In particular, we studied the consequences of the predator coupling to the alternative resource for the transient recovery dynamics. We found that disturbance strength and duration were decisive for the different transient recovery dynamics of the two trophic levels. In particular, we observed a slowed down recovery of the predator inducing a transient phase of prey release, i.e. temporarily high prey population sizes. Our results also revealed the importance of the predator coupling to an alternative resource which strongly impacted the recovery time of the predator and thus the length of the prey release phase.

Experimental methods and model
Origin and maintenance of stock cultures
The bacterium *E. coli* JM109 harboring a chromosomal green florescent protein (GFP) was used as prey organism. Using this strain allowed us to monitor *E. coli* in the food vacuoles of protists and facilitated controlling for contamination. A single clone grown on a lysogeny broth (LB) agar supplemented with 50 mg/ml kanamycin was used for establishing a pre-culture in liquid LB medium. Incubation was done in 50 ml medium in a 200 ml culture flask for 24 h on a closed rotating shaker at 25 °C. A low salt LB medium (1% tryptone, 0.5% yeast extract, 0.5% NaCl, 50 mg/ml kanamycin) was used for incubation of bacterial pre-cultures. Pre-cultures of the protist *Tetrahymena pyriformis* were established in proteose peptone yeast extract medium (1% proteose peptone, 0.15% yeast extract, 0.01 mM $FeCl_3$) at 25 °C in an incubator without shaking. These pre-cultures were cultivated axenically (i.e. growth on only dissolved nutrients without any bacteria) to avoid transfer of unwanted bacteria to the experimental cultures. *Tetrahymena pyriformis* is able to grow as a bacterivore (i.e. predating on bacterial prey) or as an osmotrophy (via direct uptake of dissolved nutrients). Prior to the experiments, pre-cultures of protists were concentrated by centrifugation ($1000g$, 10 min) and washed with experimental media twice. Both bacteria and protist pre-cultures were enumerated and diluted to the experimental concentrations with the experimental media. Enumeration techniques and all starting concentrations are described below. The *E. coli* JM109 and *Tetrahymena pyriformis* strain that were used in this work have been deposited at the public culture collection of the Department of Environmental Microbiology at the Helmholtz Centre for Environmental Research-UFZ (http://www.ufz.de/index.php?en=37703).

Experimental conditions
We used the above mentioned low-salt (in order to prevent salt damage on protists) LB medium during the experiments as the growth resource for the bacterial prey. The complex carbon source of the LB medium (i.e. yeast extract) served as an alternative resource for the predator. All experimental media were sterilized and filtered through a 0.2-μm pore sized filter. Experiments comprised 20 ml semi-continuous cultures in 50 ml sterile disposable culture flasks. Experimental cultures were always incubated at 25 °C for 24 h without shaking and all other environmental parameters were kept constant.

We found that a daily tenfold dilution prevented the collapse of the populations and resulted in an equilibrium state at which prey and predator coexist. This daily dilution went along with a replenishment of resources (i.e. LB medium) before they were depleted. It also reduced cell debris and excretion products and prevented oxygen depletion during the experiments. The remaining culture after each transfer was used for cell counts.

Experimental design

Three different treatments were applied: undisturbed (control), press disturbance and pulse disturbance. All treatments were replicated three times. All treatments were imposed by diluting the cultures with fresh medium as described below.

Undisturbed control

All replicate microcosms started with equal cell numbers of *E. coli* (3.6×10^7 cells ml^{-1}) and *Tetrahymena pyriformis* (4.2×10^4 cells ml^{-1}). Each day 2 ml from the cultures were transferred into 18 ml of fresh medium and allowed to re-grow for 24 h following this tenfold dilution.

Press disturbance

After control communities reached the equilibrium dynamics, they were exposed to the press disturbance from day 22 to 32 in separate flasks. Press disturbance was imposed as 40-fold daily dilution (simulating 4 times increased dilution rate compared to the daily constant rate) for a period of 10 days.

Pulse disturbance

Communities that had reached equilibrium dynamics were exposed to the pulse disturbance treatment on day 15. Pulse disturbance was applied as a single 2500-fold dilution (simulating a 250 times increased dilution rate). Initial cell numbers were lower than in the press experiment (i.e. 4×10^6 for bacteria and 4×10^3 for protists), but started with a similar predator: prey ratio as in the other treatments.

Sampling

A well-mixed 500 µl subsample was fixed with 0.2% Lugol's iodine solution for quantifying protists. Subsamples were diluted if the cells were too many to be counted reliably. Fixed protist cells were counted under an inverted microscope (Olympus CKX41, Olympus America Inc., Melville, NY, USA) with a Sedgewick-Rafter counting chamber (Pyser-SGI Limited, Edenbridge, UK). An additional 15 ml subsample was filtered through a 20 µm mesh filter (CellTrics, Sysmex Partec, Kobe, Japan) to remove protist cells prior to counting bacteria with a Coulter Counter (Multisizer 3, Beckman Coulter, Brea, CA, USA). Cell numbers were recorded every day.

Growth curves

Growth rates of prey and predator were determined by growing the organisms under the same experimental conditions for 24 h (i.e. without dilution). The triplicate cultures contained only prey, predator growing axenically without prey, and prey and predator together. Initial abundances of *E. coli* and *Tetrahymena* sp. were 4×10^6

and 2500 cells ml^{-1} respectively. Samples were taken with sterile syringes at 12, 14, 16, 18, 20, 22, 24 h. Protists and bacteria were counted as described above.

Modelling

We modelled the microcosm experiments as a time-discrete version of a Lotka–Volterra type predator–prey model [22]. Particularly, the model considers predator coupling to an alternative resource and the action of disturbances. Justified by experimentally determined growth curves (Additional file 1: Figure S1), we assumed a density limited prey population (P) and an exponentially growing predator (C).

$$P_{t+1} = (1 - d_t)P_t + r_P P_t \left(1 - \frac{P_t}{K_P}\right) - c_P P_t C_t \quad (1a)$$

$$C_{t+1} = (1 - d_t)C_t + r_C C_t + c_C P_t C_t \quad (1b)$$

where d_t is the dilution rate (applied once per 24 h), r_P is prey growth rate without predators, K_P is prey carrying capacity, c_P is the prey interaction coefficient describing how much prey is consumed per predator, r_C is the predator growth rate without prey and c_C is the predator interaction coefficient describing the consumption and conversion of prey to changes in C (Table 1). The model was implemented in R (version 3.1.3; [23]).

Note that the parameter r_C is important as it implicitly describes the coupling of the predator to another resource additionally to the prey population. Positive r_C imply coupling to this resource allowing the predator population to grow even in absence of prey. However, the model ignores a potential resource competition among predator and prey. Resource competition is unlikely, due to regularly strong dilution every 24 h. Dilution reduces the potential for resource competition in two ways: it removes predator and prey cells (i.e. reduces the amount of resource consumers) and it additionally renews the resource.

The model describes C and P as cells ml^{-1} and is iterated at a time step of 7.5 min, leading to 192 iterations per day. Initial tests showed that the step size was

Table 1 Parameter description and parameter values for the Eqs. (1a and 1b)

Name	Description	Value
d_t	Dilution rate	0.9 days^{-1}
r_P	Prey growth rate	0.094 (7.5 min)$^{-1}$
r_C	Predator growth rate	0.012 (7.5 min)$^{-1}$
K_P	Prey carrying capacity	4.9×10^8 cells ml^{-1}
c_P	Prey interaction coefficient	3.5×10^{-6} cells^{-1} ml
c_C	Predator interaction coefficient	1.4×10^{-11} cells^{-1} ml

sufficiently small to cover the experimental dynamics measured daily. Model results are displayed in daily time steps corresponding with experimental sampling times. For clarity, we left out the modelling time steps at a finer scale. Therefore, decline due to dilution and regrowth within the 24 h between dilutions are not visible.

To calibrate the model, initially we adjusted parameter values to the measured growth curves (see Additional file 1: Figure S1). Growth rates and prey carrying capacity were calibrated from the respective single species growth curves. Subsequently, interaction parameters were calibrated using the growth experiment with both species. We applied the Nelder–Mead optimization algorithm [24, 25] in R within reasonable wide parameter ranges. We then refined the parameter estimates by calibrating the model additionally to the control treatment. For this purpose, we applied a Latin hypercube approach on a narrow parameter space around the parameter estimates from growth curves. We then selected the parameter set that minimized the fourth power of the sum of relative distances to all cell counts in the control experiment. With the additional calibration to the control experiment we accounted for the possibility of uncontrolled changes in conditions between the separate growth and disturbance experiments.

Evaluation of results

We used the standard metric Nash–Sutcliffe efficiency (E) to quantify the general model efficiency in predicting the experimental data. E ranges between 1.0 (perfect fit) and $-\infty$. An E that is lower than 0 means that the mean value of the experimental data could be a better predictor than the model [26].

To specifically assess the differences between model and data during the first days after the start of the press or the occurrence of the pulse disturbance, we calculated the time of the species response to the disturbance by detectable abundance changes. Specifically, we defined "response time" as the time between the start of the disturbance and the day when species population size left the range of equilibrium sizes (they were calculated for the period from day 7 until disturbance start). Difference between the response times of the model and the data (average of replicates) is stated as "deviation time D_T". Deviations between the recovery times were calculated in the same manner as response time.

For the evaluation of prey release we calculated the covariance between prey and predator population sizes before, during and after disturbance. A negative covariance implies that prey population size strongly increases due to decreasing predator population size thus exhibiting prey release.

Simulation experiments

We applied the calibrated model to simulate situations that would have been difficult to directly control in the experiment. In simulation experiments, we tested the impact of (1) the duration of press disturbance, which we varied between 2 and 12 days, (2) the strength of pulse disturbance (varying between 10 and 10^6 on a 10-logarithmic scale, and (3) the strength of the predator coupling to the resource by varying parameter r_C in the range of 0.007–0.011. In these experiments, we particularly focused on speed of predator recovery, which we calculated in terms of "recovery time". Note that all source codes used in this manuscript are available upon request.

Results
Experimental population dynamics

In the control treatment, an equilibrium state appeared at which prey and predator coexisted (Fig. 1a). Under press disturbance, the prey population started to increase on day 26 reaching a higher equilibrium size than that of the control (Fig. 1b). At the end of the press disturbance, this high equilibrium population size remained for two more days and then turned back to the pre-disturbance size which was reached after full predator recovery on day 33 (at least two replicates were recovered). The predator population declined during press disturbance but started to increase during the disturbance period. After press disturbance ceased, the predator population recovered fully to its pre-disturbance size (Fig. 1b). Increasing negative covariance (*before disturbance cov* $= -0.002$; *during/after disturbance cov* $= -0.292$) indicated a phase of prey release during and after the disturbance (see "Evaluation of results").

Under pulse disturbance, the prey population increased already after one day as a consequence of the reduced predator population size (Fig. 1c). The prey population did not return back to the pre-disturbance level by the end of the experiment. The predator population continued to decline after the pulse but started to recover soon to the pre-disturbance size within 3 days (at least 2 replicates were recovered). Increasing negative covariance (*before disturbance cov* $= -0.003$; *during/after disturbance cov* $= -0.458$) indicated a phase of prey release after the pulse disturbance (see "Evaluation of results").

Modeled population dynamics

As we calibrated our model to the control treatment without disturbance, the fitted model reproduced well the non-disturbed experimental data (Fig. 1a). Also the overall response patterns to the press and pulse disturbances were captured well by the model (Fig. 1b, c). Nevertheless, the modeled population dynamics showed some slight discrepancies to the experimental

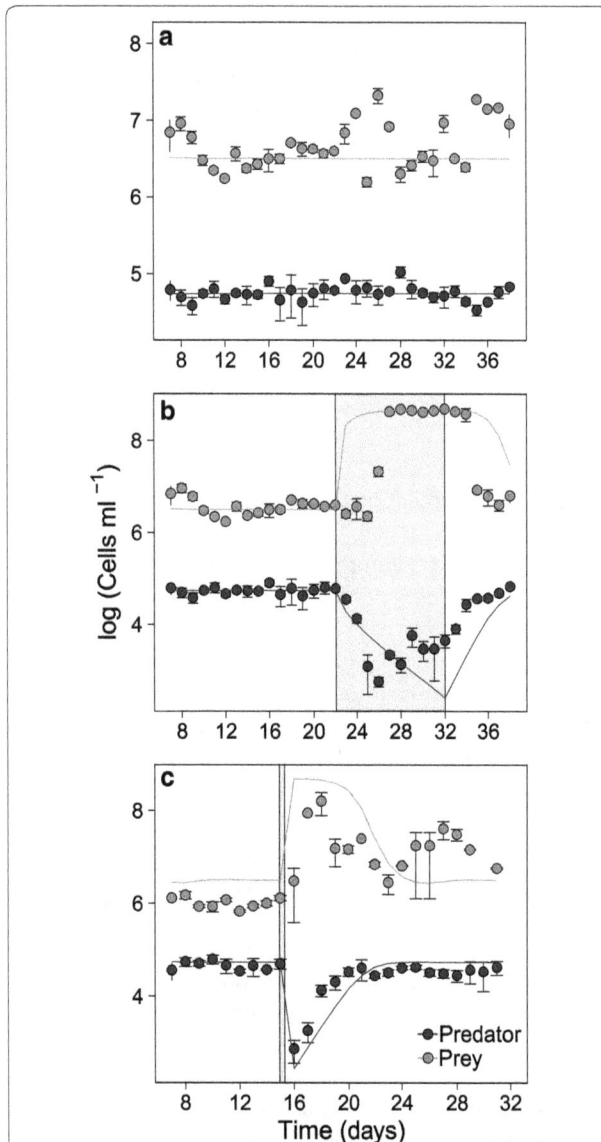

Fig. 1 Transient dynamics of predator and prey without and with disturbances. *Grey* and *dark blue filled circles* correspond to the experimentally determined mean predator and prey population dynamics respectively. *Error bars* represent ± standard deviation. *Solid lines* correspond to the model simulations (only daily time steps are shown). **a** Population dynamics without disturbance; **b** under press disturbance and **c** under pulse disturbance. Control without disturbance is with tenfold daily dilution, press disturbance corresponds to 40-fold dilution between the days 22–32 and pulse disturbance to 2500-fold dilution on the day 15. Disturbance action is shown as *grey shadows*

Specifically, some differences between modeled and experimental population dynamics occurred during the first days after the start (press) or the occurrence (pulse) of disturbance. Under press disturbance, the projected prey population showed an earlier response ($D_T = -3$) and late recovery ($D_T > 4.3$), (Fig. 1b). During the press disturbance, the experimental predator population started to increase already within the disturbance duration (around day 28), whereas the modelled population continuously declined, started to increase only after press disturbance ceased at day 32 ($D_T = -1$) and recovered later ($DT = 4.6$), (Fig. 1b).

Under pulse disturbance, the projected prey population size was slightly higher during the pre-disturbance and disturbance period. Experimental prey populations did not recover *until the end of the experiments* (see "Limitations and outlook"; Fig. 1c). Predator recovery to the equilibrium state was longer than in the experiments ($D_T = 4.3$).

Having found qualitatively similar community responses in the experiments and the simulations, we used the model to study more systematically the impact of press disturbance duration on predator (Fig. 2a) and prey (Fig. 2b), as well as pulse disturbance strength on predator (Fig. 2c) and prey (Fig. 2d).

The predator population declined stronger with both increasing press duration or pulse strength and recovered only slowly (Fig. 2a, c). Pulse and press disturbances resulted in a transient phase of decreased predator population sizes. With increasing disturbance impact, recovery times of the predator increased (Fig. 3a, b).

Note that, this was also valid for prey populations. However, it goes unnoticed on the daily sampling basis as the prey population recovered from disturbance within the 24 h sampling interval. Subsequently, it grew to higher population sizes, due to diminished predator stress (Fig. 2b, d). With increasing press duration and pulse strength, the chance increased that the prey population retained a high equilibrium population size for some time during or after the disturbance. Prey population size returned to the lower pre-disturbance equilibrium size only after significant recovery of the predator.

Impact of predator coupling to an alternative resource on predator and prey transient recovery dynamics

Changes in the coupling to the alternative resource r_C impacted predator (Fig. 2e) and prey (Fig. 2f) dynamics considerably under pulse disturbance and in a similar way also under press disturbance (see Additional file 1: Figure S2). As expected, lower values of r_C resulted in lower pre-disturbance equilibrium size of the predator (Fig. 2e) and an accordingly higher prey abundance. From these levels, disturbance reduced predator abundance according to the pulse strength. In contrast, prey grew to carrying

data. In the control treatment, predator dynamics ($E = -0.01$) were better predicted than the prey dynamics ($E = -0.32$). This is also true for the press disturbance ($E = 0.34$ *and* $E = 0.11$, respectively) and even more pronounced for the pulse disturbance ($E = 0.61$ *and* $E = -0.44$, respectively).

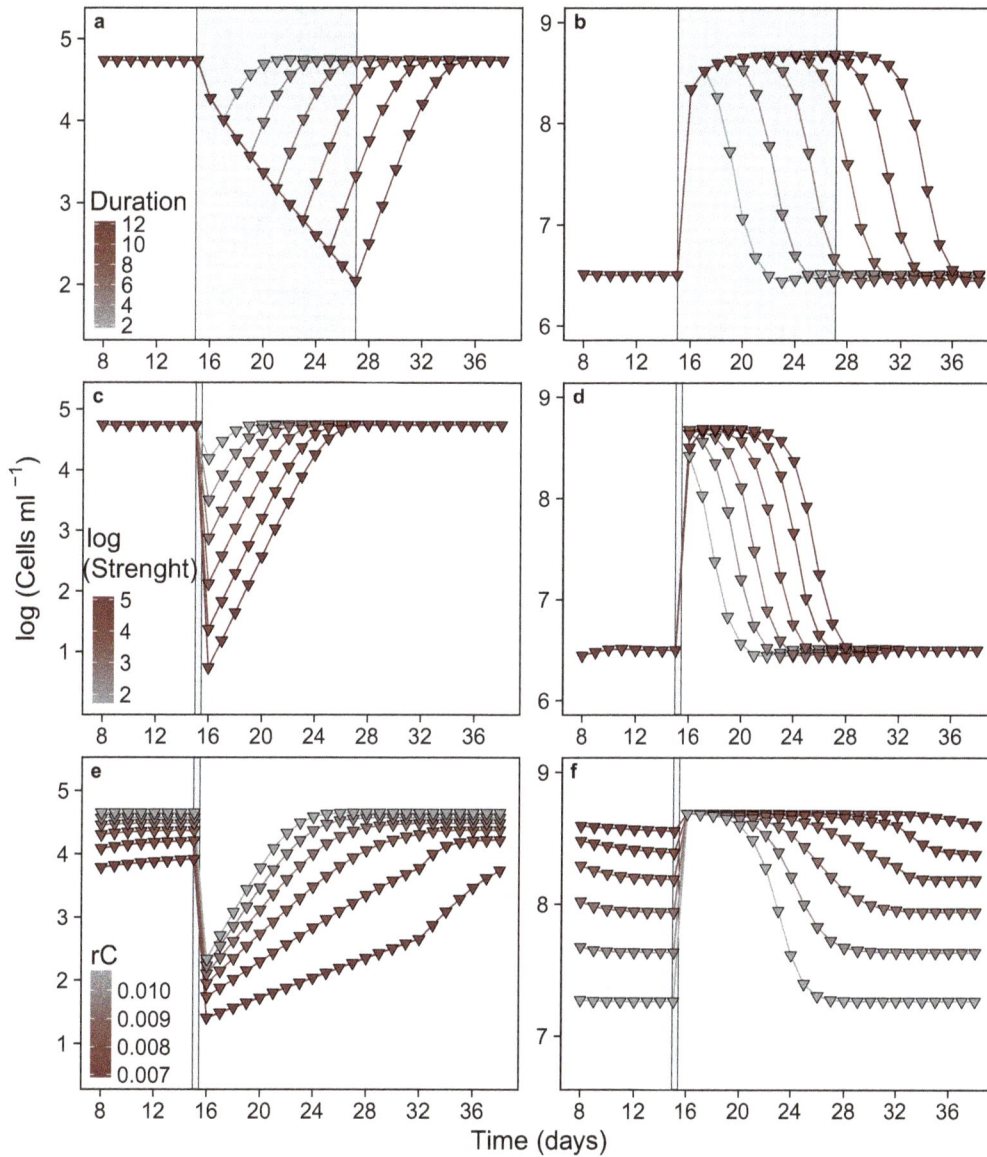

Fig. 2 Impact of disturbance duration, strength and predator coupling to an alternative resource on transient dynamics. **a, b** Impact of press disturbance duration (varied from 2 to 12 days) projected by the model simulations for predator (**a**) and prey (**b**). Strength of disturbance was kept constant (40-fold). Disturbance has started on day 15. *Color gradient* shows the shortest (*grey*) to the longest (*dark red*) disturbance duration. **c, d** Impact of pulse disturbance strength (varied from 50 to 100,000-fold) projected by the model simulations for predator (**c**) and prey (**d**). *Color gradient* shows the lowest (*grey*) to the highest (*dark red*) disturbance strength. **e, f** Impact of predator coupling r_C to an alternative resource besides prey (varied from 0.007 to 0.011) under pulse disturbance (2500-fold dilution) projected by the model simulations for predator (**e**) and prey a (**f**). *Color gradient* shows the lowest (*grey*) to the highest (*dark red*) r_C values. Disturbance action is shown as *grey shadows*

capacity (maximum size that the density dependent prey population can reach) within 24 h, independent of its pre-disturbance abundance (Fig. 2f). Recovery time of the predator extended significantly with decreasing r_C (Fig. 3c). Therefore, the prey population could retain its carrying capacity for a longer time (Fig. 2f).

Discussion

We found strong impacts of the strength and duration of disturbances on the transient dynamics and recovery time of predator and prey, and strong differences among the dynamics of the two species due to their position in the food web. In particular, our results revealed a slowed

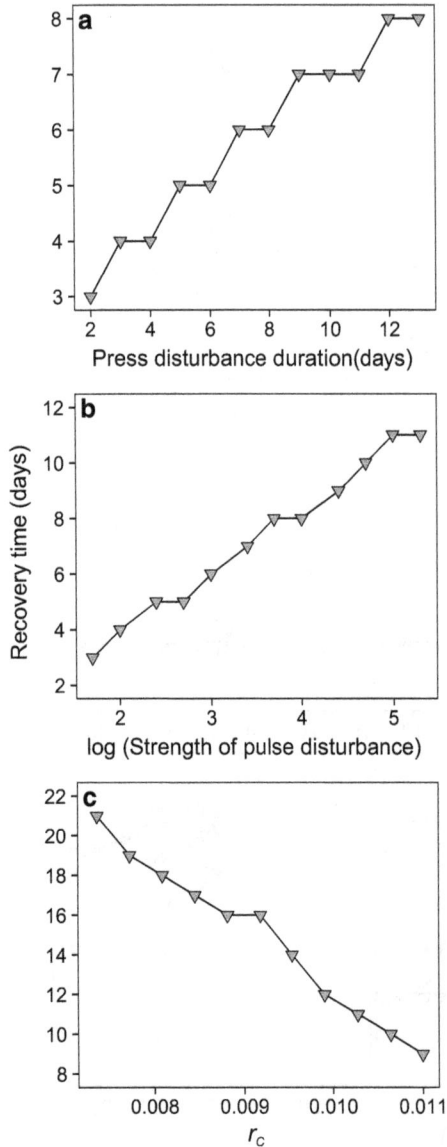

Fig. 3 Recovery time of predator depending on the disturbance duration, strength and resource coupling. We explicitly focused on predator recovery and defined the time a predator population needed to reach the pre-disturbance population size again as "recovery time". Recovery time is calculated as the duration (days) from the end of disturbance to the return to the pre-disturbance population size. **a** The dependence of recovery time of predator on the duration of press disturbance, **b** the strength of pulse disturbance and **c** changing resource coupling r_c resulting from the model simulations

Transient recovery dynamics of predator and prey may result in prey release

After disturbance ceased, the predator population recovered to pre-disturbance size (Fig. 2a, c). The respective recovery time was strongly related to the disturbance duration and strength (Fig. 3a, b). This finding is highly relevant, because prolonged recovery time, during which population size is low, comes along with increased extinction risk [27]. Extinction of top predators may cause radical changes in ecosystems by altering community structures [28, 29].

We found similar structural changes in our protist-bacteria system. The prey population size considerably increased during and after the disturbances due to missing predation pressure (Fig. 2b, d). This is a clear sign of prey release [16, 30]. Effectively, disturbance had uncoupled the two interacting species, such that the prey population was no longer relevantly affected by its predator. Prey release is common in systems with substantial disturbance on predators, e.g. by hunting [31]. For example, it was previously observed that the prey population release following the hydrological disturbance in a freshwater ecosystem was due to the reduced abundance of large sized predators [16]. A similar pattern has been also observed in an island ecosystem following a hurricane which reduced the abundance of top predators and caused herbivore outbreak [15]. We found that even if disturbance is affecting both species with equal mortality, as in our study, this can initiate prey release. The duration of this prey release depended on both the duration and the strength of the disturbance (Fig. 2b, d). Thus, even if a species is heavily impacted by disturbance (such as the bacterial prey), it might still benefit due to diminished competitor or enemy stress.

Predator coupling to an alternative resource is important for predator recovery and prey release

The use of an alternative resource is a known phenomenon for the studied protist. *Tetrahymena* species are able to grow on dissolved carbon sources and even fail to reduce the density of bacteria offered to them [32]. Foraging may be flexible due to specific predator traits such as absolute time or effort needed for grazing and relative intake rates, which, in turn impact the transient dynamics of the communities [21]. We found that a strong coupling to the alternative resource allowed the predator to reach higher pre-disturbance equilibrium sizes and accelerated the predator's recovery after the disturbances (Fig. 2e). Accordingly, weak couplings are advantageous for the prey (Fig. 2f) and may result in prey release as well. These results support previous findings on the importance of alternative resources for food web stability [33].

recovery of the predator from the disturbance inducing a temporary phase of prey release. The predator's coupling to an alternative resource was strongly impacting its own recovery time and thus also the length of the prey release phase. These general findings are discussed in the following in more detail.

Limitations and outlook

Despite its simplicity, our simulation model well reflects the transient dynamics of both predator and prey under pulse and press disturbance. Although this simplicity greatly facilitates a general understanding of the mechanisms, it has also drawbacks coming along with some mismatches between experimental data and model results.

As explained in the results section, the experimental prey population took longer to increase than indicated by the model (Fig. 1b, c) and reached lower values after pulse disturbance (Fig. 1c). Also, the experimental predator population already started to increase, while press disturbance was still impacting the community (Fig. 1b). These responses indicated a weaker impact of disturbance on the predator than expected from the model. We therefore tested the impact of an alternative resource across a range of coupling strengths as this could attenuate the impact of disturbances on the predator. We found that coupling of the predator to an alternative resource did clearly reduce its recovery time (Fig. 3c) but could not reproduce an increase of the predator population already during press disturbance (see Additional file 1: Figure S2). Stronger consumption of an alternative resource could be possible during a phase of increased dilution rates along with very low and high prey abundance. For future work, we suggest to relax the assumption of a constant coupling and to test coupling strengths dependent on prey density.

Another mismatch is that in contrast to model projection, the experimental prey population after pulse disturbance (Fig. 1c) did not completely return to the pre-disturbance equilibrium, but remained slightly elevated. Prey adaptation mechanisms such as cell aggregation and biofilm formation may cause this deviation and might provoke an alternative system state triggered by the disturbance [34–36].

It should also be taken into account that our simple Lotka–Volterra type model ignores possible predator satiation effects (Holling Type II and Type III non-linear functional responses) and assumes a linear functional response (Holling Type I without saturation). This is because the good fit of the L–V model to the experimentally measured predator and prey growth curves (Additional file 1: Figure S1C) indicates that predator's linear functional response describes the empirical data well and therefore density-dependent predation in form of non-linear functional responses is unlikely. However, given the discrepancies, especially during the prey release phase, one should investigate in future the applicability of non-linear functional responses. These investigations can be combined with the above described density dependent couplings to an alternative resource.

Conclusions

By combining experimental and modelling approaches we found that the interplay of disturbance attributes and food web structure determines the transient recovery dynamics of interacting species. This can lead to diverging population growth with one trophic level suffering and the other one profiting even if disturbance induces the same mortality. Most importantly, coupling of the predator to alternative resources may stabilize the community dynamics. These findings are essential for understanding how through changing disturbance attributes or creation of alternative resources (additional couplings) the transient food web dynamics can be changed to the benefit or harm of a species. These factors should therefore be taken into account in future food web studies. Taking a closer look at the impact of disturbances on species and communities and the resulting transient recovery dynamics might turn out to be pivotal in establishing intervention tools for conservation biology, biological control and epidemiology.

Authors' contributions
CK, KJ, AS and AC designed research; CK made the lab experiments, AS performed the simulations; AC and HH supervised the lab experiments; KJ supervised the theoretical part; CK, AS, KJ, HH and AC interpreted the results and wrote the paper. All authors read and approved the final manuscript.

Author details
[1] Department of Environmental Microbiology, Helmholtz Centre for Environmental Research-UFZ, Permoserstraße 15, 04318 Leipzig, Germany. [2] Department of Ecological Modelling, Helmholtz Centre for Environmental Research-UFZ, Permoserstraße 15, 04318 Leipzig, Germany. [3] Centre for Integrative Biodiversity Research (iDiv) Halle-Jena-Leipzig, Deutscher Platz 5e, 04103 Leipzig, Germany. [4] Present Address: Swedish Species Information Centre, Swedish University of Agricultural Sciences, P.O. Box 7007, 75007 Uppsala, Sweden.

Acknowledgements
We acknowledge Verena Jaschik and Jana Hoffkamp for their technical assistance in the laboratory. We thank Prof. Martin Schlegel (University Leipzig, Germany) and Prof. Kornelia Smalla (JKI Braunschweig, Germany) for kindly providing us the ciliate and the E. coli strain used in this study.

Competing interests
The authors declare that they have no competing interests.

Funding
This work was funded by the Helmholtz Association via the integrated project "Emerging Ecosystems" for the research topic "Land Use, Biodiversity, and Ecosystem Services" within the research program "Terrestrial Environment".

References

1. Turner MG. Disturbance and landscape dynamics in a changing world. Ecology. 2010;91:2833–49.
2. White PS, Jentsch A. The search for generality in studies of disturbance and ecosystem dynamics. Prog Bot. 2001;62:399–450.
3. Johst K, Gutt J, Wissel C, Grimm V. Diversity and disturbances in the antarctic megabenthos: feasible versus theoretical disturbance ranges. Ecosystems. 2006;9:1145–55.
4. White PS, Pickett S. Natural disturbance and patch dynamics. Cambridge: Academic Press; 1985.
5. Shade A, Read JS, Welkie DG, Kratz TK, Wu CH, McMahon KD. Resistance, resilience and recovery: aquatic bacterial dynamics after water column disturbance. Environ Microbiol. 2011;13:2752–67.
6. Shade A, Peter H, Allison SD, Baho DL, Berga M, Bürgmann H, et al. Fundamentals of microbial community resistance and resilience. Front Microbiol. 2012;3:417.
7. Benedetti-Cecchi L. Variance in ecological consumer–resource interactions. Nature. 2000;407:370–4.
8. Shade A, Read JS, Youngblut ND, Fierer N, Knight R, Kratz TK, et al. Lake microbial communities are resilient after a whole-ecosystem disturbance. ISME J. 2012;6:2153–67.
9. Fraterrigo JM, Rusak JA. Disturbance-driven changes in the variability of ecological patterns and processes. Ecol Lett. 2008;11:756–70.
10. dos Santos FAS, Johst K, Grimm V. Neutral communities may lead to decreasing diversity-disturbance relationships: insights from a generic simulation model. Ecol Lett. 2011;14:653–60.
11. Gallet R, Alizon S, Comte P, Gutierrez A, Depaulis F, van Baalen M, et al. Predation and disturbance interact to shape prey species diversity. Am Nat. 2007;170:143–54.
12. Murphy GEP, Romanuk TN. A meta-analysis of community response predictability to anthropogenic disturbances. Am Nat. 2012;180:316–27.
13. Wootton JT. Effects of disturbance on species diversity: a multitrophic perspective. Am Nat. 1998;152:803–25.
14. Dulvy NK, Freckleton RP, Polunin NVC. Coral reef cascades and the indirect effects of predator removal by exploitation. Ecol Lett. 2004;7:410–6.
15. Spiller DA, Schoener TW. Alteration of island food-web dynamics following major disturbance by hurricanes. Ecology. 2007;88:37–41.
16. Dorn NJ, Cook MI. Hydrological disturbance diminishes predator control in wetlands. Ecology. 2015;96:2984–93.
17. Liu M, Bjørnlund L, Rønn R, Christensen S, Ekelund F. Disturbance promotes non-indigenous bacterial invasion in soil microcosms: analysis of the roles of resource availability and community structure. PLoS ONE. 2012;7:e45306.
18. Hastings A. Transient dynamics and persistence of ecological systems. Ecol Lett. 2001;4:215–20.
19. Steiner CF, Klausmeier CA, Litchman E. Transient dynamics and the destabilizing effects of prey heterogeneity. Ecology. 2012;93:632–44.
20. van Baalen M, Křivan V, van Rijn PC, Sabelis MW. Alternative food, switching predators, and the persistence of predator-prey systems. Am Nat. 2001;157:512–24.
21. Abrams PA. Implications of flexible foraging for interspecific interactions: lessons from simple models. Funct Ecol. 2010;24:7–17.
22. Begon M, Townsend CR, Harper JL. Ecology: from individuals to ecosystems. Malden: Blackwell Pub; 2006.
23. R Core Team. R: a language and environment for statistical computing [Internet]. Vienna: R Foundation for Statistical Computing; 2015. http://www.R-project.org/.
24. Nelder JA, Mead R. A simplex method for function minimization. Comput J. 1965;7:308–13.
25. Bolker BM. Ecological models and data in R. New Jersey: Princeton University Press; 2008.
26. Krause P, Boyle DP, Bäse F. Comparison of different efficiency criteria for hydrological model assessment. Adv Geosci. 2005;5:89–97.
27. Caughley G. Directions in conservation biology. J Anim Ecol. 1994;63:215.
28. Johnson CN, Isaac JL, Fisher DO. Rarity of a top predator triggers continent-wide collapse of mammal prey: dingoes and marsupials in Australia. Proc R Soc Lond B Biol Sci. 2007;274:341–6.
29. Boersma KS, Bogan MT, Henrichs BA, Lytle DA. Top predator removals have consistent effects on large species despite high environmental variability. Oikos. 2014;123:807–16.
30. Müller CB, Brodeur J. Intraguild predation in biological control and conservation biology. Biol Control. 2002;25:216–23.
31. Ritchie EG, Johnson CN. Predator interactions, mesopredator release and biodiversity conservation. Ecol Lett. 2009;12:982–98.
32. Fox JW, Barreto C. Surprising competitive coexistence in a classic model system. Community Ecol. 2006;7:143–54.
33. Post DM, Conners ME, Goldberg DS. Prey preference by a top predator and the stability of linked food chains. Ecology. 2000;81:8–14.
34. Yoshida T, Jones LE, Ellner SP, Fussmann GF, Hairston NG. Rapid evolution drives ecological dynamics in a predator–prey system. Nature. 2003;424:303–6.
35. Liess A, Diehl S. Effects of enrichment on protist abundances and bacterial composition in simple microbial communities. Oikos. 2006;114:15–26.
36. Friman V-P, Laakso J, Koivu-Orava M, Hiltunen T. Pulsed-resource dynamics increase the asymmetry of antagonistic coevolution between a predatory protist and a prey bacterium. J Evol Biol. 2011;24:2563–73.
37. Karakoç C, Singer A, Johst K, Harms H, Chatzinotas H. Data from: transient recovery dynamics of trophically interacting species under press and pulse disturbances. Dryad Digit Repos. 2017. doi:10.5061/dryad.1gq66.

3

Hunting as a management tool? Cougar-human conflict is positively related to trophy hunting

Kristine J. Teichman[1,2*†], Bogdan Cristescu[3†] and Chris T. Darimont[1,4,5]

Abstract

Background: Overexploitation and persecution of large carnivores resulting from conflict with humans comprise major causes of declines worldwide. Although little is known about the interplay between these mortality types, hunting of predators remains a common management strategy aimed at reducing predator-human conflict. Emerging theory and data, however, caution that such policy can alter the age structure of populations, triggering increased conflict in which conflict-prone juveniles are involved.

Results: Using a 30-year dataset on human-caused cougar (*Puma concolor*) kills in British Columbia (BC), Canada, we examined relationships between hunter-caused and conflict-associated mortality. Individuals that were killed via conflict with humans were younger than hunted cougars. Accounting for human density and habitat productivity, human hunting pressure during or before the year of conflict comprised the most important variables. Both were associated with increased male cougar-human conflict. Moreover, in each of five regions assessed, conflict was higher with increased human hunting pressure for at least one cougar sex.

Conclusion: Although only providing correlative evidence, such patterns over large geographic and temporal scales suggest that alternative approaches to conflict mitigation might yield more effective outcomes for humans as well as cougar populations and the individuals within populations.

Keywords: British Columbia, Mountain lion, Predator-human coexistence, Puma, *Puma concolor*, Skull size, Trophy hunting, Wildlife

Background

Exploitation and persecution related to conflict with humans form major causes of predator declines worldwide [1–4]. Killing takes several forms and its ecological and evolutionary effects might be more severe than the number of removed predators suggests [5]. Expansion of human activities into previously undisturbed areas enables increased killing through facilitated human access; roads, cut lines and trails associated with extractive industries facilitate hunting of predators during and/or after resource extraction [6, 7]. As human populations expand, the likelihood of wildlife-human conflict also increases [8].

When conflicts involve large mammalian predators that pose a perceived or real threat to humans and property, a common outcome is the lethal removal of the predator by management agencies or sometimes by land owners, for example in response to predation on livestock [9]. In addition, conflict is often managed through increasing human-caused killing of carnivores, under the premise that human hunting can reduce conflict incidence over depredation or decrease predation on wild ungulates sought by hunters (hereafter, 'human hunting hypothesis'; e.g., [10–12]).

In the case of predator-human conflict over depredation, Treves and Naughton-Treves [13] suggested that

*Correspondence: kristine.teichman@ubc.ca
†Kristine J Teichman and Bogdan Cristescu contributed equally to this work
[1] Department of Geography, University of Victoria, PO Box 3060, STN CSC, Victoria, BC V8W 3R4, Canada
Full list of author information is available at the end of the article

carnivore killing by hunters may actually promote conflict. The process is thought to operate via shifts in age composition to younger age animals, which might depredate more because of higher encounter rates with livestock. This process is thought to occur via the increased mobility of juvenile age classes of carnivores caused by decline in adult male territory tenure [14]. Young individuals become locally more abundant and thereby have increased chance of encountering livestock—and/or young animals might be bolder, more curious or lacking experience in interactions with people [15] or in capturing wild prey effectively [16]. Collectively these factors suggest that younger animals are more conflict-prone (hereafter, 'young animal hypothesis'). Moreover, hunting, culling or other lethal control targeted at specific individuals (e.g. those involved in livestock predation) may reduce conflict ('problem individuals hypothesis'; e.g., [16]), which has been challenged by the assertion that dispersing individuals often quickly recolonize conflict areas, offering only temporary relief [17].

To confront these hypotheses, we examined a long-term dataset on human hunting of cougars and conflict involving cougars in BC, Canada. Cougar-human conflict and cougar hunting are relatively widespread and common, the latter attracting both local BC hunters as well as foreign hunters for guided hunts. We used this system to test whether: (1) cougars killed by hunters would be larger than those that came into conflict with people (young animal hypothesis); and (2) human hunting mortality and conflict incidence would be related (problem animal and human hunting hypotheses).

Methods
Cougar data
We used a 30-year dataset (1979–2008) on recorded cougar mortality in BC, Canada provided by the BC Ministry of Environment, wherein all records had an associated date. We used cougar kill records resulting from conflict and legal hunting events. For analyses involving age of conflict and legally hunted cougars [(1) above] we used only those records with associated spatial data, sex and skull sizes. The other analyses [(2) above] were carried out using the larger dataset of spatially-referenced conflict and legal hunting mortalities of cougars with known sex, irrespective of whether skull size had been recorded. Only 96 illegal kills were recorded during 1979–2008, of which 35 had associated skull length and width data and these were not used in analyses. We consider this a minimum estimate because evaluations of the frequency of illegal cougar kills have not been performed. We do not expect illegal killing to vary across regions. Additional spatially-referenced mortality records of cougars with known sex (356, of which 139 had associated skull

information) had unclear or unrecorded cause of death and were not used in analyses.

Spatial data included universal transverse mercator (UTM) coordinates and we considered only conflict and legal hunting records occurring within the 5 of 8 total 'development regions' of BC (region size mean ± SE, $72,173 \pm 19,388$ km^2) in which mortality was highest (Cariboo, Kootenay, Lower Mainland South-West (SW), Thompson Okanagan and Vancouver Island). After plotting kill locations by region in ArcGIS v.10.3 (ESRI, Redlands, USA) for validation and discarding records occurring outside the 5 regions or in water, as well as a small number of erroneous records (e.g., skull width > skull length), the final dataset for cougar age analysis consisted of 3665 records. The data included records of kills by BC resident hunters and non-resident guided hunters ($n = 3219$) as well as conflict-related cougar deaths ($n = 449$). 'Conflict' was defined as any incident of cougar road mortality, predation on livestock, perceived risk to people such as cougars sighted in urban areas, or recorded attack on humans. More male ($n = 2240$) than female ($n = 1428$) mortality records occurred in the data. The larger dataset for analysis of cougar conflict in relation to human hunting levels included 8788 records. The data were dominated by hunting mortalities ($n = 7550$), with conflict-related kills being less frequent ($n = 1238$). The dataset had more male ($n = 5348$) than female records ($n = 3440$).

Skull size data (length and width in mm) were collected by BC Ministry of Environment personnel as a proxy for age. These variables are positively correlated [18] with the skull growth continuing long into adulthood [19]. Skull size has been used as a proxy for age/body size in other large felids, such as African lion [20], leopard [21] and jaguar [22]. Because skull length and width were highly correlated for males (Pearson $r = 0.761$, $df = 2239$, $P < 0.001$) and females ($r = 0.669$, $df = 1427$, $P < 0.001$), we used an index known as the total skull length (or total skull size) for all analyses. This index is the sum of length and width [22] and is the standard age/body/trophy size metric used by the Boone and Crockett Club and the International Council for Game and Wildlife Conservation when assessing cougar and jaguar trophies [23].

Statistical analyses
To assess if skull sizes varied in relation to different human-caused mortality types, we first assessed if the variable was normally distributed with Shapiro–Wilk tests. Separate assessments were carried out for each sex and region. For males and females in all regions, the skull size variable was not normally distributed. Therefore we used two-sample Wilcoxon rank-sum (Mann–Whitney) tests to compare mean skull size for conflict and hunter

kills. Separate testing was performed for each sex and region for a total of 10 tests (2 sexes × 5 regions).

We used time series analysis to test factor combinations hypothesized a priori to influence annual conflict frequency across time (Additional file 1: Table S1). Newey-West Heteroskedasticity and Autocorrelation (HAC) standard errors were computed in multiple linear regression to account for potential variability and temporal autocorrelation in the models' error terms. Conflict incidence (dependent variable) was standardized per 10,000 km^2 and square root-transformed prior to modelling to reduce skewness. Predictor variables included human density (D), human hunting pressure (annual number of cougars hunted) in the year of conflict (H_{t0}) and the Normalized Difference Vegetation Index (NDVI; a proxy for plant and prey productivity) in the year of conflict (N_{t0}). A squared term was included for human density (D^2) to account for possible thresholds in human density beyond which conflict would decrease because of an assumed limitation to cougar habitat. Yearly lag 1 and 2 terms were used for human hunting pressure (H_{t1}; H_{t2}) and NDVI (N_{t1}; N_{t2}) to incorporate potential influences of hunting and habitat productivity in the periods preceding conflict. Human density (per 10,000 km^2) was calculated for each year by dividing annual census counts by region size (details in Additional file 2). Human density calculation for the Vancouver Island region included a small part of the mainland coast as constrained by data availability. Human hunting pressure was standardized per 10,000 km^2 and included hunting by residents and non-residents of BC. Because habitat quality and prey availability can influence large carnivore-human conflict [24, 25], but such data over our broad temporal and spatial extents were not available, we used NDVI as a habitat productivity surrogate [26–28]. These data came from the National Oceanic and Atmospheric Administration (NOAA) Climate Data Records (CDR), which derived NDVI from surface reflectance data acquired by the advanced very high resolution radiometer (AVHRR) sensor ([29]; details in Additional file 2). Highly correlated variables ($r > |0.8|$) were not included together in the same model structure..

We evaluated candidate models using Akaike's Information Criterion for small sample sizes (AICc) [30]. We estimated relative importance of variables by applying multi-model inference to rank variables in the supported model set ($\Delta AICc \leq 7$) by their summed AICc weights [31]. We used the proportion of variance explained (R^2) to evaluate model performance. For all models that received support we plotted residuals against fitted values and inspected for patterns in the residual distribution. We used Stata v.14.1 (StataCorp, College Station, USA) and an alpha level of 0.10 for all statistical analyses.

The Newey-West HAC standard errors were computed in Stata using the hacreg command [32].

Results

Skull size comparisons between hunter- and conflict-killed cougars

At the provincial level, conflict-killed male cougar skulls were smaller than those of hunter-killed animals (Two-sample Wilcoxon rank-sum $z = -5.376$, $df = 2239$, $P < 0.001$). Skull sizes differed between kill types for males in 4 of the 5 BC regions, similarly larger for hunter-killed than for conflict-killed males for Cariboo (Two-sample Wilcoxon rank-sum $z = -1.959$, $df = 329$, $P = 0.050$), Lower Mainland SW (Two-sample Wilcoxon rank-sum $z = -2.195$, $df = 113$, $P = 0.028$), Thompson Okanagan (Two-sample Wilcoxon rank-sum $z = -2.210$, $df = 763$, $P = 0.027$) and Vancouver Island (Two-sample Wilcoxon rank-sum $z = -2.762$, $df = 571$, $P = 0.006$) (Fig. 1a).

At the provincial level, skull sizes of females were similarly smaller among conflict animals compared with hunter-killed individuals (Two-sample Wilcoxon rank-sum $z = -3.464$, $df = 1427$, $P < 0.001$). Skull sizes likewise differed between kill types in 2 of the 5 BC regions (Lower Mainland SW; Two-sample Wilcoxon rank-sum $z = -1.701$, $df = 114$, $P = 0.089$; Thompson Okanagan; Two-sample Wilcoxon rank-sum $z = -4.311$, $df = 520$, $P < 0.001$) (Fig. 1b).

Predictors of cougar-human conflict

Regional models (see Additional file 3) that received substantial support explained roughly half of the variation in cougar-human conflict for males (R^2: mean = 0.504; range = 0.258–0.816; all $P < 0.10$) as well as females (R^2: mean = 0.507; range = 0.124–0.772; all $P < 0.10$). For both sexes, models that received substantial support were of intermediate or low complexity (with 1–4 parameters, including the intercept; Table 1). Only for males in the Lower Mainland SW did the intercept-only model receive substantial support, but two candidate models were superior. All supported models ($\Delta AICc \leq 7$) [33] are listed in Additional file 4: Tables S2–S6, provided their $\Delta AICc$ was smaller than that of the corresponding null model.

Human hunting pressure in both current (Figs. 2, 3) and lagged periods (Fig. 2) had the most relative importance for predicting cougar-human conflict for male cougars across the five regions. Human hunting was positively associated with conflict involving this cougar sex. Variables for human hunting during the conflict year or hunting lagged occurred in all but one male model that received substantial support and which had AICc less than the null model's AICc (Table 1).

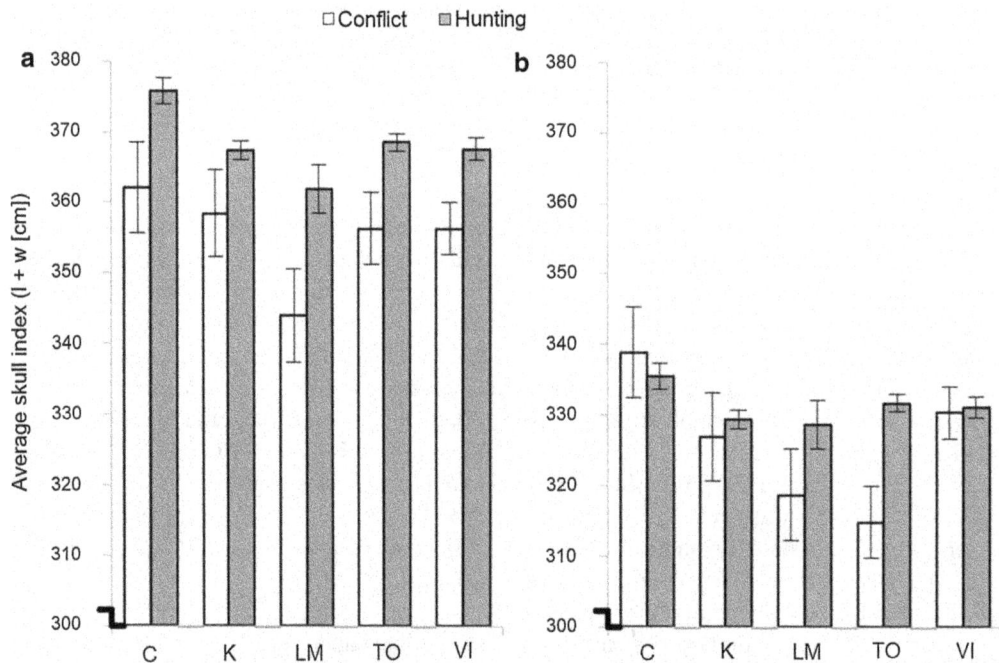

Fig. 1 Average skull sizes of cougars killed in five regions of British Columbia, Canada, as a result of conflict and human hunting. Data include kill records with associated geographic coordinates and age (skull size) information for **a** males and **b** females. BC regions are: *C* Cariboo, *K* Kootenay, *LM* Lower Mainland SW, *TO* Thompson Okanagan and *VI* Vancouver Island. *Error bars* represent ± 1 SE. Note broken *Y axis*

Table 1 Models for assessing temporal patterns of cougar-human conflict in British Columbia, Canada that received substantial support (ΔAICc < 2)

Region	Sex	Model description	ΔAICc	w_{AICc}	R^2
Cariboo	Male	$D + D^2 + H_{t0}$	0.0	0.33	0.557
		$D + D^2$	1.0	0.20	0.456
		H_{t0}	1.1	0.19	0.369
	Female	$D + D^2$	0.0	0.64	0.599
Kootenay	Male	$N_{t0} + H_{t0}$	0.0	0.82	0.816
	Female	$D + D^2$	0.0	0.48	0.736
		$N_{t0} + D + D^2$	0.3	0.42	0.772
Lower Mainland SW	Male	$H_{t1} + H_{t2}$	0.0	0.19	0.258
		$H_{t0} + H_{t1} + H_{t2}$	0.6	0.14	0.347
	Female	$D + D^2$	0.0	0.46	0.334
Thompson Okanagan	Male	$D + D^2 + H_{t0}$	0.0	0.51	0.590
	Female	$H_{t0} + H_{t1} + H_{t2}$	0.0	0.30	0.406
		H_{t0}	1.8	0.12	0.124
		$D + D^2$	1.8	0.12	0.236
Vancouver Island	Male	$H_{t1} + H_{t2}$	0.0	0.35	0.539
		$H_{t0} + H_{t1} + H_{t2}$	0.2	0.32	0.602
	Female	H_{t0}	0.0	0.50	0.668
		$N_{t0} + H_{t0}$	1.6	0.23	0.688

D human density, H_{t0} human hunting pressure, H_{t1} Human hunting pressure (lag 1), H_{t2} human hunting pressure (lag 2), N_{t0} NDVI, N_{t1} NDVI (lag 1), N_{t2} NDVI (lag 2)

Human hunting pressure was also the most important factor associated with cougar-human conflict for female cougars in 2 of 5 BC regions. Only for one model (female cougars, Thompson-Okanagan) was increased human hunting (lag 2) associated with decreased conflict.

Overall, increased human hunting was related to greater conflict for 16 of 17 models that included hunting variables with estimates that did not overlap zero and that received substantial support (Table 2; Additional file 4: Tables S7–S11).

Human density was the key variable associated with conflict for female cougars in 3 BC regions (Fig. 2) and was also important for male cougar-human conflict in 1 BC region (Table 2). Years with intermediary human densities were generally associated with conflict (Additional file 4: Tables S9, S10). For both cougar sexes, NDVI was the least important variable tested in relation to conflict (Fig. 2), but three substantially supported models revealed conflict increases in years when habitat productivity was low (Table 2).

Discussion

With expanding human populations and influence, conflict between carnivores and humans is expected to increase, which requires evidence-informed approaches to conflict mitigation. A long-term data set on human-caused cougar mortality allowed us to confront

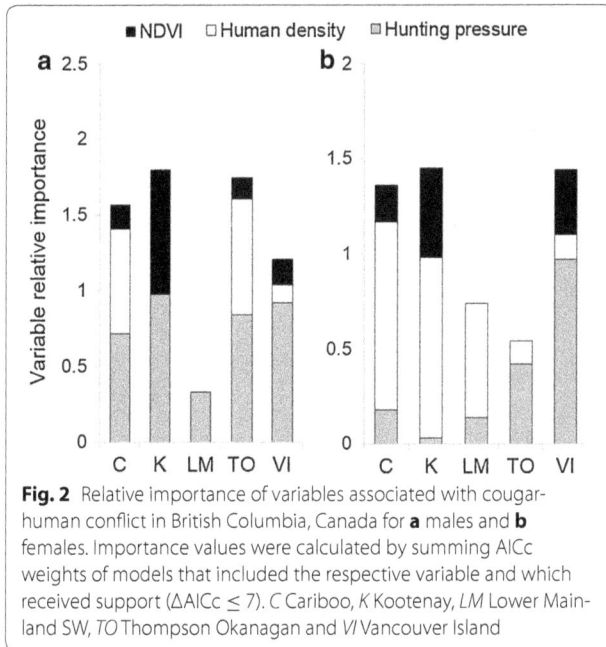

Fig. 2 Relative importance of variables associated with cougar-human conflict in British Columbia, Canada for **a** males and **b** females. Importance values were calculated by summing AICc weights of models that included the respective variable and which received support (ΔAICc ≤ 7). *C* Cariboo, *K* Kootenay, *LM* Lower Mainland SW, *TO* Thompson Okanagan and *VI* Vancouver Island

fundamental hypotheses on the relationship between human hunting, cougar-human conflict and cougar population demography, including testing of the commonly accepted but under-examined assumption that hunting of large carnivores could result in decreased conflict incidence (see [34] for an overview and call for inquiry into the relationship between hunting of carnivores and conflict).

As we expected, we found support for the young animal hypothesis in most comparisons, with individuals that came into conflict with humans younger compared to those hunted. Human encroachment into cougar habitat increases conflict potential [35–37] and young animals are more likely to occur in areas used by people

than other age classes [38]. Dispersing juveniles are more likely to come into conflict on travel routes through fragmented habitats and high risk areas including human inhabited areas, roads [39] and ranches [24, 40]. In addition, food resources may be limited while dispersers attempt to establish home ranges [41]. As a result, when available, cougars may attack livestock [42] (however, see [43] for an alternative documentation of old cougars being disproportionately involved in livestock predation). Finally, hunters might be more likely to forgo killing small individuals for trophies, particularly if they are treed by trained hounds, although this has not been examined.

The manner by which carnivore populations respond to regulated hunting depends on social structure, reproductive strategies and dispersal patterns [14]. Human hunting of old individuals can increase immigration of juveniles from neighboring areas [14, 44], which could result in increased conflict. We therefore hypothesized that increased human hunting pressure would be associated with increased conflict via social disruption and younger population age structure (problem animal and human hunting hypotheses). We demonstrated that high hunting-related mortality in the same or preceding time period is positively associated with cougar-human conflict for at least one sex in all five regions tested (Table 2; Figs. 2, 3), with the most consistent pattern (both sexes: regression $P < 0.10$) for Thompson-Okanagan and Vancouver Island. While Thompson-Okanagan is an inland region, Vancouver Island is a large land mass off the British Columbia mainland known to be the world's 'hotspot' of cougar-human conflict [45]. Our results corroborate and extend recent findings on impacts of human hunting on cougar complaints and depredations in Washington State [46]. In British Columbia, male cougars appeared most susceptible to conflict if hunted more intensively and conflict records involving males were almost double

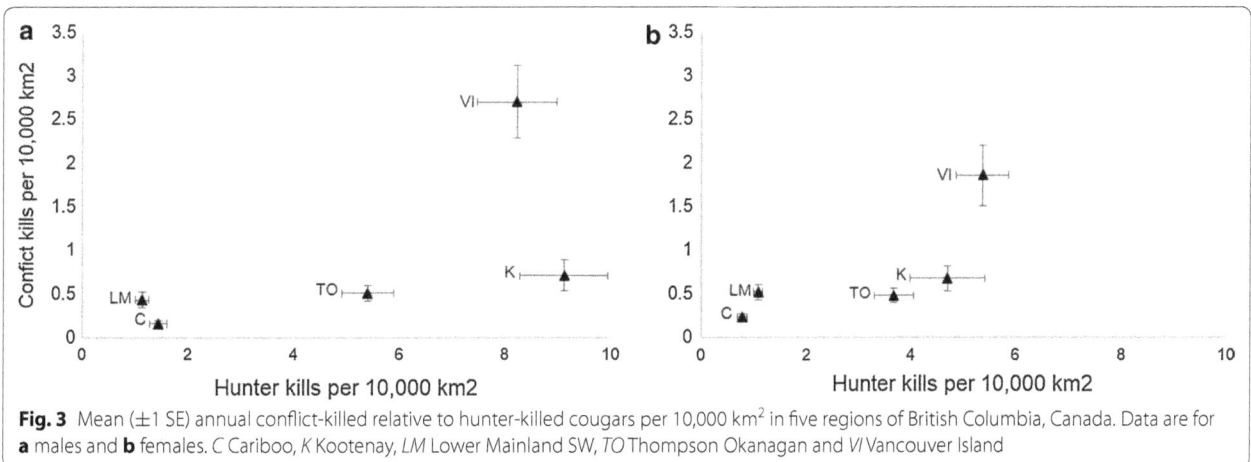

Fig. 3 Mean (±1 SE) annual conflict-killed relative to hunter-killed cougars per 10,000 km² in five regions of British Columbia, Canada. Data are for **a** males and **b** females. *C* Cariboo, *K* Kootenay, *LM* Lower Mainland SW, *TO* Thompson Okanagan and *VI* Vancouver Island

Table 2 Direction (+ positive, − negative) and confidence interval overlap with zero for parameter estimates from substantially supported ΔAICc models for cougar-human conflict in British Columbia, Canada

Region	Sex	N_{t0}	N_{t1}	N_{t2}	D	D^2	H_{t0}	H_{t1}	H_{t2}
Cariboo	Male				++	−−	+*+*		
	Female				+	−			
Kootenay	Male	−*					+*		
	Female	−*			−*−*	+*+*			
Lower Mainland SW	Male						+*	+*+*	−−
	Female				+*	−*			
Thompson Okanagan	Male				+*	−*	+*		
	Female				+*	−*	+*+*	−	−*
Vancouver Island	Male						+*	+*+*	+*+*
	Female	−*					+*+*		

Estimates for which confidence intervals did not overlap zero have an asterisk. No reporting of coefficients refers to the specific variable(s) not being included in supported models

D human density, H_{t0} human hunting pressure, H_{t1} Human hunting pressure (lag 1), H_{t2} human hunting pressure (lag 2), N_{t0} NDVI, N_{t1} NDVI (lag 1), N_{t2} NDVI (lag 2)

in number than those involving females. The latter findings are in accordance with Linnell et al.'s conclusion that male large carnivores are most likely to get into conflict with humans [16], a proposition also more recently supported by research on cheetah-human [47] and jaguar-human conflicts [48]. One mechanism that might explain why males of hunted cougar populations are involved more frequently in conflicts than females might be the altered male spatial organization under greater hunting pressure [49].

Human densities were associated with male cougar-human conflict in only one BC region, whereas conflict with females appeared related to variation in human density. Females might use suboptimal areas with human development by means of spatially avoiding male-caused mortality risk for themselves and their offspring, possibly resulting in increased conflict for females in connection to human densities, as we detected. Selection of areas close to human development by females with offspring presumably to avoid males has been recently documented for cougars in California [50] and grizzly bears in Alberta [51]. Thompson-Okanagan was the only region where human density was related with conflict for both sexes, with conflicts most likely at intermediary densities of people. Such intermediate densities are typically found in exurban or suburban areas and are thought to have high levels of cougar-human conflict in California [52]. Despite high human populations in Lower Mainland SW, human density in this region did not influence frequency of conflict involving males. The documented decreases in conflict associated with decreased human hunting of males in this region suggest that, similar to other carnivores [53], cougar populations can persist in regions with high human densities as long as human hunting pressure is low.

We found limited relationship with NDVI, our proxy for habitat productivity. Decreased productivity was hypothesized to be associated with increased cougar-human conflict. Conversely, a positive relation between conflict and NDVI might have been expected due to increased productivity resulting in increased reproductive output [54], with the indirect effect of increased sub-adult dispersal and greater conflict potential. Kootenay was the only region where decreased productivity was associated with increased conflict for both males and females. This region comprises substantial high elevation mountain ranges compared to the other regions and habitat productivity in the Kootenay is possibly an important limiting factor for cougars and their prey. Future monitoring of the associations between habitat productivity and carnivore-human conflict should not be neglected, given increased variability in vegetation conditions/NDVI associated with climate change, which might have implications for future predator-human conflicts that have yet to be explored. When possible, finer scale prey availability metrics should be incorporated, because prey use differences among cougar sexes [55] could influence conflict incidence. Furthermore, it is important to recognize that inferences from this study should be placed in the context of the relative coarseness of covariate data utilized, which is to be expected when focusing on broad spatiotemporal extents such as the one we considered. Our results showed that human-related variables had the strongest association with conflict. We acknowledge that the patterns of association we reveal do not necessarily imply causation. Our results, however, are generally consistent with the hypothesis that high hunter mortality leads to young animals becoming involved in conflict. Unlike natural agents of mortality (other predators, competitors, disease), hunters typically target adult individuals. The ability of resident males to maintain territories means that sub-adults are more likely to come into conflict, likely because of their movements during dispersal in search for vacant territories [56]. Human hunting can disrupt social structure leading to increased juvenile immigration from surrounding source populations [14] and result in younger age structure [57, 58] exacerbating conflicts between cougars and humans. With increasing human populations, interactions between predators and humans are expected to become more common, underlining the need for research into patterns and mechanisms of conflict, conflict prevention and non-traditional management strategies to facilitate coexistence.

Conclusions

Wildlife managers often prescribe hunting of carnivores to reduce competition with hunters for prey and to minimize conflicts with humans and their property [8]. If lethal control such as through human hunting is to facilitate coexistence between wildlife and humans, control must minimize wildlife-human conflict or increase tolerance of the public towards wildlife, without compromising wildlife population viability [13]. In some situations lethal management focused on targeted individuals associated with conflict (e.g., individuals that injure or kill people in predatory attacks) offers one route to address large carnivore-human conflicts. However, we showed that overall increased human hunting in fact can be associated with increased conflict, especially for males. Although our results are only correlative, we caution against the universal use of hunting as a tool for managing conflict with large predators.

Additional files

Additional file 1: Table S1. Hypotheses for frequency of cougar-human conflict.

Additional file 2. Information on NDVI and human density data.

Additional file 3. Information on regional models of cougar-human conflict.

Additional file 4: Tables S2–S11. List of all cougar-human conflict models that received support (ΔAICc $\leq$ 7). Estimated coefficients from substantially supported (ΔAICc < 2) models are also listed.

Abbreviations

AICc: Akaike's information criterion for small sample sizes; AVHRR: advanced very high resolution radiometer; BC: British Columbia; CDR: climate data records; D: human density; HAC: heteroskedasticity and autocorrelation; H_{t0}: human hunting pressure; H_{t1}: human hunting pressure (lag 1); H_{t2}: human hunting pressure (lag 2); NDVI: normalized difference vegetation index; NOAA: national oceanic and atmospheric administration; N_{t0}: NDVI; N_{t1}: NDVI (lag 1); N_{t2}: NDVI (lag 2); SW: south–west; UTM: universal transverse mercator; ΔAICc: difference between the model AICc and the lowest AICc for the model set.

Authors' contributions

KJT and BC conceived and designed the study and CTD helped to refine it. KJT obtained the cougar mortality dataset from the British Columbia Ministry of Environment. KJT and BC performed the statistical analyses. KJT, BC and CTD wrote the manuscript. All authors reviewed and approved the final manuscript.

Author details

[1] Department of Geography, University of Victoria, PO Box 3060, STN CSC, Victoria, BC V8W 3R4, Canada. [2] Biology Department, University of British Columbia, 3333 University Way, Kelowna, BC V1V 1V7, Canada. [3] Department of Biological Sciences, University of Cape Town, Private Bag X3, Rondebosch 7701, South Africa. [4] Raincoast Conservation Foundation, P.O. Box 77, Bella Bella, BC V0T 1B0, Canada. [5] Hakai Institute, P.O. Box 309, Heriot Bay, BC V0P 1H0, Canada.

Acknowledgements

We thank Tony Hamilton for providing the Ministry of Environment cougar mortality data for British Columbia, Canada.

Competing interests

The authors declare that they have no competing interests.

References

1. Woodroffe R, Ginsberg J. Edge effects and the extinction of populations inside protected areas. Science. 1998;280:2126–8.
2. Schipper J, Chanson JS, Chiozza F, Cox NA, Hoffmann M, Katariya V, Lamoreux J, Rodrigues ASL, Stuart SN, Temple HJ, et al. The status of the world's land and marine mammals: diversity, threat, and knowledge. Science. 2008;322:225–30.
3. Fa JE, Brown D. Impacts of hunting on mammals in African tropical moist forests: a review and synthesis. Mamm Rev. 2009;39:231–64.
4. Ripple WJ, Estes JA, Beschta RL, Wilmers CC, Ritchie EG, Hebblewhite M, Berger J, Elmhagen B, Letnic M, Nelson MP, Schmitz OJ, Smith DW, Wallach AD, Wirsing AJ. Status and ecological effects of the world's largest carnivores. Science. 2014. doi:10.1126/science.1241484.
5. Krofel M, Treves A, Ripple WJ, Chapron G, López-Bao JV. Hunted carnivores at outsized risk. Science. 2015;30:518–9.
6. Brinkman TJ, Chapin T, Kofinas G, Person DK. Linking hunter knowledge with forest change to understand changing deer harvest opportunities in intensively logged landscape. Ecol Soc. 2009;14. http://www.ecologyand-society.org/vol14/iss1/art36/.
7. Festa-Bianchet M, Ray JC, Boutin S, Cote SD, Gunn A. Conservation of caribou (Rangifer tarandus) in Canada: an uncertain future. Can J Zool. 2011;89:419–34.
8. Treves A, Karanth KU. Human-carnivore conflict and perspectives on carnivore management worldwide. Conserv Biol. 2003;17:1491–9.
9. Treves A, Wallace RB, Naughton-Treves L, Morales A. Co-managing human-wildlife conflicts: a review. Hum Dim Wildl. 2006;11:383–96.
10. Linnell JDC, Andersen R, Kvam T, Andren H, Liberg O, Odden J, Moa PF. Home range size and choice of management strategy for lynx in Scandinavia. Environ Manage. 2001;27:869–79.
11. Schwartz CC, Swenson JE, Miller SD. Large carnivores, moose, and humans: a changing paradigm of predator management in the 21st century. Alces. 2003;39:41–63.

12. Sidorovich VE, Tikhomirova LL, Jedrzejewska B. Wolf *Canis lupus* numbers, diet and damage to livestock in relation to hunting and ungulate abundance in northeastern Belarus during 1999–2000. Wildl. Biol. 2003;9:103–11.

13. Treves A, Naughton-Treves L. Evaluating lethal control in the management of human-wildlife conflict. In: Woodroffe R, Thirgood S, Rabinowitz A, editors. People and wildlife: conflict or coexistence?. Cambridge: Cambridge University Press; 2005.

14. Robinson HS, Wielgus RB, Cooley HS, Cooley SW. Sink populations in carnivore management: cougar demography and immigration in a hunted population. Ecol Appl. 2008;18:1028–37.

15. Elfström M, Zedrosser A, Støen O-G, Swenson JE. Ultimate and proximate mechanisms underlying the occurrence of bears close to human settlements: review and management implications. Mamm Rev. 2014;44:5–18.

16. Linnell JDC, Odden J, Smith ME, Aanes R, Swenson JE. Large carnivores that kill livestock: Do "problem individuals" really exist? Wildl Soc Bull. 1999;27:698–705.

17. Conner MM, Jaeger MM, Weller TJ, McCullough DR. Effect of coyote removal on sheep depredation in Northern California. J Wildl Manage. 1998;62:690–9.

18. Gay SW, Best TL. Age-related variation in skulls of the puma (*Puma concolor*). J Mammal. 1996;77:191–8.

19. Segura V, Prevosti F, Cassini G. Cranial ontogeny in the Puma lineage, *Puma concolor, Herpailurus yagouaroundi,* and *Acinonyx jubatus* (Carnivora: Felidae): a three-dimensional geometric morphometric approach. Zool J Linnean Soc. 2013;169:235–50.

20. Smuts GL, Robinson GA, Whyte IJ. Comparative growth of wild male and female lions (*Panthera leo*). J Zool. 1980;190:365–73.

21. Lukarevsky V, Malkhasyan A, Askerov E. Biology and ecology of the leopard in the Caucasus. CAT News Special Issue 2—Caucasus Leopard. 2007;9–14.

22. Hoogesteijn R, Mondolfi E. Body mass and skull measurements in four jaguar populations and observations on their prey base. Bull Florida Museum Nat Hist. 1996;39:195–219.

23. Whitehead GK. The game-trophies of the world—international formula for the measurement and evaluation of trophies. Hamburg and Berlin: Verlag Paul Parey; 1981.

24. Polisar J, Matix I, Scognamillo D, Farrell L, Sunquist ME, Eisenberg JF. Jaguars, pumas, their prey base, and cattle ranching: ecological interpretations of a management problem. Biol Conserv. 2003;109:297–310.

25. Packer C, Ikanda D, Kissui B, Kushnir H. Conservation biology: lion attacks on humans in Tanzania. Nature. 2005;436:927–8.

26. Pettorelli N, Pelletier F, von Hardenberg A, Festa-Bianchet M, Côté SD. Early onset of vegetation growth vs. rapid green-up: impacts on juvenile mountain ungulates. Ecology. 2007;88:381–90.

27. Hamel S, Garel M, Festa-Bianchet M, Gaillard J-M, Côté SD. Spring normalized difference vegetation index (NDVI) predicts annual variation in timing of peak faecal crude protein in mountain ungulates. J Appl Ecol. 2009;46:582–9.

28. Pettorelli N, Ryan S, Mueller T, Bunnefeld N, Jedrzejewska B, Lima M, Kausrud K. The normalized difference vegetation index (NDVI): unforeseen successes in animal ecology. Clim Res. 2011;46:15–27.

29. United States geological survey. Earth explorer. vegetation monitoring—NOAA CDR NDVI. http://earthexplorer.usgs.gov/ Accessed 20 July 2016.

30. Burnham KP, Anderson DR. Model selection and multimodel inference: a practical information-theoretic approach. 2nd ed. New York: Springer-Verlag; 2002.

31. Symonds MRE, Moussalli A. A brief guide to model selection, multimodel inference and model averaging in behavioural ecology using Akaike's information criterion. Behav Ecol Sociobiol. 2011;65:13–21.

32. Wang Q, Wu N. Long-run covariance and its applications in cointegration regression. The Stata J. 2012;12:515–42.

33. Grueber CE, Nakagawa S, Lewis RJ, Jamieson IG. Multimodel inference in ecology and evolution: challenges and solutions. J Evol Biol. 2010;24:699–711.

34. Treves A. Hunting for large carnivore conservation. J Appl Ecol. 2009;46:1350–6.

35. Beier P. Determining minimum habitat areas and habitat corridors for cougars. Conserv Biol. 1993;7:94–108.

36. Torres SG, Mansfield TM, Foley JE, Ludo T, Branches A. Mountain lion and human activity in California: testing speculations. Wildl Soc Bull. 1996;24:451–60.

37. Weaver JL, Paquet PC, Ruggiero LF. Resilience and conservation of large carnivores in the Rocky Mountains. Conserv Biol. 1996;10:964–76.

38. Kertson BN, Spencer RD, Grue CE. Demographic influences on cougar residential use and interactions with people in western Washington. J Mammal. 2013;94:269–81.

39. Teichman KJ, Cristescu B, Nielsen SE. Does sex matter? Temporal and spatial patterns of cougar-human conflict in British Columbia. PLoS ONE. 2013;8:e74663.

40. Michalski F, Boulhosa RLP, Faria A, Peres CA. Human-wildlife conflicts in a fragmented Amazonian forest landscape: determinants of large felid depredation on livestock. Anim Conserv. 2006;9:179–88.

41. Hornocker M. Pressing business. In: Hornocker M, Negri S, editors. Cougar ecology and conservation. Chicago: University of Chicago Press; 2010. p. 235–47.

42. Cunningham SC, Gustavson CR, Ballard WB. Diet selection of mountain lions in southeastern Arizona. J Range Manage. 1999;52:202–7.

43. Fairaizl SD, Stiver SJ. A profile of depredating mountain lions—proceedings of the seventeenth vertebrate pest conference. Davis: University of California Press; 1996.

44. Wielgus RB, Sarrazin F, Ferriere R, Clobert J. Estimating effects of adult male mortality on grizzly bear population growth and persistence using matrix models. Biol Conserv. 2001;98:293–303.

45. Beier P. Cougar attacks on humans in the United-States and Canada. Wildl Soc Bull. 1991;19:403–12.

46. Peebles KA, Wielgus RB, Maletzke BT, Swanson ME. Effects of remedial sport hunting on cougar complaints and livestock depredations. PLoS ONE. 2013;8:e79713.

47. Marker LL, Dickman AJ, Leo RM, Mills MGL, MacDonald DW. Demography of the Namibian cheetah *Acinonyx jubatus jubatus*. Biol Conserv. 2003;114:413–25.

48. Conde DA, Colchero F, Zarza H, Christensen NL Jr, Sexton JO, Manterola C, Chávez C, Rivera A, Azuara D, Ceballos G. Sex matters: modeling male and female habitat differences for jaguar conservation. Biol Conserv. 2010;143:1980–8.

49. Maletzke BT, Wielgus R, Koehler GM, Swanson M, Cooley H, Alldredge JR. Effects of hunting on cougar spatial organization. Ecol Evol. 2014;4:2178–85.

50. Benson JF, Sikich JA, Riley SPD. Individual and population level resource selection patterns of mountain lions preying on mule deer along an urban-wildland gradient. PLoS ONE. 2016;11:e0158006.

51. Cristescu B, Stenhouse GB, Symbaluk M, Nielsen SE, Boyce MS. Wildlife habitat selection on landscapes with industrial disturbance. Env Conserv. 2016. doi:10.1017/S0376892916000217.

52. Burdett CL, Crooks KR, Theobald DM, Wilson KR, Boydston EE, Lyren LM, Fisher RN, Winston Vickers T, Morrison SA, Boyce WM. Interfacing models of wildlife habitat and human development to predict the future distribution of puma habitat. Ecosphere. 2010;1:4.

53. Linnell JDC, Swenson JE, Anderson R. Predators and people: conservation of large carnivores is possible at high human densities if management policy is favorable. Anim Conserv. 2001;4:345–9.

54. Balme GA, Batchelor A, De Woronin Britz N, Seymour G, Grover M, Hes L, MacDonald DW, Hunter LTB. Reproductive success of female leopards *Panthera pardus*: the importance of top-down processes. Mamm Rev. 2012;43:221–37.

55. White KR, Koehler GM, Maletzke BT, Wielgus RB. Differential prey use by male and female cougars in Washington. J Wildl Manage. 2011;75:1115–20.

56. Thompson DJ, Jenks JA. Dispersal movements of subadult cougars from the black hills: the notions of range expansion and recolonization. Ecosphere. 2010;1:1–10.

57. Lambert CM, Wielgus RB, Robinson HR, Cruickshank HS, Clarke R, Almack J. Cougar population dynamics and viability in the Pacific Northwest. J Wildl Manage. 2006;70:246–54.

58. Cooley HS, Wielgus RB, Koehler GM, Robinson HS, Maletzke BT. Does hunting regulate cougar populations? A test of the compensatory mortality hypothesis. Ecology. 2009;90:2913–21.

4

Feeding preferences of the Asian elephant (*Elephas maximus*) in Nepal

Raj Kumar Koirala[1,2]*, David Raubenheimer[3], Achyut Aryal[3,4,5], Mitra Lal Pathak[6] and Weihong Ji[1]

Abstract

Background: Nepal provides habitat for approximately 100–125 wild Asian elephants (*Elephas maximus*). Although a small proportion of the world population of this species, this group is important for maintaining the genetic diversity of elephants and conservation of biodiversity in this region. Knowledge of foraging patterns of these animals, which is important for understanding their habitat requirements and for assessing their habitat condition, is lacking for the main areas populated by elephants in Nepal. This study investigates the feeding preferences of the Asian elephant in Parsa Wildlife Reserve (PWR) and Chitwan National Park (CNP), Nepal.

Result: Fifty-seven species of plants in 28 families were found to be eaten by Asian elephants, including 13 species of grasses, five shrubs, two climbers, one herb and 36 species of trees. The species that contributed the greatest proportion of the elephant's diet were *Spatholobus parviflorus* (20.2%), *Saccharum spontaneum* (7.1%), *Shorea robusta* (6.3), *Mallotus philippensis* (5.7%), *Garuga pinnata* (4.3%). *Saccharum bengalensis* (4.2%), *Cymbopogan* spp (3.7%), *Litsea monopetala* (3.6) and *Phoenix humilis* (2.9%). The preference index (PI) showed that browsed species were preferred during the dry season, while browsed species and grasses were both important food sources during the rainy season. Elephants targeted leaves and twigs more than other parts of plants ($P < 0.05$).

Conclusion: This study presents useful information on foraging patterns and baseline data for elephant habitat management in the PWR and CNP in the south central region of Nepal.

Keywords: Browse. elephant habitat, Feeding sign, Food preferences, Micro-histological analysis

Background

Elephants are among the internationally endangered large mammals [1]. The habitat of Asian elephants (*Elephas maximus)* has been decreasing throughout their range, due primarily to habitat destruction and fragmentation resulting from human land use practices [2, 3]. Even though elephant populations have decreased, in general, the local density of elephants has increased due to habitat loss [4]. This has caused resource competition among elephants [5], and increased human–elephant conflict [6]. Asian elephants are generalised herbivores utilising a variety of plant species [2, 7]. Large herbivores such as elephants require extensive home ranges to satisfy their high food demand [8]. Reduction in food availability due to loss of habitat has created challenges for elephant conservation in the many regions in Asia.

Although the dietary requirements of Asian elephants have been studied, the majority of these studies [2, 5, 9, 10] have dealt with the documentation of food plant species, the rate of consumption and seasonal comparative diet overlap between sympatric elephants and rhinos [11, 12]. However, details regarding food choice and seasonal diet composition remain unknown. Such information is important for Asian elephant conservation in terms of habitat management and human–elephant conflict mitigation.

Nepal provides important habitat for Asian elephants. Historically habitat in the Terai range was continuous. Currently, elephants are found only in four regions of the country, eastern, central, western and far-western. In central Nepal, Parsa Wildlife Reserve (PWR) is the main elephant habitat. However, elephants were found

*Correspondence: raj.koirala68@gmail.com
[1] Human Wildlife Interaction Research Group, Institute of Natural and Mathematical Sciences, Massey University, Private Bag 102 904, Albany, Auckland 0745, New Zealand
Full list of author information is available at the end of the article

to migrate between PWR and Chitwan National Park (CNP) since the middle of the 1990s [13]. The migration of elephants between these sites was thought to be primarily due to the reduction of water availability in the Bara Forest near PWR resulting in reduced food availability and aggravated competition with livestock [13]. Currently, all the four isolated population of elephants in Nepal are in the lowland Terai region. These widespread and fragmentary distributed elephants strongly prefer floodplain communities, and there is a significant shift from browse to grass-dominated vegetation between seasons in Bardia National Park [12, 14]. However, the diet has not been studied for other elephant populations of the country.

This study aims to investigate the food preferences and seasonal changes in foraging patterns of the Asian elephant in the PWR, CNP and adjoining forests. We predict a climate-related reduction in grass productivity in the Parsa area will correspond with a reduction of grass in the elephant diet during the dry season. Information obtained from this study will aid elephant conservation in respect to the restoration of their habitats, and will thereby contribute towards minimising human–elephant conflict.

Methods

The study was carried out at the Parsa Wildlife Reserve and part of adjoining reserve forest (Bara forest) in the north and Chitwan National Park and part of its buffer Zone forests. Permission for the study was acquired from the Department of National Park and Wildlife Conservation, the government of Nepal. Parsa Wildlife Reserve is the largest wildlife reserve in Nepal (Fig. 1), consisting of 499 km² sub-tropical forests in the south-central lowland Terai ecoregion of Nepal. The PWR is located in the Churia hills, the outermost foothills of the Himalayas [15], which are a part of the Bhabar District. The PWR is typically dry with average rainfall between 300 and 450 mm during the summer months [13, 16]. The typical vegetation of this reserve and the adjoining Bara forest is tropical and subtropical forest types with Sal (*Shorea robusta*) forest about 90% of the vegetation. Chirpine (*Pinus roxburghii*) grows in the Churia hills. Khair (*Acacia catechu*), Sisso (*Dalbergia sisso*) and Silk cotton (*Bombax ceiba*) trees occur along water channel. Sabai grass (*Eulaliopsis binata*) grows well on the southern face of the Churia Hills [17, 18]. Chitwan National Park was established in 1973 as the first national park in Nepal and was listed as a World Heritage Site in 1984. The CNP spans 932 km² and is situated in the sub-tropical lowlands of the Inner Terai, in the Chitwan district of south-central Nepal (Fig. 1). Elevation ranges from

approximately 100 m in lowland river valleys to 815 m on Churia Hill ridgetops. In the north-west of this protected area, the Narayani and Rapti rivers separate the park from human settlements [19]. The buffer zone has mostly agriculture fields and human settlements along with community forests. The typical vegetation of CNP and its buffer zone forests is Himalayan subtropical broadleaf forests with primarily Sal (*Shorea robusta*) trees covering about 70% of the national park area. On northern slopes, Sal associated with smaller flowering tree and shrub species such as *Terminalia bellirica*, *Dalbergia sissoo*, *Dillenia indica*, *Garuga pinnata* and climbers such as *Bauhinia vahlii* and *Spatholobus parviflorus* [17, 18, 20].

Both the PWR and CNP are prime habitats for wild Asian elephants and both parks are adjacent to Valmiki tiger reserve in India (Fig. 1). These three trans-boundary, contiguous protected areas cover a 3549 km² mixed-habitat zone containing large tracts of grasslands and humid deciduous forests, which provide suitable habitat for a large number of megaherbivores and big cats such as Asian elephants, endangered tigers (*Panthera tigris*) and greater one-horned rhinos (*Rhinoceros unicornis*).

Elephant feeding data collection

Opportunistic direct feeding observations and the observation of elephant feeding sign on food trails (elephant feeding routes) were the methods used in the present study to determine diet selection of elephants residing in different areas and travelling on different migration routes [9, 21]. The feeding routes observed to be taken by elephants were followed by field researchers, and all plant species showing signs of being eaten by elephants were recorded. Evidence of feeding sign included elephant footprints, fresh dung piles nearby to browsed foliage, and the identifying characteristics of plant damage caused by elephant browse, such as debarkation, branch breaking and uprooting. The following data were recorded to determine the feeding preferences of Asian elephants: (1) plant species browsed, (2) parts of the plant eaten (leaves, branches and/or bark), (3) habitat type and (4) global positioning system (GPS) coordinates of sample sites (Fig. 1). The relative frequency (percentage) of feeding sign was calculated to yield a feeding sign score. Feeding sign was ranked according to the intensity of browsing, the proportion of bark, stem and foliage removed and/or the area of grass eaten.

Elephant dietary analysis from dung samples

Samples of elephant dung encountered during a total 24 days of field survey in the wet season (June–September

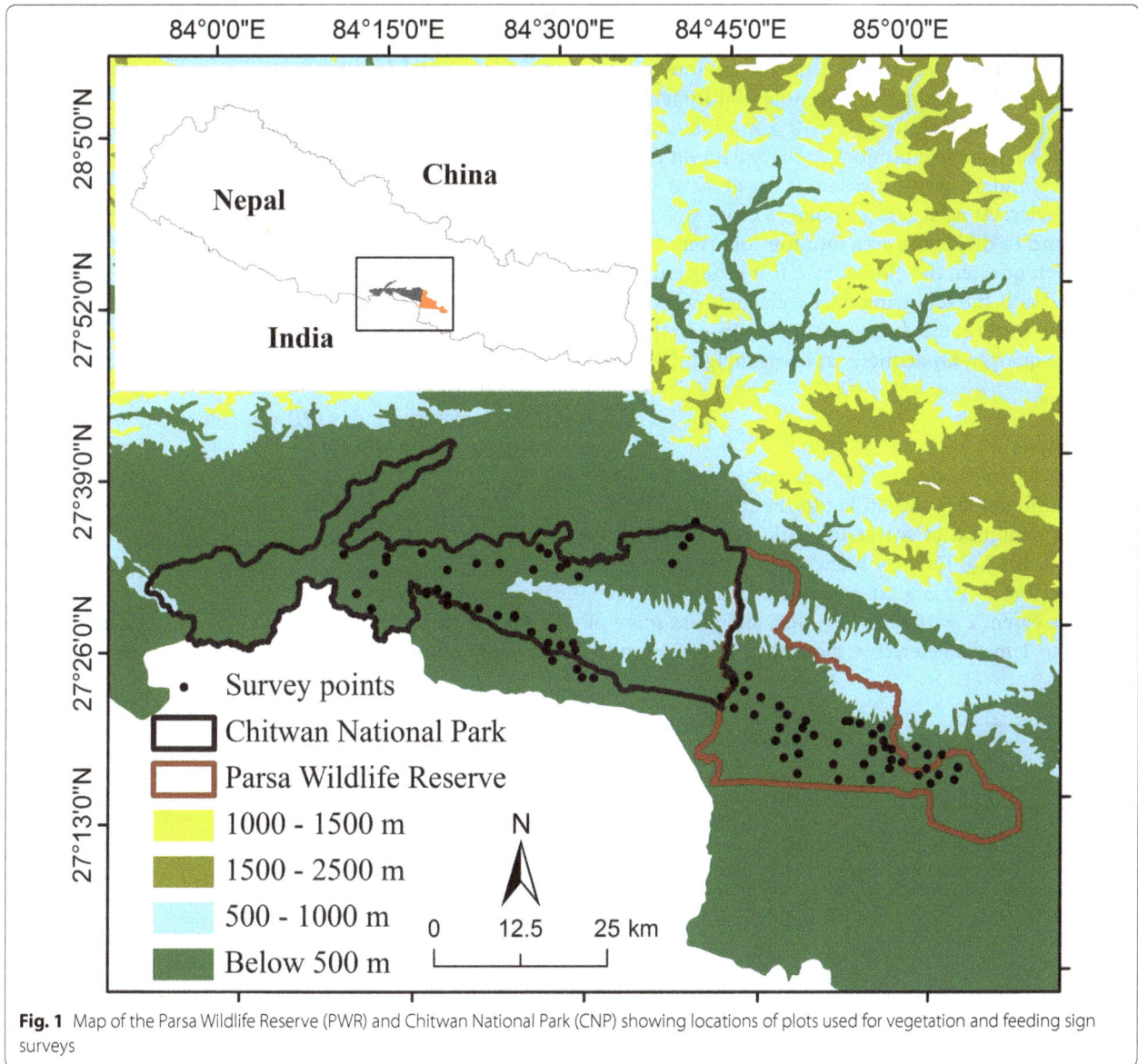

Fig. 1 Map of the Parsa Wildlife Reserve (PWR) and Chitwan National Park (CNP) showing locations of plots used for vegetation and feeding sign surveys

2012) and the dry season (February–April 2013) were collected. Visual examination of deposited elephant dung piles was performed to identify the presence of macro plant fragments. Micro-plant fragments were identified through micro-histological analysis [22–24]. This dual methodology is widely used for estimating the diet composition of herbivores [25]. Fragments of probable food species were collected for the preparation of reference slides. The collection was made as per methods used in the previously published literature describing elephant food plants [11, 12]. A total of 20, non-overlapping random fragments were isolated on each dung slide and were compared with a reference slide for epidermal derivatives. Microphotographs were taken using

a 100 × 4× lens and an Am Scope MT130 1.3 megapixel USB2.0 microscope eyepiece digital camera.

Food availability survey

To assess the food preferences of elephants, we carried out vegetation surveys using the point-centred quarter technique [26] to obtain data on the relative abundance of different plant species. A total of 30 transects of 2 km length each, one each per habitat type, were created for this survey. To compare the availability of food plants within and outside protected areas, 20 of these transects were placed in the protected areas, while the remaining ten transects were located in habitats outside national parks. Each transect was surveyed twice, once in the wet season (August/September) and once in the dry season (March/April). A total of 10 sample points were assigned to each transect at 200 m interval for the purpose of gathering data on potential forage trees. Also, 10 quadrats measuring 1 m × 1 m each were created near each sample points to collect data on density, frequency and visual estimation of cover % of dietary grass species. At each sample point, a cross was laid on the ground to divide the area into four quarters (Fig. 2). From each quarter, the closest tree from the centre was identified and the following data collected: (1) the species of the tree, (2) the distance from the tree to the centre of the quarter, (3) diameter at breast height (DBH) of the tree.

Data analysis

Elephant feeding sign survey was conducted by scoring the different signs according to the intensity of damage. The definition of scores for bark was: $1 \leq 0.5$ m^2; $2 = 0.5–1$ m^2; $3 \geq 1$ m^2; for branch score: $1 =$ up to 5 cm diameter; $2 = 5–20$; $3 > 20$; while foliage score: $1 \leq 10\%$ of foliage eaten; $2 = 10–40\%$, $3 \geq 40\%$ and grass score of $1 =$ up to 1 m diameter; $2 =$ more than 2 m; $3 =$ more

than 5 m. Total feeding score was calculated by multiplying the frequency of each plant species showing feeding signs with total feeding sign score of that species. Total feeding score of each species was multiplied by 100 and divided by the total feeding score of all species to calculate an index equivalent to utilisation percent. The importance value index (IVI) of a plant species in each habitat was calculated by adding the relative frequency, density and dominance (basal area) for trees. The relative frequency, density and cover for grass and herbaceous species was used as an index of availability of a species in the study area [21]. The density of browse was calculated using the distance from the tree to the centre of the quarter following Mitchell [27].

The preference index (PI) was calculated using the following equation [21, 28]:

Preference index (PI)

$$= \frac{\text{Utilization percentage}}{\text{Percentage availability in the environment}} \quad (1)$$

A PI score >1 indicates food that was utilised proportionately more than its occurrence in the environment,

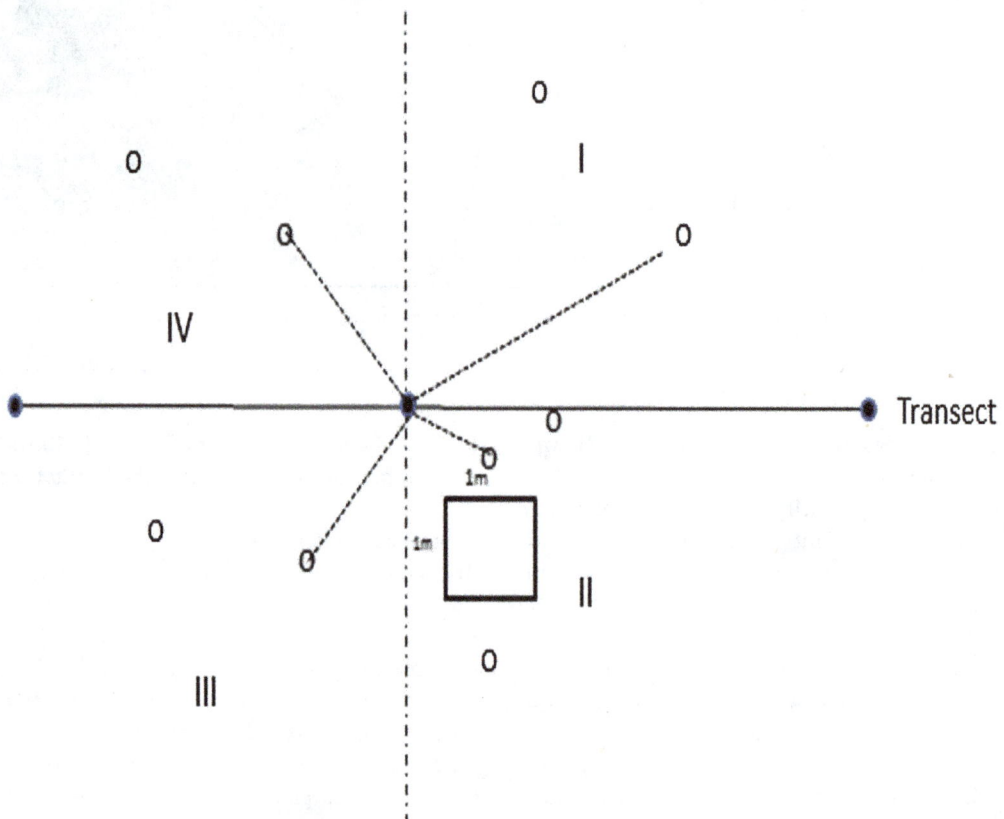

Fig. 2 Sample points along a transect with the nearest trees in each quarter indicated by *dash lines* and a grass of 1 m × 1 m near each sample points

and a PI score <1 indicates a plant food that was used proportionately less than its presence in the environment.

The Chi-square test was used to test for differences in feeding preferences between plant parts, seasons and sites differences in vegetation density; Pearson correlation was used to determine the correlation between forage availability and preference. Simpson's diversity index was used to estimate the vegetation diversity, and the independent sample t test was applied to test for seasonal dietary intake differences in monocot and dicot plants. All tests were performed using Excel and IBM SPSS statistical version 22.

Results

Elephant foraging patterns

In total, 57 species of plants (13 grass, five shrubs, two climbers, one herb and 36 tree species) belonging to 28 taxonomic families were eaten by Asian elephants. In the Parsa area, 40 species (10 grass, four shrub, two climber, one herb and 23 tree species) were consumed, and in the Chitwan area 37 species (nine grass, three shrub, one climber and 24 tree species) were utilised; the utilisation pattern suggests that 76% of all identified food species were consumed during the wet season, with only 24% consumed during the dry season (Additional file 1: Appendix). The foliage (leaves and twigs) of both grasses and browsed trees were selected more than the stems, bark, roots and fruits during the wet season in both Parsa and Chitwan ($\chi^2 = 10.72$, df = 6, P < 0.05), whereas debarkation and uprooting were more common in the dry season ($\chi^2 = 5.24$, df = 4, P < 0.05).

Dietary analysis from dung samples

Microscopic analysis of 36 dung samples collected during two seasons showed a higher dicot-to-monocot ratio in the dry season compared to the wet season. The average dicot-to-monocot ratio was 1:0.57 in the dry season, whereas the ratio was 1:1.11 in the rainy season. The observations from the feeding sign survey and the micro-histological analysis revealed that dicots were consumed more during the dry season (t = −4.27, df = 10, P = 0.002). There was no significant difference in the presence of dicot and monocot plants in elephant diet during the rainy season (t = 1.59, df = 58, P = 0.117).

Regional food availability

There was no difference in the types of plants availability in and outside the two sites (P ≥ 0.05). However, species diversity was slightly lower in CNP (Simpson's diversity index, D = 0.097) than in the PWR (D = 0.091). Similarly, in both study sites and seasons, food species densities and frequencies recorded were significantly different.

There was a significant relationship of grass and browse abundance in dry and wet season in Parsa and Chitwan, indicating an association between these factors ($\chi^2 = 8.92$, df = 1, P = 0.002). Higher densities of each browse species were recorded in the PWR (mean density, 25.00/ha) than in the CNP (mean density, 20.4/ha). Seasonally the wet season mean density of each browse species in Chitwan and Parsa were 23.2 and 15.4/ha respectively. In the dry season, the mean density of browse in Chitwan was 16.3 and in Parsa 20.0/ha. There was significant difference in the frequency of grasses ($\chi2 = 20$, df = 1, P < 0.001) in the dry season in both parks with higher frequencies of grass species recorded in the dry season in CNP (mean frequency 3.45/q; mean density, 115.7/m^2) than in PWR (mean frequency, 1.57/q; mean density, 22.85/m^2). The mean grass frequency and density in the wet season in Chitwan was 4.5/q and 160 individuals/m^2, respectively. In Parsa, the mean grass frequency and density was 2.0/q and 131 individuals/m^2. There was a negative correlation between the availability of individual plant food species in the habitat and their utilisation by elephants (r = −0.244, P = 0.02).

Plant species preferences

Elephants showed a positive PI score for 26 out of the 57 utilised plant species (Fig. 3). Elephant browse that had relatively high PI scores ranged from 1.04 (*Bombax ceiba*) to 9.2 (*Ficus racemosa*). Similarly, vine PI scores ranged from 0.02 (*Bauhinia vahlii*) to 9.32 (*Spatholobus parviflorus*). Shrubs that had relatively high PI scores were *Hypericum uralum* (1.18) and the palm *Phoenix humilis* (2.91). Grass PI scores ranged from 1.28 (*Saccharum bengalensis*) to 5.51 (*Thysanolaena maxima*). Species that were highly abundant, which may have led to lower PI scores, included *Shorea robusta*, *Dillenia pentagyna*, *Hemarthria compressa*, *Imperata cylindrica* and *Cymbopogon* spp.

Overall, in both sites, elephants showed the strongest preferences for common species such as *Spatholobus parviflorus*, *Saccharum spontaneum*, *Phoenix humilis*, *Saccharum bengalensis*, *Mallotus philippensis*, and *Phragmites karka*. In addition to these species, elephants in the Chitwan area showed a strong preference for *Cleistocalyx operculata* and *Bridelia retusa*, while Parsa area elephants showed a strong preference for *Litsea monopetala*, *Thysanolaena maxima*, *Sterculia villosa*, *Equisetum debile*, *Bambusa* spp. and *Hypericum uralum*. The availability of these species in the two parks varies. Amongst these 26 most preferred species, 17 species were preferred more by elephants in Parsa, while the remainder (nine species) were preferred relatively more by Chitwan elephants (Fig. 3).

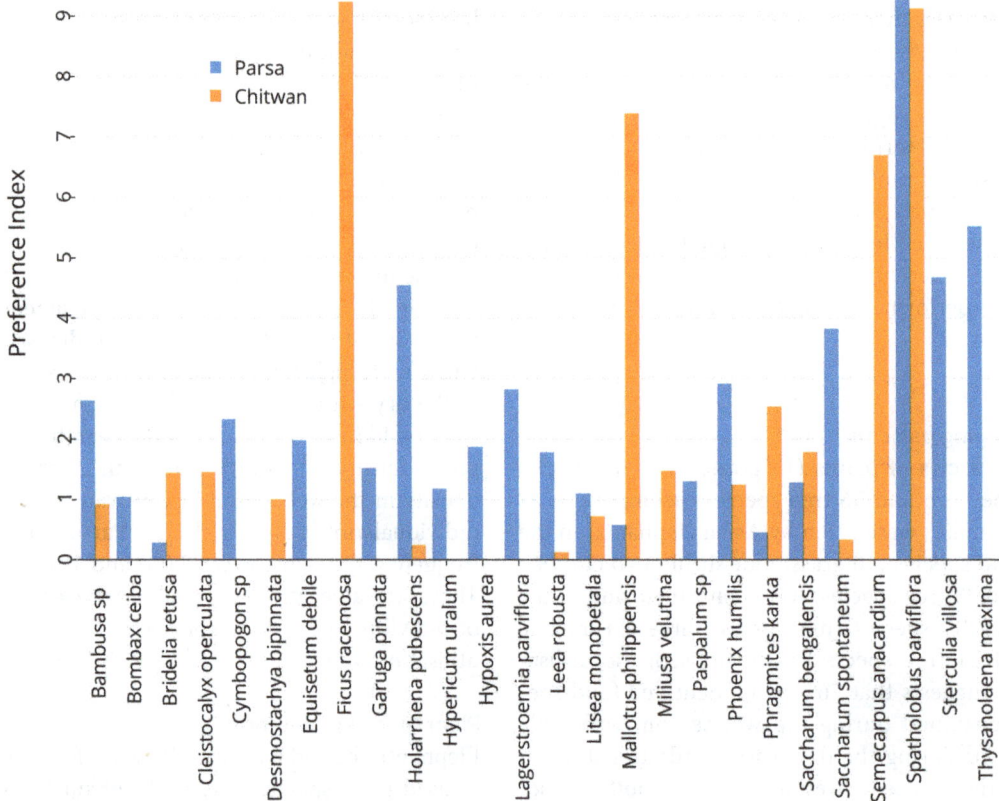

Fig. 3 Preference indices (PI) for the most prevalent plant species found in the diet of wild Asian elephants in the Chitwan National Park (CNP) and the Parsa Wildlife Reserve (PWR) in both the rainy and dry seasons

Discussion

Nepal has lost over 80% of its elephant habitat to human settlement [5]. As a result, the resident elephant population, estimated to number between 109 and 142 individuals, is presently restricted to four isolated areas [29]. Available diet and nutritional preference are the two most important factors that drive elephant movements, and that generate conflict with humans, especially when available elephant habitat is shrinking [30]. Reduction in grass, especially in the dry seasons may result in elephant migration. Human–elephant conflicts may arise mainly due to elephant migration [31]. Thus, knowledge of elephant foraging patterns and seasonal food availability is important for mitigation of human–elephant conflict. The management of grass species in the dry season is crucial. In areas like Parsa where there is an environmental constraint in retaining surface water, some potential habitats could be irrigated during the dry season to maintain grass productivity.

The present study recorded 57 plant species within 28 families that were foraged by Asian elephants in the PWR and CNP. In a similar study, Sukumar [2] reported 112 species of plants in the elephant's diet in southern India,

and Chen et al. [9] reported 106 plant species in the diets of elephants in Shangyong National Natural Reserve in Xishuangbanna, the People's Republic of China were catalogued. The wide range of results between studies may be due to differences in the number and diversity of plant species available. Divergent results may also be partly due to differences in sampling methods; variances in forest condition (disturbed versus undisturbed), composition, and sampling area could also have contributed to divergent results.

Elephants are mixed feeders, and there is seasonal variation in their food selection [8]. In the present study, we found that browse flora and grasses were both eaten by elephants during the wet season, while browse vegetation dominated the dry season diet. Indeed, it seems that the proportion of dicot and monocot species in the diet of elephants varies across different home ranges. In southern India, elephants are known to rely heavily on graminoids (grasses, sedges and rushes) in the wet season and almost exclusively on woody plants during the dry season [32]. Similar patterns of seasonal variation in feeding have been reported by Pradhan et al. [12] in Bardia National Park in Nepal, and also for African elephant in Uganda [33]. In Nilgiri Biosphere Reserve, southern India, grasses dominate the elephant diet in all seasons, while browse flora forms an important portion of their diet only during the dry season [10]. Likewise, in the foothills of the Himalayas, browse forms the majority of the diet in dry seasons [34]. In similar studies, browse dominated the diet of elephants all year in the rainforests of Malaysia [35], north-eastern India [36] and in the state of Bihar, Central India [37].

Results of the present study are comparable to the data obtained in the above-mentioned studies in terms of dry-season diet. This browse-dominated dry season diet could be due to the lower average grass biomass available when the dry season causes a reduction in grass cover. It could also be due to the need to meet specific nutrient requirements, for example, the high levels of essential minerals in the hard wood of browse plants [12]. However, our study revealed a slightly different trend in the wet season, when a similar proportion of grass and browse were found in the elephant diets. This could be due to the migration patterns of elephants in Nepal: at the onset of the rainy monsoon season, elephants move from Chitwan to Parsa and towards upper slopes [13]. As the monsoon develops, elephants migrate from grass-rich lower elevations south to the foothills of the Churia range for occasional resting. In the Churia foothills, elephants have more opportunities to eat foods other than grasses, as these foothills are rich in preferred woody species.

In the present study, we noted a difference in feeding preference for stems, leaves and twigs, bark and other parts of woody plant species. Foliage (leaves and twigs) of both grass and browse flora were eaten more than other parts of plants in the wet season, while bark dominated the dry season diet. The use of bark from various tree species by elephants might relate to macronutrient balancing [38], and for gaining moisture and mineral supplements [39] that would otherwise have been unavailable during the dry season. The current study aligns with the findings of Pradhan et al. [12] in Bardia National Park, Nepal, where bark consumption dominated the diet of elephants in the dry season. Differences in forest structure, methodologies used and spatial and temporal availability of different groups of plants could explain the variance in PI between the two studies, which are both based on elephant populations in Nepal.

Spatially and temporally, PI can vary between species. In the present study, widely abundant foods such as *Shorea robusta*, *Mallotus philippensis*, *Imperata cylindrica* and *Saccharum bengalensis* were avoided by elephants in some seasons and locations, despite their high availability (Additional file 1: Appendix). Therefore, it is important to examine independently the PI scores of species that are of high availability (or rare) to determine whether the score

could be due to the methodological limitations of this index alone [40], or could involve other factors. The PI scores derived from Parsa and Chitwan could be obtained from multiple rather than single factors [41]. Factors such as seasonal availability [42], palatability [43], nutritive value and plant tissue toxicity are all important influences on the selection of food plants by elephants [35].

Although in both the PWR and the CNP, elephants prefer common plants such as *Spatholobus parviflorus*, in fact *Saccarum spontaneum*, *Phoenix humilis*, *Saccharum bengalensis*, *Mallotus philippensis* and *Phragmites karka*, there are some less common species such as *Acacia catechu*, *Bombax ceiba*, *Bamboosa* spp and *Ficus* spp that are important food for elephants. In the present study, feeding patterns observed in both areas revealed that Parsa elephants ate a more diverse, species-rich diet than did Chitwan's Asian elephant population. The Parsa area has a higher number of elephants, possibly suggesting that nutrition is superior in PWR due to greater dietary diversity. However, further study on habitat preference in all seasons is needed to further investigate this. In addition, the present study has also yielded new data supporting previously unrecorded Asian elephant preferences for *Thysanolaena maxima*, *Sterculia villosa*, *Equisetum debile*, *Semecarpus anacardium* and *Hypericum uralum*.

Conclusion

Asian elephants have a diverse diet including monocot and dicot plants. Their diet in the dry season (February–April) contained a higher proportion of dicots compared to that of the wet season (June–September). There was a negative correlation between availability of plants and preference by elephants, suggesting food selection by elephants is not passively driven by relative availability, but related to specific preferences [44]. Further studies are needed to understand this feeding selectivity and its implications for the elephants. The current study provides baseline information about different types of natural food available in the Parsa and Chitwan regions of Nepal, and their relative importance in the diets of elephants in and around the PWR and CNP. This information is important for realising successful outcomes for the conservation of Asian elephants and improved seasonal management for the long-term protection of this endangered species and its shrinking habitat.

Additional file

Additional file 1: Appendix Species, family, type of plant and plant parts consumed, and preference index for the majority of plants consumed by wild Asian elephants. A preference index score >1 indicates a food that was utilised proportionately more than its occurrence in the environment, and food with a preference index score <1 was utilised proportionately less than its occurrence in the environment.

Authors' contributions
RKK, DR, and WJ designed the study; RKK, MP collected the data; RKK, AA analysed the data. RKK wrote the manuscript, and all authors contributed to the final version of the paper. All authors read and approved the final manuscript.

Author details
[1] Human Wildlife Interaction Research Group, Institute of Natural and Mathematical Sciences, Massey University, Private Bag 102 904, Albany, Auckland 0745, New Zealand. [2] Institute of Forestry, Tribhuvan University, Pokhara, Nepal. [3] The Charles Perkins Centre and School of Life and Environmental Sciences, University of Sydney, Sydney, NSW, Australia. [4] Department of Forestry & Resource Management, Toi Ohomai Institute of Technology , Rotorua 3046, New Zealand. [5] Waste Management NZ Ltd, Auckland, New Zealand. [6] Department of Plant Resources, National Herbarium and Plant Laboratories, Godawari, Nepal.

Acknowledgements
We thank the Department of National Park and Wildlife Conservation government of Nepal for granting permission and field support to conduct this field research. We also thank Institute of Forestry, Pokhara, Nepal for providing logistic support. DR is an Adjunct Professor at the New Zealand Institute for Advanced Study. We thank to Rufford Small Grant Foundation, UK; Chester Zoo, UK; Keidanren Nature Conservation Fund (KNCF) for their financial support of this study. We are grateful to Dr Margaret Sheridan for her interest and contribution towards supporting this work.

Competing interests
The authors declare that they have no competing interests.

Funding
Rufford Small Grant Foundation, UK; Chester Zoo, UK; Keidanren Nature Conservation Fund (KNCF).

References
1. Choudhury A, Lahiri Choudhury DK, Desai A, Duckworth JW, Easa PS, Johnsingh AJT, Fernando P, Hedges S, Gunawardena M, Kurt F, Karanth U, Lister A, Menon V, Riddle H, Rübel A, Wikramanayake E. (IUCN SSC Asian Elephant Specialist Group). 2008. *Elephas maximus*. The IUCN Red List of Threatened Species 2008: e.T7140A12828813. http://dx.doi.org/10.2305/IUCN.UK.2008.RLTS.T7140A12828813.en.
2. Sukumar R. Ecology of the Asian elephant in southern India. Feeding habits and crop raiding patterns. J Trop Ecol. 1990;6(1):33–53.
3. Owen-Smith RN. Megaherbivores: the influence of very large body size on ecology. Cambridge university press; 1992.
4. Croze H, Hillman AKK, Lang EM. Elephants and their habitats: how do they tolerate each other. In: Fowler CW, Smith TD, editors. Dynamics of large mammal populations. New York: Wiley; 1981. p. 68–95.
5. Joshi R, Singh R. Asian elephants are losing their seasonal traditional movement tracks: a decade of study in and around the Rajaji National Park, India. Gajah. 2007;27:15–26.
6. Sukumar R. Elephant days and nights. New York: Oxford University Press; 1994.
7. Dierenfeld ES. Nutrition. In: Fowler ME, Mikota SK, editors. Biology, medicine, and surgery of elephants. Ames: Blackwell Publishing; 2006. p. 57–65.
8. Sukumar R. The Asian elephant: ecology and management. Cambridge: Cambridge University Press; 1989.
9. Chen J, Deng X, Zhang L, Bai Z. Diet composition and foraging ecology of Asian elephants in Shangyong, Xishuangbanna, China. Acta Ecologica Sinica. 2006;26(2):309–16.
10. Baskaran N, Balasubramanian M, Swaminathan S, Desai AA. Feeding ecology of the Asian elephant Elephas maximus Linnaeus in the Nilgiri Biosphere Reserve, southern India. J Bombay Nat Hist Soc. 2010;107(1):3.
11. Steinheim G, Wegge P, Fjellstad JI, Jnawali SR, Weladji RB. Dry season diets and habitat use of sympatric Asian elephants (*Elephas maximus*)) and greater one-horned rhinoceros (*Rhinocerus unicornis*) in Nepal. J Zool. 2005;265(04):377–85.
12. Pradhan N, Wegge P, Moe SR, Shrestha AK. Feeding ecology of two endangered sympatric megaherbivores: Asian elephant *Elephas maximus* and greater one-horned rhinoceros *Rhinoceros unicornis* in lowland Nepal. Wildlife Biol. 2008;14(1):147–54.
13. Ten Velde PF. A status reports on Nepal's wild elephant population. Kathmandu: WWF Nepal Program; 1997.
14. Pradhan NMB. An ecological study of a re-colonizing population of Asian elephants (*Elephas maximus*) in lowland Nepal. Akershus: Norwegian University of Life Sciences, Department of Ecology and Natural Resource Management; 2007.
15. Thapa K, Shrestha R, Karki J, Thapa GJ, Subedi N, Pradhan NMB, Kelly MJ. Leopard *Panthera pardus fusca* density in the seasonally dry, subtropical forest in the Bhabhar of Terai Arc, Nepal. Adv Ecol. 2014;286949. doi:10.1155/2014/286949.
16. MOSTE. Agro climatic atlas of Nepal, Report, Department of Hydrology and Meteorology, Ministry of Science, Technology and Environment (MoSTE), Government of Nepal (GoN) CGIAR-CCAFS Regional(IGP), Program unit, International water management Institute: New Delhi; 2013.
17. Bhuju UR, Shakya PR, Basnet TB, Shrestha S. Nepal biodiversity resource book. Protected areas, Ramsar sites, and World Heritage sites. 2007.
18. Majupuria TC, Kumar R. Wildlife National Parks and Reserves of Nepal. ISBN 974-89833-5-8. Bangkok: Devi, Saharanpur and Tecpress Books; 1998. p. 245–8.
19. Aryal A, Brunton D, Pandit R, Shrestha TK, Koirala RK, Lord J, Thapa YB, Adhikari B, Ji W, Raubenheimer D. Biological diversity and management regimes of the northern Barandabhar Forest Corridor: an essential habitat for ecological connectivity in Nepal. Trop Conserv Sci. 2012;5(1):38–49.
20. Dinerstein E, Wemmer CM. Fruits Rhinoceros eat: dispersal of Trewia nudiflora (*Euphorbiaceae*) in lowland Nepal. Ecology. 1988;69(6):1768–74.
21. Afework Biru Y B. Food habits of the African elephant (*Loxodonta africana*) in Babile Elephant Sanctuary, Ethiopia. Trop Ecol. 2012;53(1):43–52.
22. Holechek JL, Gross BD. Training needed for quantifying simulated diets from fragmented range plants. J Range Manag. 1982;35(5):644–7.
23. Metcalfe CR. Anatomy of the monocotyledons. Gramineae. Oxford: Oxford University Press; 1990.
24. Aryal A, Brunton D, Ji W, Yadav HK, Adhikari B, Raubenheimer D. Diet and habitat use of Hispid hare *Caprolagus hispidus* in Shuklaphanta Wildlife Reserve. Nepal. Mammal Study. 2012;37(2):147–54.
25. Shrestha R, Wegge P. Determining the composition of herbivore diets in the trans-himalayan rangelands: a comparison of field methods. Rangel Ecol Manag. 2006;59(5):512–8.
26. Bryant DM, Ducey MJ, Innes JC, Lee TD, Eckert RT, Zarin DJ. Forest community analysis and the point-centered quarter method. Plant Ecol. 2005;175(2):193–203.
27. Mitchell K. Quantitative analysis by the point-centered quarter method. 2010. arXiv:1010.3303.
28. Fritz H, De Garine-Wichatitsky M, Letessier G. Habitat use by sympatric wild and domestic herbivores in an African savanna woodland: the influence of cattle spatial behaviour. J Appl Ecol. 1996;1:589–98.
29. Pradhan NM, Williams AC, Dhakal M. Current status of Asian elephants in Nepal. Gajah. 2011;35:87–92.
30. Rode KD, Chiyo PI, Chapman CA, McDowell LR. Nutritional ecology of elephants in Kibale National Park, Uganda, and its relationship with crop-raiding behaviour. J Trop Ecol. 2006;22(04):441–9.
31. Koirala RK, Ji W, Aryal A, Rothman J, Raubenheimer D. Dispersal and ranging patterns of the Asian Elephant (*Elephas maximus*) in relation to their interactions with humans in Nepal. Ethol Ecol Evol. 2016;28(2):221–31.
32. Sukumar R. A brief review of the status, distribution and biology of wild Asian elephants, (*Elephas maximus*). Intern Zoo Yearb. 2006;40:1–8. doi:10.1111/j.1748-1090.2006.00001.x.
33. Field CR, Ross IC. The savanna ecology of Kidepo Valley National park. Afr J Ecol. 1976;14(1):1–15.
34. Lahkar BP, Das JP, Nath NK, Dey S, Brahma N, Sarma PK. A study of habitat utilization patterns of Asian elephant Elephas maximus and current status of human–elephant conflict in Manas National Park within Chirang-Ripu Elephant Reserve, Assam. Guwahati: Aaranyak; 2007.

35. Olivier RCD. On the ecology of Asian Elephant. PhD Thesis. Cambridge: University of Cambridge; 1978.

36. Sukumar R. The living elephants: evolutionary ecology, behaviour, and conservation. Oxford: Oxford University Press; 2003.

37. Daniel JC, Desai AA, Sivaganesan N, Datye HS, Rameshkumar S, Baskara N, Balasubramaniam M, Swaminathan S. Ecology of the Asian elephant. Final Report 1987–1994. Bombay: Bombay Natural History Society; 1995.

38. Laws RM, Parker ISC, Johnstone RCB. Elephant and their habitats: the ecology of elephants in North Bunyoro. London: Oxford University Press; 1975.

39. Bax PN, Sheldrick DL. Some preliminary observations on the food of elephant in the Tsavo Royal National Park (East) of Kenya. Afr J Ecol. 1963;1(1):40–51.

40. Johnson DH. The comparison of usage and availability measurements for evaluating resource preference. Ecology. 1980;61(1):65–71.

41. Ishwaran N. Elephant and woody-plant relationships in Gal Oya, Sri Lanka. Biol Conserv. 1983;26(3):255–70.

42. Mwalyosi RB. The dynamics ecology of Acacia tortilis woodland in Lake Manyara National Park, Tanzania. Afr J Ecol. 1990;28(3):189–99.

43. Caister LE, Shields WM, Gosser A. Female tannin avoidance: a possible explanation for habitat and dietary segregation of giraffes (Giraffa camelopardalis peralta) in Niger. Afr J Ecol. 2003;41(3):201–10.

44. Raubenheimer D. Toward a quantitative nutritional ecology: the right-angled mixture triangle. Ecol Monogr. 2011;81(3):407–27.

5

Siberian flying squirrels do not anticipate future resource abundance

Vesa Selonen[1*] and Ralf Wistbacka[2]

Abstract

Background: One way to cope with irregularly occurring resources is to adjust reproduction according to the anticipated future resource availability. In support of this hypothesis, few rodent species have been observed to produce, after the first litter born in spring, summer litters in anticipation of autumn's seed mast. This kind of behaviour could eliminate or decrease the lag in population density normally present in consumer dynamics. We focus on possible anticipation of future food availability in Siberian flying squirrels, *Pteromys volans*. We utilise long-term data set on flying squirrel reproduction spanning over 20 years with individuals living in nest-boxes in two study areas located in western Finland. In winter and early spring, flying squirrels depend on catkin mast of deciduous trees. Thus, the temporal availability of food resource for Siberian flying squirrels is similar to other mast-dependent rodent species in which anticipatory reproduction has been observed.

Results: We show that production of summer litters was not related to food levels in the following autumn and winter. Instead, food levels before reproduction, in the preceding winter and spring, were related to production of summer litters. In addition, the amount of precipitation in the preceding winter was found to be related to the production of summer litters.

Conclusions: Our results support the conclusion that Siberian flying squirrels do not anticipate the mast. Instead, increased reproductive effort in female flying squirrels is an opportunistic event, seized if the resource situation allows.

Keywords: Resource pulse, Masting, Demographic responses

Background

One way to cope with irregularity of resource availability is to adjust reproduction according to the anticipated future resource availability [1–4]. This would be particularly useful in resource pulse systems, where resource levels fluctuate remarkably over time [5]. Due to the unpredictable nature of resource pulses animals may be doomed to boom and bust dynamics with dramatic population decline when the resource pulse is over [5]. Anticipation of the resource pulse [1] or anticipation of resource crash [6] could eliminate the lag in population density normally present in consumer dynamics.

A common cause for fluctuation in recourse levels in forest communities is masting by trees, synchronous production of large seed crops, which dramatically affects the whole forest community [5, 7, 8]. For example, densities of seed predators often peak in spring-summer following the resource pulse from the previous autumn [7, 9, 10]. To optimize reproduction with masting events, it is suggested that in European and North American red squirrels [1, 10], chipmunks [11] and fat dormice [12] a mother may increase reproductive output in summers before mast autumns. However, the role of this behaviour in population dynamics of the species remains uncertain [13, 14]. It also remains unclear how general the anticipation behaviour might be for rodents living in forest communities [15].

In this study we test whether Siberian flying squirrels, *Pteromys volans* (hereafter flying squirrels), which depend upon resource pulses of catkins from deciduous trees [16, 17], are able to anticipate current year's resources in fall by increasing reproductive output in

*Correspondence: vessel@utu.fi
[1] Department of Biology, Section of Ecology, University of Turku, 20014 Turku, Finland
Full list of author information is available at the end of the article

summer. Earlier red squirrel and chipmunk studies have indicated that summer litters are produced in anticipation, whereas spring litters are less affected by future food conditions [10, 13, 14]. Thus, we focus our analysis on production of summer litters in flying squirrels.

We predict that (1) if flying squirrels anticipate the abundance of food resources available to juveniles in the winter of their first year, the production of summer litters is related to resource levels in the following autumn and winter. If flying squirrels do not anticipate the resource availability (2) the production of summer litters is related to the resource abundance in the preceding winter and spring. In addition to the food availability, climate is an important determinant of animal reproduction [18]. Thus, we also analyse whether (3) temperature and precipitation preceding reproduction affects the production of summer litters in flying squirrels.

Methods

Study species and its food

The Siberian flying squirrel is a nocturnal, arboreal rodent, which nests in tree cavities, nest-boxes and dreys in spruce-dominated boreal forests. The flying squirrel feeds in deciduous trees that occur within spruce forests, birch, *Betula pubescens/pendula*, alder, *Alnus incana/glutinosa*, and aspen, *Populus tremula*, being the only used deciduous trees in our study areas [19–21; own observation]. During winter and early spring, when flying squirrel mating, pregnancy and parturition of spring litters occur, birch and alder catkins are the main food of flying squirrels [19, 22; Fig. 1]. Birch catkins form the main part of the winter diet (80% of used food, based on faecal diet analysis; [20]), but only alder catkins are stored and are preferred over birch based on analysis of use versus availability [20–22; Fig. 2]. Catkins begin development during the previous summer, and flying squirrels may

start to consume them in autumn [20], continuing to do so during the following winter and early spring (Fig. 1). Catkins flower in spring and birch catkins, which are not stored, are not available when reproductive decisions for summer litters are made. How long alder catkins can be stored in caches is not known, such storage prolongs the time period that catkins are edible, as it prevents catkins from flowering. Catkin production varies considerably between years [16, 17]; see 23] for frequency of pulses in our study areas. Catkin production increases when the previous summer has been warm, however, trees seldom manage to produce mast for two successive years [17, 23]. After the opening of leaves, on average in the beginning of May in our study areas, leaves form major part of late spring and summer diet, together with flower buds of conifers [20, 21]. However, during pregnancy and parturition of spring litters, females may still use catkins, since females are in oestrus and mating occurs starting from mid-March. Spring litters are born in late April. After the spring litter, the second (summer) litter may be born in June, gestation starting in May [24]. Females seem to be territorial, living in separate on average 7 ha home ranges, but males live in overlapping on average 60 ha home ranges encompassing several males and females [25].

Study areas and data gathering

The studies on flying squirrels were done with individuals living in nest-boxes in two study areas located in western Finland: Luoto (63°49′N, 22°49′E) and Vaasa (63°3′N, 22°41′E), located about 100 km from each other. We do not know any obvious behavioural or reproductive differences between individuals living in nest-boxes, dreys, or natural cavities [26], nor in predator community between sites. The entrance-hole diameter of nest-boxes was 4.5 cm. This entrance-hole size prevents main predators

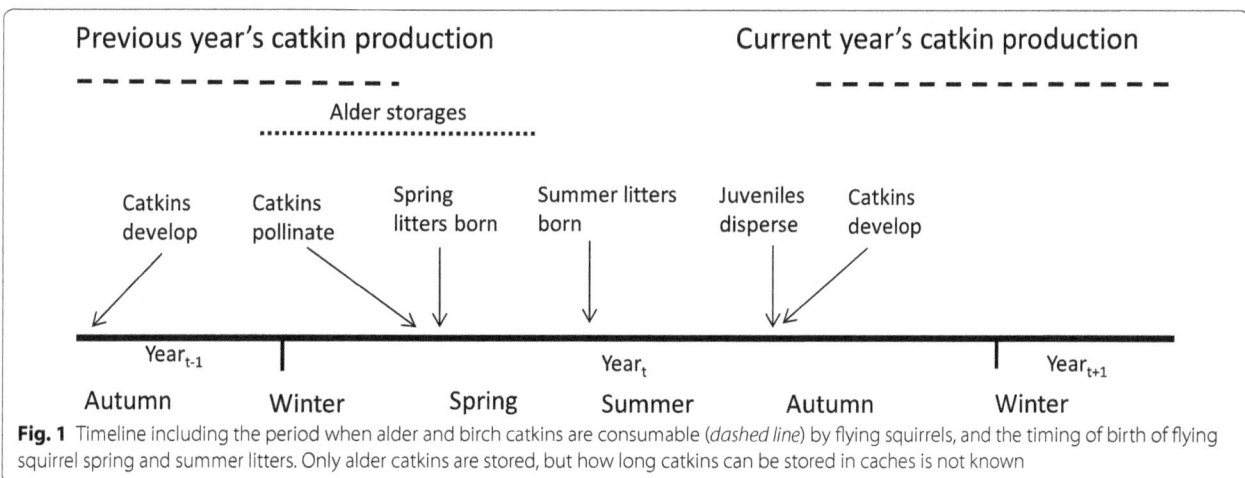

Fig. 1 Timeline including the period when alder and birch catkins are consumable (*dashed line*) by flying squirrels, and the timing of birth of flying squirrel spring and summer litters. Only alder catkins are stored, but how long catkins can be stored in caches is not known

Fig. 2 An example of alder catkin storage made by a flying squirrel. Catkins are cached typically on branches of spruces high up in trees, as in this case, and also sometimes in cavities and nest-boxes. An individual can make several different caches. ©Pertti and Risto Sulkava

(the pine marten, *Martes martes*, and large owls) entering the nest box. Nest-boxes were made from a piece of aspen or spruce trunk, so that they resembled natural cavities. Natural cavities were rare in our study areas (on average 0.1 cavities per hectare based on 742 spruce forest hectares surveyed within our study areas).

In Luoto flying squirrels were studied during 1993–2014 within an area of 44 km^2, where between 300 and 400 nest-boxes were placed for flying squirrels. The main forest types in Luoto are shoreline spruce-dominated mixed forests, clear-cuts, and cultivated Scots pine forests. In Vaasa the marking of flying squirrels started in 1992. The Vaasa study area was 25 km^2 and is covered by spruce forest patches, clear-cuts, and agricultural fields (for more information see [26, 27]. 200–400 nest-boxes were placed within the Vaasa study area to be used by flying squirrels during the study period.

We placed flying squirrel nest-boxes in forest patches of various sizes in sets of 2–4 nest-boxes per site, on average two nest-boxes per mature spruce forest hectare. Box occupancy percentage by the flying squirrel was low (25% nest box occupancy), that is, in most cases a nest box was empty when checked. Flying squirrels were captured by hand in nest-boxes, sexed, weighed, and marked with ear-tags (Hauptner 73850, Hauptner, Germany). The nest-boxes were checked during two sessions in June and August. The latter session was for locating summer litter

juveniles. All boxes were checked in spring, but on our second (August) nest box session we focused only on nest box sites occupied by females during the spring (June) nest-box check.

We calculated the number of summer litters occurring in both study areas each year (Table 1), spring litter production of flying squirrels is analysed in [23]. For analysis of summer litters we only included cases where the female was witnessed to successfully produce the spring litter juveniles close to weaning age. In some cases we observed only late born litters without knowledge whether the mother had produced the spring litter. These cases were omitted from the data, because we did not know whether we missed the spring litter (it could be in drey, i.e. a twig nest) or whether the female failed to produce the spring litter. Number of omitted litters was on average 1.1 ± 0.9 litters in Luoto and 1.3 ± 1.5 litters in Vaasa per year. The occurrence of these omitted litters was positively correlated with the number of summer litters born to mothers with observed spring litter each year (estimate 0.15 ± 0.07; $F_{40} = 4.80$, p = 0.03). During the first nest-box checking session (mean date 14th June) litters had not been weaned (mean weight 59 ± 11 g). The summer litters were on average 56 ± 12 g during the second nest box checking session on average in 18th of August (during this time spring litter juveniles are around 100 g; adult body mass is usually 100–150 g).

Table 1 Data for spring litters (n = 640) and summer litters (n = 93) within the two flying squirrel study areas

	Vaasa (mean ± SD)	Luoto (mean ± SD)
Spring litters		
No of litters[a]	n = 404, 18 ± 11 per year	n = 236, 12 ± 6 per year
Litter size[b]	2.5 ± 0.72	2.5 ± 0.86
Body mass[c]	58 ± 11 g	60 ± 10 g
Summer litters		
No of litters	n = 70, 3.3 ± 3.5 per year, min 0, max 12	n = 23, 1.2 ± 1.1 per year, min 0, max 4
Litter size[b]	2.3 ± 0.8	2.6 ± 0.7
Body mass[c]	54 ± 11 g	59 ± 12 g
Years studied	1992–2014	1993–2014[d]

[a] Number of sites with spring litters and checked to locate the possible summer litter

[b] Mothers with summer litter: 2.48 ± 0.65, n = 88 and Mothers without summer litter: 2.52 ± 0.8, n = 547

[c] Body mass at capture on average 14th of June for spring litter and 18th of August for summer litter

[d] Luoto: years 2007 and 2008 omitted due to lack of data

Food abundance indices

We used estimates from an annual birch-catkin survey conducted by the Finnish Forest Research Institute [28] to estimate food available to flying squirrels each year. These data describe nation-wide pollen availability in Finland. Birch catkins were counted in winter from seed-crop observation stands. The catkin data originated from 15 permanent research stands, where catkins were counted from 30 to 50 birches per stand. Observations were made repeatedly from the same individual trees each year [28]. At our Vaasa study site, a seed-crop observation stand was located within the study area. The closest seed-crop observation stand to our Luoto study site was the Vaasa observation stand located 90 km away. Thus, we used Vaasa indices for both of our study areas, since according to previous analysis of this catkin data, correlation between two sampling sites at this distance is high (r ≈ 0.7), because catkin production of deciduous trees is spatially correlated at scales of up to few hundred kilometers in Finland [16]. Although the food index for Luoto is less accurate than for Vaasa, it describes the yearly variation in catkin production in the area. Both study areas located in coastal area with very similar weather conditions.

For alder there was no catkin count data, but as a proxy we used aerial pollen estimates that correlate with catkin production [16]. Pollen data was collected by the aerobiology unit at University of Turku. Pollen samples were collected from 10 different locations in Finland with EU standard methods and Burkard samplers. The data consisted of accumulated sums of average daily counts of airborne pollen in 1 m³ of air during spring (16; http://www.siitepoly.fi/en/). Similarly as above for birch catkin data, we used Vaasa sampling site for both of our study areas.

Alder pollen and birch catkin data are correlated, albeit not very strongly ($r^2 = 0.31$ for years 1992–2014 in our dataset).

Weather indices

We used weather information from the closest weather station maintained by the Finnish Meteorological Institute to both study areas. For Vaasa the closest weather station was located within our study area, and for Luoto it was 10 km southeast of the study area. Weather recording stations were at the same altitude with study areas.

We used monthly average weather indices from November prior to gestation to June following lactation. We selected the following periods: For winter weather, we used average temperature and the amount of precipitation in December–January (early winter) and the average temperature and amount of precipitation in February–March (late winter) in our analysis. For spring weather, instead of monthly average temperatures, we used (1) the start date of the growing season, that is, the date after which the average daily temperature in spring was permanently above +5 °C. Additionally, we used (2) growing degree days in April and May (the sum of degrees that in daily average temperature were above 5 °C in a given month). These indices were assumed to describe spring conditions better than mere temperature, although we also tested the effect of temperature in April–May. Temperature permanently above +5 °C is determined to indicate start of growing season by Finnish Meteorological Institute (http://en.ilmatieteenlaitos.fi/seasons-in-finland) and has been observed, for example, to well correspond to birch bud burst in Finland [29]. Lastly, we used precipitation in April–May and temperature and precipitation in June (summer) in our analyses.

Analysis

Despite the obvious correlations between different weather and resource data, the explanatory variables were relatively independent from each other. We did not allow the variables, past birch and start of growing season, in the same model. This resulted in low collinearity between variables (Variance inflation factor values <2, Proc Reg, SAS 9.3).

To analyse the effects of different food and weather variables on occurrence of summer litters, we used multi-model inference based on Akaike's information criterion (AIC, smaller values being better). We used AICc values designed for a small sample size and did not include more than three explanatory variables at a time to the model to avoid over-parameterisation. This was done because, in this analysis, the sampling unit was a year. If there was no single clear best fit model or parameter, we used model averaging, using cut-off ΔAIC of 10 and including all models where the term of interest appeared [30]. From the results of model averaging, we considered a parameter to be important in explaining squirrel reproduction if its coefficient and associated 95% confidence interval did not include zero (the obtained results were the same, if we used generalized linear models, analysis not shown). We built models with binomial distribution with GLIM-MIX (SAS), using the events/trial option, such that the 'event' was the number of summer litters observed and the 'trial' was the total number of sites that had a spring litter and that were inspected for a possible summer litter in each study area each year. The explanatory variables were future (current years' autumn and winter following lactation and weaning) and past (previous winter and spring preceding gestation) catkin production of birch or pollen estimate of alder and aspen (proxy for catkin production) and above described temperature and precipitation estimates before reproduction. The study area was selected as a class variable in the model.

To gain further information on recourse availability/female condition before production of summer litters, we compared body mass of spring litters born to mothers with summer litters and spring litters born to mothers without summer litters. If spring litters were large when observed (born earlier and/or grown faster) that indicates good resource situation before reproduction [23]. For this analysis we only used litters weighed during the same day each year (body mass was calculated as an average for a litter). In addition, we tested whether the age of mother affected its likelihood to have a summer litter. For this analysis we used only females ear-tagged as juveniles, so that the exact age of individual was known. Whether or not a female was observed to produce one or two litters a year was a dependent variable (binomial distribution). The age of the mother as well as the study area were

selected as explanatory variables. The ID of the mother was a random variable using Kenward-Roger method to determine degrees of freedom. Finally, with binomial model we tested whether or not a female had summer litter was related to the size of its spring litter. In this model individual ID and year were random variables; study area was included as class variable. The above analyses were done with generalized linear mixed models in GLIM-MIX, SAS.

Results

We had data for 547 females with only a spring litter and 93 females with both summer and spring litters (total 733 litters; Table 1). Thus, about 15% of mothers were observed to produce summer litters (Figs. 3, 4). Litter sizes were quite similar between summer and spring litters (Table 1) and the size of spring litter was not related to likelihood to produce a summer litter ($F_{1,323.8} = 1.3$, $p = 0.24$; Table 1). The mother's age (age range 1–6 years) was not related to the likelihood of producing summer litters (n = 111 cases; $F_{1,24} = 0.32$, $p = 0.58$).

Summer litters were not produced in anticipation of the future resource availability in the autumn and winter of a juvenile's first year. Instead, alder catkin production during winter and spring before reproduction was significantly related to occurrence of summer litters (coefficient 1.5, 95% CI 1.1 and 2.1; Fig. 4). The top models explaining the occurrence of summer litters included also birch catkins before reproduction and early winter rain (Fig. 3; Table 2). The effect of early winter rain was significant in the model (coefficient 2.1, 95% CI 1.0 and 3.1) but birch had no obvious effect (coefficient 0.29, 95% CI −0.02 and 0.6). Future alder and birch estimates clearly lowered the model fit (Table 2; increase in ΔAIC, alder: 26, birch: 29). Future alder pollen production had a significant, but negative, effect (coefficient −0.30, 95% CI −0.05 and −0.65; Fig. 4), because a mast year with a high number of summer litters is typically followed by a low resource year (Fig. 3). Body mass of juveniles in spring litters born to mothers with summer litters was on average 5 ± 6 g larger than body mass of juveniles in spring litters born to mothers without summer litters (difference to expected 0 g difference: $t_{45} = 5.4$, $p < 0.0001$). In other words, if a female produced summer litter, her spring juveniles were born earlier or grew faster compared to juveniles of females who did not produce summer litters.

Discussion

We observed that flying squirrels reproductive investment did not anticipate future resource availability. Instead, food levels before reproduction explained increased reproductive effort, in the form of summer litters. In addition, females who produced summer

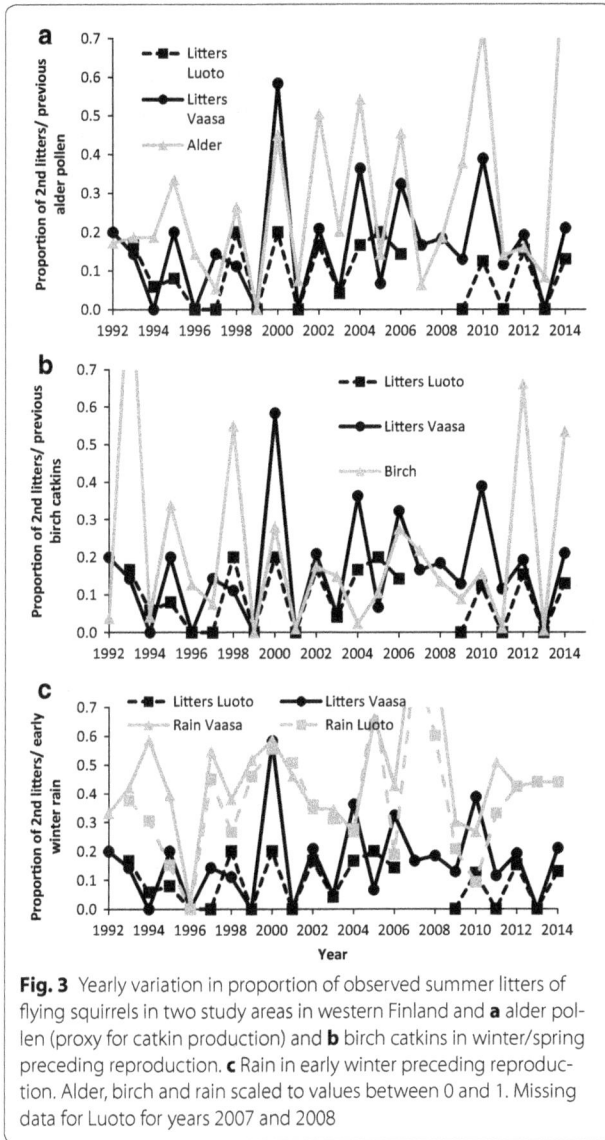

Fig. 3 Yearly variation in proportion of observed summer litters of flying squirrels in two study areas in western Finland and **a** alder pollen (proxy for catkin production) and **b** birch catkins in winter/spring preceding reproduction. **c** Rain in early winter preceding reproduction. Alder, birch and rain scaled to values between 0 and 1. Missing data for Luoto for years 2007 and 2008

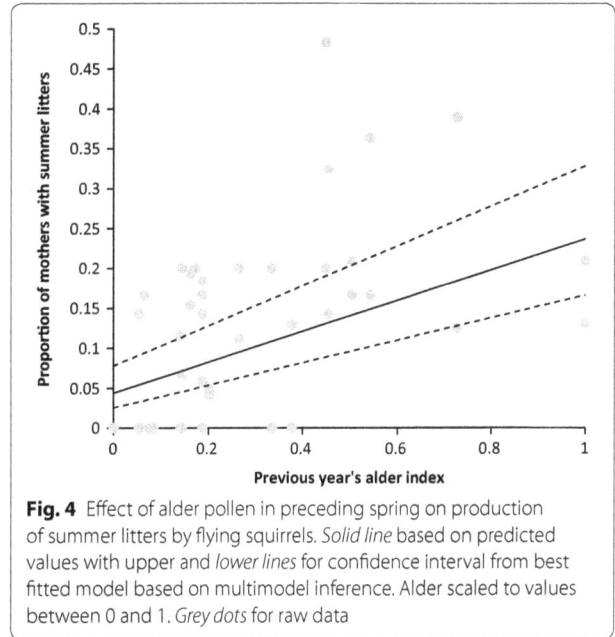

Fig. 4 Effect of alder pollen in preceding spring on production of summer litters by flying squirrels. *Solid line* based on predicted values with upper and *lower lines* for confidence interval from best fitted model based on multimodel inference. Alder scaled to values between 0 and 1. *Grey dots* for raw data

Table 2 Ranking of the best candidate models to explain occurrence of flying squirrel summer litters in Vaasa and Luoto study areas between 1992 and 2014

Model[a]	AICc	ΔAICc	AICc weight
Alder_previous + Rdecjan	117.3	0	0.26
Alder_previous + Rdecjan + birch_previous	117.5	0.8	0.20
Alder_previous + Rdecjan + Tfebmar	119.3	2.0	0.09
Alder_previous + Rdecjan + aspen_previous	119.4	2.1	0.09
Alder_previous + Rdecjan + DdaysApril	119.5	2.2	0.09
Alder_previous + Rdecjan + DdaysMay	119.8	2.5	0.07
Alder_previous + Rdecjan + TJune	119.9	2.6	0.07
Alder_current + Rdecjan	143.5	26.2	0

The best model for both the future and the past food availability are shown. The AICc value, as well as the change in AICc (ΔAICc) and relative weight of support (AIC weight) are shown for each model. Models with cumulative $Wi = 0.90$ presented

[a] Variable names: *T* temperature in given month; *R* rain in given month; *decjan* December–January; *febmar* February–March; *Ddays* degree days; *Aspen* Aspen pollen estimate; *Alder* alder pollen estimate; *Birch* birch catkin estimate; *previous* pollen/catkin estimate available preceding gestation; *current* current years' pollen/catkin estimate available after lactation and weaning. Study area was included in all models, and it had a significant effect (coefficient and c.l. > 0), since the proportion of summer litters was low in Luoto likely due to a lower density of nest-boxes in Luoto than in Vaasa

litters had managed to produce spring litters earlier than females who only produced spring litters. This further supports the conclusion that production of summer litters is related to the condition of a female before reproduction.

Our results support the hypothesis that reproductive decisions are determined by the condition of females at the time of reproduction. This kind of behaviour is typical in, for example, income breeding species, like grazers depending on spring plant growth [31] or insectivorous birds [32]. Foraging behaviour of flying squirrels differs from the behaviour of these species since, in winter and spring flying squirrels depend on food that has already developed during the previous autumn, i.e. an example of a capital breeder strategy. It seems likely

that storages of alder catkins are important for fuelling reproduction of flying squirrels in summer, since the alder was more clearly related to reproduction than was birch catkin production, an important, but not cached, winter food.

After successfully weaning the spring litter, reproducing again during the same summer seems to require good environmental conditions. The observed relationship between food resources and production of summer litters was clear (Figs. 3a, 4). However, the proportion of mothers with summer litters may be slightly underestimated, since it is possible that we missed a few summer litters, if some females moved from nest boxes to dreys (twig nest). In particular in the Luoto study area, the low number of summer litters is likely due to a lower nest-box density in this study area than in the Vaasa study area [26], which lowers the likelihood of finding summer litters. Nevertheless, both study areas gave similar support for the effect of past alder catkin availability on production of summer litters in flying squirrels.

Weather was also linked to production of summer litters in flying squirrels. Surprisingly, precipitation in winter prior to gestation, not the temperature in spring or summer, was linked to the occurrence of summer litters. It remains unclear what is behind this observed correlation, and in the time-series of the data (Fig. 3c) the relationship was not very clear. However, the lack of sufficient soil moisture is an important stress factor for deciduous trees [33], and it is possible that dry or snowless winter conditions affect moisture conditions and consequently flowering buds or leaves in spring and summer, which provide food for flying squirrels. Indeed, the quality of summer food is a likely candidate that affects summer reproduction of the species. Unfortunately, we were unable to directly study this, but leaf growth is tightly linked to weather conditions during the time period included in our analysis. In any case, the effect of weather on production of summer litters needs further study due to correlative nature of our analysis.

Our results from flying squirrels provide an example of forest-dependent rodent species not able to anticipate a mast. This result is in contrast to observations in some other studies on rodents [1, 10, 11, 34, 35]. For example, North American red squirrels [1] are likely more dependent on cached food than Siberian flying squirrels. North American red squirrels clip new spruce cones containing seeds each autumn and cache them in a larger hoard called a midden [36, 37]. The dependency on middens [38] might increase the adaptive reasons to anticipate the mast in North American red squirrels. However, anticipation is suggested to also occur in forest rodents other than North American red squirrels [10, 11, 34]. The adaptive reasons to anticipate the mast should occur also in flying squirrels as the production of food consumed by flying squirrels is quite similar to that of, for example, Eurasian red squirrels [10]. Flying squirrels start to consume catkins in autumn and continue to do so during the following winter and early spring, when the catkins flower. Thus, if a female could anticipate the coming mast, its offspring would face the winter with optimal resource availability. In addition, variance in birch and alder catkin production [23] is comparable to variation in spruce cone production used by red squirrels [1, 14]. Furthermore, the Siberian flying squirrel are entirely dependent on trees, and very seldom move on the ground (North American flying squirrels, *Glaucomys* spp., move regularly on the ground, e.g. when they harvest truffles). In winter the only foods available for flying squirrels are catkins and buds. However, during summer food other than catkins seems to be sufficient for reproduction as some summer litters were also produced following poor catkin winters. Thus, mast conditions do not appear to be essential for the production of summer flying squirrel litters.

For species observed to anticipate mast, it has previously been speculated that buds that eventually develop into cones/seeds are used to predict the future resource availability [1, 36]. For example, the edible dormouse, *Glis glis*, has been suggested to use the flower buds of the European beech, *Fagus sylvatica*, in spring as a sign of mast [12, 35]. The dormice gain energy from eating these buds, and it has also been observed that food supplementation in spring increases the summer production of this species ([12, 35]; however, for North American red squirrels see [1, 39]). Similarly, in flying squirrels, abundant food resources in the spring were positively correlated with the production of summer litters. However, for dormice the situation is different, as increased energy from flower buds also correlates with a future good seed situation that will benefit the offspring the next autumn and winter [40, 41]. With flying squirrels, the juveniles from summer litters will face the winter without catkins, since mast is generally followed by poor investment in reproduction by trees [42]. This may be problematic, since the survival rate of rodent juveniles is generally highest in mast conditions [43].

Conclusions

We observed that Siberian flying squirrels do not anticipate the coming mast, but instead adjust their reproductive decision based on current and past food availability. For flying squirrels, an increased reproductive effort is simply a consequence of favourable environmental conditions, which allow females to increase offspring production. The reproductive strategy of Siberian flying squirrels appears to be an opportunistic strategy, depending on the current resource availability, without possibilities to anticipate the future conditions the offspring will face when they mature.

Authors' contributions
RW collected data, VS analysed data and wrote the manuscript. Both authors read and approved the final manuscript.

Author details
[1] Department of Biology, Section of Ecology, University of Turku, 20014 Turku, Finland. [2] Department of Biology, University of Oulu, 90014 Oulu, Finland.

Acknowledgements
We thank all of the field workers, Timo Hyrsky, Rune Jakobsson, Antero Mäkelä and Markus Sundell, who have assisted during data gathering. Annika Saarto kindly provided the catkin and pollen data. Leigh Ann Lindholm gave valuable comments on an earlier version of the manuscript.

Competing interests
The authors declare that they have no competing interests.

Funding
The study was financially supported by Oskar Öflunds stiftelse (to RW), Societas Pro Fauna et Flora Fennica (to RW), Vuokon luonnonsuojelusäätiö (to RW), the Academy of Finland (Grant number 259562 to VS).

References
1. Boutin S, Wauters LA, McAdam AG, Humphries MM, Tosi G, Dhondt AA. Anticipatory reproduction and population growth in seed predators. Science. 2006;314:1928–30.
2. Valone TJ. Are animals capable of Bayesian updating? An empirical review. Oikos. 2006;112:252–9.
3. Raby CR, Alexis DM, Dickinson A, Clayton NS. Planning for the future by western scrub-jays. Nature. 2007;445:919–21.
4. Railsback SF, Harvey BC. Trait-mediated trophic interactions: is foraging theory keeping up? Trends Ecol Evol. 2013;28:119–25.
5. Ostfeld RS, Keesing F. Pulsed resources and community dynamics of consumers in terrestrial ecosystems. Trends Ecol Evol. 2000;15:232–7.
6. Lindström E. Reproductive effort in the red fox, *Vulpes vulpes*, and future supply of a fluctuating prey. Oikos. 1988;52:115–9.
7. Jędrzejewska B, Jędrzejewski W. Predation in vertebrate communities: the Bialowieza Primeval Forest as a case study. Berlin: Springer-Verlag; 1998.
8. Lobo N, Millar JS. Indirect and mitigated effects of pulsed resources on the population dynamics of a northern rodent. J Anim Ecol. 2013;82:814–25.
9. Bogdziewicz M, Zwolak R, Crone EE. How do vertebrates respond to mast seeding? Oikos. 2016;125:300–7.
10. Wauters LA, Githiru M, Bertolino S, Molinari A, Tosi G, Lens L. Demography of alpine red squirrel populations in relation to fluctuations in seed crop size. Ecography. 2008;31:104–14.
11. Bergeron P, Réale D, Humphries MM, Garant D. Anticipation and tracking of pulsed resources drive population dynamics in eastern chipmunks. Ecology. 2011;92:2027–34.
12. Lebl K, Kürbisch K, Bieber C, Ruf T. Energy or information? the role of seed availability for reproductive decisions in edible dormice. J Comp Physiol B. 2010;180:447–56.
13. Williams CT, Lane JE, Humphries MM, McAdam AG, Boutin S. Reproductive phenology of a food-hoarding mast-seed consumer: resource- and density-dependent benefits of early breeding in red squirrels. Oecologia. 2014;174:777–88.
14. Selonen V, Varjonen R, Korpimäki E. Immediate or lagged responses of a red squirrel population to pulsed resources. Oecologia. 2015;177:401–11.
15. White TCR. Mast seeding and mammal breeding: can a bonanza food supply be anticipated? N Z J Zool. 2007;34:179–83.
16. Ranta H, Hokkanen T, Linkosalo T, Laukkanen L, Bondenstam K, Oksanen A. Male flowering of birch: spatial synchronization, year-to-year variation and relation of catkin numbers and airborne pollen counts. Forest Ecol Manage. 2008;255:643–50.
17. Ranta H, Oksanen A, Hokkanen T, Bondestam K, Heino S. Masting by *Betula*-species; applying the resource budget model to north European data sets. Int J Biometeorol. 2005;49:146–51.
18. White TCR. The role of food, weather and climate in limiting the abundance of animals. Biol Rev. 2008;83:227–48.
19. Mäkelä A. Liito-oravan, Pteromys volans L. Ravintobiologiasta (feeding biology of flying squirrel, in Finnish). Master's Thesis, University of Oulu, Finland. 1981.
20. Mäkelä A. Liito-oravan (*Pteromys volans* L.) ravintokohteet eri vuodenaikoina ulosteanalyysin perusteella (diet of flying squirrel, in Finnish). Helsinki: WWF Finland Reports 8; 1996. p. 54–8.
21. Hanski IK, Mönkkönen M, Reunanen P, Stevens PC. Ecology of the Eurasian flying squirrel (*Pteromys volans*) in Finland. In: Goldingay R, Scheibe J, editors. Biology of gliding mammals. Furth (Germany): Filander; 2000. p. 67–86.
22. Sulkava P, Sulkava R. Liito-oravan ravinnosta ja ruokailutavoista Keski-Suomessa (feeding habits of flying squirrel; in Finnish). Luonnon tutkija. 1993;97:136–8.
23. Selonen V, Wistbacka R, Korpimäki E. Food abundance and weather modify reproduction of two arboreal squirrel species. J Mammal. 2016;97:1376–84.
24. Hanski IK, Selonen V. Female-biased natal dispersal in the Siberian flying squirrel. Behav Ecol. 2009;20:60–7.
25. Selonen V, Painter JN, Rantala S, Hanski IK. Mating system and reproductive success in the Siberian flying squirrel. J Mammal. 2013;94:1266–73.
26. Selonen V, Hanski IK, Wistbacka R. Communal nesting is explained by subsequent mating rather than kinship or thermoregulation in the Siberian flying squirrel. Behav Ecol Socio. 2014;68:971–80.
27. Lampila S, Wistbacka A, Mäkelä A, Orell M. Survival and population growth rate of the threatened Siberian flying squirrel (*Pteromys volans*) in a fragmented forest landscape. Ecoscience. 2009;16:66–74.
28. Hokkanen T. Seed crops and seed crop forecasts for a number of tree species. In: Mälkönen E, Babich NA, Krutov VI, Markova IA, editors. Forest regeneration in the northern parts of Europe. Proceedings of the Finnish–Russian forest regeneration seminar in Vuokatti, Finland, Sept 28th–Oct 2nd, 1998. Metsäntutkimuslaitoksen tiedonantoja—The Finnish Forest Research Institute, Research Papers. 2000;790: 87–97.
29. Rousi M, Heinonen J. Temperature sum accumulation effects on within-population variation and long-term trends in date of bud burst of European white birch (*Betula pendula*). Tree Physiol. 2007;27:1019–25.
30. Burnham KP, Anderson DR. Model selection and multimodel inference: a practical information-theoretic approach. 2nd ed. New York: Springer; 2002.
31. Post E, Forchhammer MC. Climate change reduces reproductive success of an arctic herbivore through trophic mismatch. Phil Trans Roy Soc Lond B. 2008;363:2369–75.
32. Eeva T, Veistola S, Lehikoinen E. Timing of breeding in subarctic passerines in relation to food availability. Can J Zool. 2000;78:67–78.
33. Possen BJHM, Oksanen E, Rousi M, Ruhanen H, Ahonen V, Tervahauta A, Heinonen J, Heiskanen J, Kärenlampi S, Vapaavuori EM. Adaptability of birch (*Betula pendula* Roth) and aspen (*Populus tremula* L.) genotypes to different soil moisture conditions. Forest Ecol Manage. 2011;262:1387–99.
34. Marcello GJ, Wilder SM, Meikle DB. Population dynamics of a generalist rodent in relation to variability in pulsed food resources in a fragmented landscape. J Anim Ecol. 2008;77:41–6.
35. Kager T, Fietz J. Food availability in spring influences reproductive output in the seed-preying edible dormouse (*Glis glis*). Can J Zool. 2009;87:555–65.
36. Smith CC. The coevolution of pine squirrels (*Tamiasciurus*) and conifers. Ecol Monogr. 1970;40:349–71.
37. Fletcher QE, Boutin S, Lane JE, LaMontagne JM, McAdam AG, Krebs CJ, Humphries MM. The functional response of a hoarding seed predator to mast seeding. Ecology. 2010;91:2673–83.

38. LaMontagne JM, Williams CT, Donald JL, Humphries MM, McAdam AG, Boutin S. Linking intraspecific variation in territory size, cone supply, and survival of North American red squirrels. J Mammal. 2013;94:1048–58.

39. Larsen KW, Becker CD, Boutin S, Blower M. Effects of hoard manipulations on life history and reproductive success of female red squirrels (*Tamiasciurus hudsonicus*). J Mammal. 1997;78:192–203.

40. Bieber C. Population dynamics, sexual activity, and reproduction failure in the fat dormouse (*Myoxus glis*). J Zool. 1998;244:223–9.

41. Pilastro A, Tavecchia G, Marin G. Long living and reproduction skipping in the fat dormouse. Ecology. 2003;84:1784–92.

42. Kelly D, Sork VL. Mast seeding in perennial plants: why, how, where? Annu Rev Ecol Syst. 2002;33:427–47.

43. Yang LH, Edwards KF, Byrnes JE, Bastow JL, Wright AN, Spence KO. A meta-analysis of resource pulse–consumer interactions. Ecol Monogr. 2010;80:125–51.

6

Population density of the western burrowing owl (*Athene cunicularia hypugaea*) in Mexican prairie dog (*Cynomys mexicanus*) colonies in northeastern Mexico

Gabriel Ruiz Ayma[1], Alina Olalla Kerstupp[1], Alberto Macías Duarte[2], Antonio Guzmán Velasco[1] and José I. González Rojas[1*]

Abstract

Background: The western burrowing owl (*Athene cunicularia hypugaea*) occurs throughout western North America in various habitats such as desert, short-grass prairie and shrub-steppe, among others, where the main threat for this species is habitat loss. Range-wide declines have prompted a need for reliable estimates of its populations in Mexico, where the size of resident and migratory populations remain unknown.

Results: Our objective was to estimate the abundance and density of breeding western burrowing owl populations in Mexican prairie dog (*Cynomys mexicanus*) colonies in two sites located within the Chihuahuan Desert ecoregion in the states of Nuevo Leon and San Luis Potosi, Mexico. Line transect surveys were conducted from February to April of 2010 and 2011. Fifty 60 ha transects were analyzed using distance sampling to estimate owl and Mexican prairie dog populations. We estimated a population of 2026 owls (95 % CI 1756–2336) in 2010 and 2015 owls (95 % CI 1573–2317) in 2011 across 50 Mexican prairie dog colonies (20,529 ha).

Conclusions: The results represent the first systematic attempt to provide reliable evidence related to the size of the adult owl populations, within the largest and best preserved Mexican prairie dog colonies in Mexico.

Keywords: Chihuahuan Desert, Distance sampling, Grassland, Mexican prairie dog, Mexico, Population, Western burrowing owl

Background

Rigorous estimates of regional population size are critical for the development and assessment of avian conservation strategies, particularly for species undergoing shifts in their distribution and range. The western burrowing owl (*Athene cunicularia hypugaea*) (Fig. 1), a species with special conservation status throughout much of its range, has experienced range-wide shifts from southern Canada to northern Mexico [1]. Western burrowing owls belong to a grassland bird guild that is threatened by habitat loss [2].

The species uses open habitats such as grasslands, deserts and areas of disturbance [3]. These owls also prefer areas with discontinuous vegetation and low growth shrubs, allowing them to increase visibility for hunting, vigilance against predators and caring of burrows [4, 5].

Published data from owl populations vary within the range of distribution in North America. For example, in the 1990's population estimates of this species in Canada and the United States of America (USA) ranged from as low as 2000–20,000 to as high as 20,000–200,000 individuals [6]. In Canada, the populations have declined abruptly and even disappeared from British Columbia and Manitoba [7]. Previous reports indicate a wide variation of population trends ranging from stable in some

*Correspondence: jose.gonzalezr@uanl.mx
[1] Facultad de Ciencias Biologicas, Universidad Autonoma de Nuevo Leon, Ave. Universidad s/n. Cd. Universitaria, 66455 San Nicolas de los Garza, Nuevo Leon, Mexico
Full list of author information is available at the end of the article

Fig. 1 Western burrowing owl in the colony of Mexican prairie dog, in Chihuahuan Desert

areas in the USA and Canada, to reduced, extirpated or increasing in others [2, 7–16].

Local density estimates a range of 13–31 owls/km^2 in Canada (Manitoba, Alberta, Saskatchewan and British Columbia) and the USA (Arizona, California, Colorado, Idaho, Iowa, Kansas, Minnesota, Montana, Nebraska, New Mexico, North and South Dakota, Oklahoma, Oregon, Texas, Utah and Washington) during the western burrowing owl breeding season indicating variation in the density estimates [1, 7, 9, 14, 17–26]. In Mexico, the federal government classifies the western burrowing owl under the category of special protection [27]. Habits of the western burrowing owl such as summer diet, prey selection, movement of juveniles, selection of nesting sites and threats remain poorly known. Densities estimated during the breeding season in 2002 in Mexico include 14.1 owls/km^2 near Mexicali [28], 3.2 pairs/km^2 in Yaqui-Mayo Valley, Sonora, 4.5 pairs/km^2 in Valle del Fuerte, Sinaloa, and 4.7 pairs/km^2 in Valle de Culiacan, Sinaloa [29]. Winter season density estimates in central Mexico include 11 owls/km^2 in Guanajuato [30] and 5.2 owls/km^2 in Nuevo Leon [31].

The western burrowing owl has been strongly associated with two species of prairie dogs in Mexico, the Mexican prairie dog (*Cynomys mexicanus*) and black-tailed prairie dog (*C. ludovicianus*) (Fig. 2) [9, 32–35]. Both of these species are federally listed in Mexico as endangered and threatened, respectively [27]. The black-tailed prairie dog is distributed from Saskatchewan in Canada to southern Montana and Nebraska in the United States to northern Chihuahua and Sonora in Mexico where the colonies are fragmented and isolated. The habitat occupied by the species of prairie dog is herbs, grasses and shrubs. Currently, the regions supporting black-tailed

prairie dog colonies cover 18,500 ha [36]. The Mexican prairie dog is endemic of central and northern of Mexico within the states of Coahuila, Nuevo Leon, Zacatecas and San Luis Potosi, in colonies covering approximately 25,000 ha. These two species of dogs have lost more than 80 % of their original range [37]. Mexican prairie dog colonies provide burrows and foraging opportunities for breeding burrowing owls, which apparently keep the prairie dog population stable, despite disturbance and loss of habitat in prairie dog colonies caused by expanding agricultural and cattle grazing activities [38], the use of pesticides, collisions with vehicles, diseases, predators, and urbanization [1, 2, 7, 11, 32, 39–44].

Based on the problems and the lack of knowledge mentioned above, in this study we estimate the abundance and density of western burrowing owls in colonies of Mexican prairie dog in northeastern Mexico. Density/abundances of western burrowing owls and their association with Mexican prairie dog colonies provide relevant conservation information to ensure the long-term persistence of both species. In addition, this study can be integrated across North America to establish baseline range-wide population estimate(s) to improve our understanding of the recent range-wide shifts in owl populations.

Methods

Study area

Our study sites were located in Nuevo Leon (NL) and San Luis Potosi (SLP) within the Chihuahuan Desert ecoregion [45] (Fig. 2) that is part of the physiographic region known as the Mexican Plateau within the Mexican states of Coahuila, Zacatecas, NL and SLP. The semi-arid climate features temperatures ranging from 6 to 25 °C with an annual average of 16 °C [46]. Average precipitation totals 427 mm [47].

Previously, studies in NL have been conducted in the areas known as Llano de la Soledad (23°53′N, 100°42′W) and Compromiso (23°53′N, 100°42′W). These areas maintain the largest Mexican prairie dog populations, including those at Martha (25°0′N, 100°40′W), Concha (25°1′N, 100°35′W), and Hediondilla (24°57′N, 100°42′W). Western burrowing owls of SLP were studied in Llano del Manantial (24°7′N, 100°55′W) and Gallo (24°12′N, 100°54′W) in the municipality of Vanegas.

The Llano de la Soledad has been provided with several conservation designations by the NL government such as State Natural Protected Area [48], and Important Site for Bird Conservation [49]. This site hosts several vulnerable, endemic and migratory species [50, 51]. The dominant vegetation in Mexican prairie dog colonies is characterized by halophytic grassland and consists largely of *Muhlenbergia villiflora*, *Muhlenbergia repens*, *Pleuraphis*

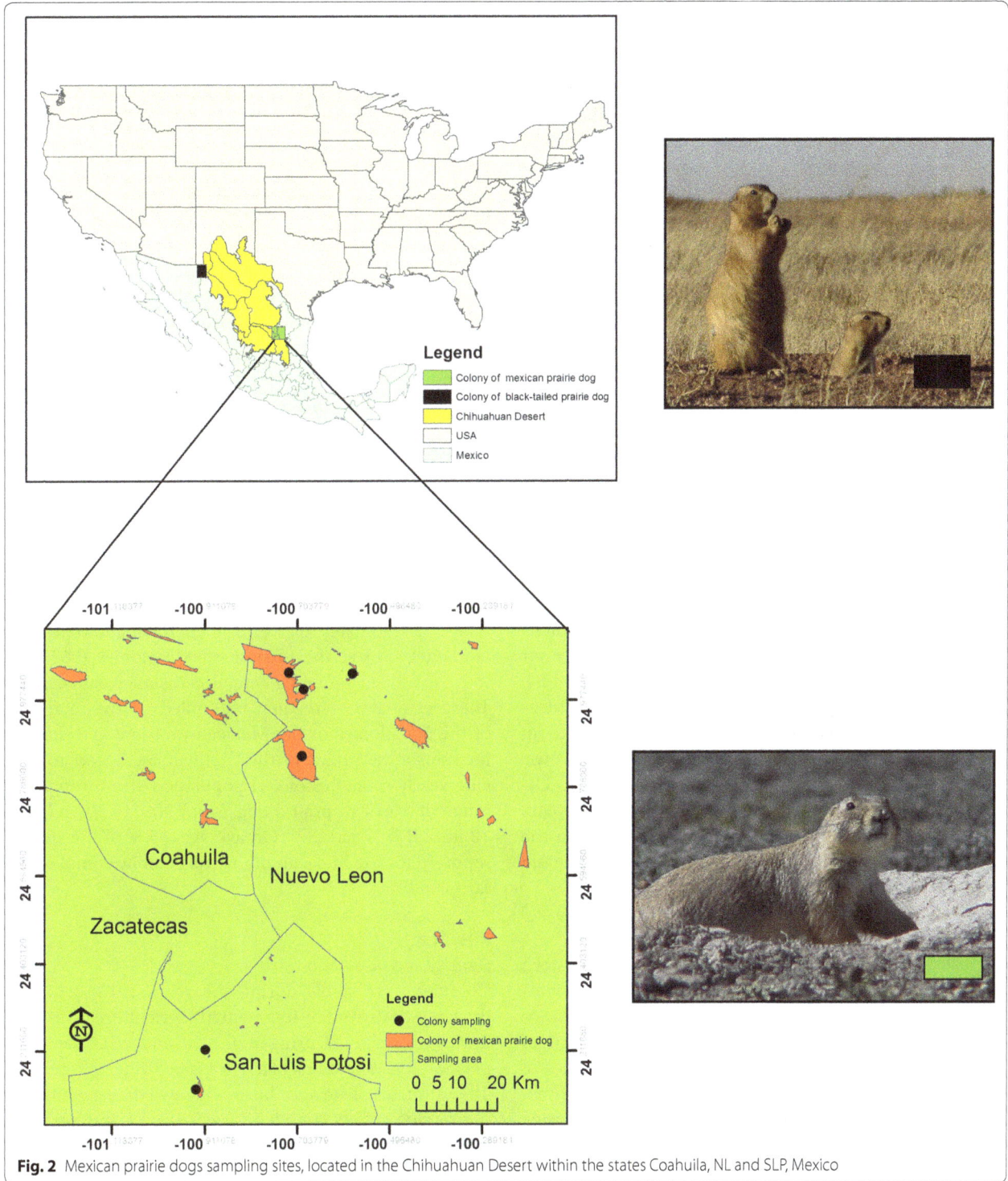

Fig. 2 Mexican prairie dogs sampling sites, located in the Chihuahuan Desert within the states Coahuila, NL and SLP, Mexico

mutica, Sporobolus airoides, Frankenia gypsophila and *Dalea gypsophila.* Other coexisting plant communities include microphyllus vascular plants and rosette shrubs [31, 52–56].

From 50 colonies of prairie dogs existing in NL and SLP, nine were selected for sampling. These colonies were selected based on the following characteristics: spatial continuity of the community and a lack of fragmentation, conservation status of the site, vegetation type that was homogeneous enough to contain at least one complete transect. The sampled colonies covered about 55 % of the area available for all colonies of Mexican prairie dogs in the southern Chihuahuan Desert. Sampling was conducted between February and April in both 2010 and 2011. The transect line method was used [57]. Fifty transects (each 2 km long × 0.3 km wide and ≥0.5 km apart from each other) were traveled using the remote sampling method by the observer as described below to estimate the density of adult owls [58]. The number (n) of transect routes for each area was: Soledad ($n = 28$) and Compromiso (15), Marta (2) and Concha (2) in NL; Manantial (2) and Gallo (1) in SLP. We walked each transect at a constant rate using a global positioning system (GPS) to ensure a straight survey line. Owls were detected visually or with binoculars. Then, the perpendicular distance from the transect line route was measured using a laser rangefinder (15–815 m, Leica Rangemaster 900, Optics Planet, Inc. Northbrook, IL, USA). To meet the assumptions of distance sampling, only adult owls were recorded on the ground outside the burrows or without movement [58, 59]. If the bird under observation moved because of the presence of the observer, registering the perpendicular distance was performed at the original site without the observer leaving their sighting transect travel line. Those adult owls flying with an unknown initial location were not documented. To reduce bias and avoid an overestimation of population density, only adult owls were recorded. Considering the extreme desert climate, personal observations made during previous years and different criteria of previous authors related to the activities of owls, the field observations were conducted from 0600 to 1200 h [20, 23, 60, 61].

Data analysis

We used program *DISTANCE ver. 6.0* to obtain western burrowing owl density estimates from distance sampling [62]. *DISTANCE* calculates density and abundance using modeling detection probability as a function of the perpendicular distance to the transect in a series of monotonic models. Several standard detection functions (uniform, half-normal, or hazard-rate) with cosine series adjustment were evaluated using the Akaike information criterion (AIC). We used the AIC to select the model with the most parsimonious detection function in *DISTANCE* [58, 59, 63]. We pooled all data to estimate a single detection function (probability of detection, $g(x)$, at a given distance (x) from the transect) because we did not anticipate effects of environmental features on detection, such as *age* (adult) and factor *STATE* (levels: NL and SLP). We considered serial adjustments of one to three

parameters. We did not truncate the data because the frequencies of long distances observations were better maintained in this manner [4].

The estimator of density ($\hat{D}$) is given by the expression:

$$\hat{D} = \frac{\hat{n}\hat{f}(0)}{2L}$$

where $\hat{f}(0)$ is the probability density function of detection distances from the line evaluated at zero distance, calculated in *DISTANCE* as the average number of individuals per detection [62]. The standard error of density SE ($\hat{D}$), assuming a Poisson distribution of counts, can be approximated using the delta method as follows [58]:

$$SE(\hat{D}) = \hat{D}\sqrt{\frac{1}{n} + \frac{Var(\hat{f}(0))}{(\hat{f}(0))^2}},$$

where $Var(\hat{f}(0)) = (SE(\hat{f}(0)))^2$ also is a direct output of *DISTANCE*. The component cluster size was omitted from the above formulas because virtually all detections were individual records. Estimates of density and their standard errors were used to test statistical differences in density between states and years using a Wald test [64]. Values are presented as mean ± SE.

Overall estimates of western burrowing owl density (and their SE) at the nine sampled colonies (9620 ha) were obtained by pooling detection distance data by year. These estimates were then multiplied by the total area of the 50 colonies of the Mexican prairie dog described for the southern part of the Chihuahuan Desert to provide yearly estimates of owl population size through the range of Mexican prairie dog: 38 in NL (19,802 ha) and 12 in SLP (727 ha) [37]. On average, 55 % of the surface reported for the Mexican prairie dog colony complex was sampled [37].

Results

Density and population size

Colonies were stable during the years 2010–2011 and were not destroyed or fragmented (agriculture, livestock) during this time. During the 2010 and 2011 sampling periods, 235 detections of at least one owl were recorded. The estimates of western burrowing owl density in the 50 prairie dog colony complex were 9.8 ± 1.0 ind/km^2 (CV 0.107) in 2010 and 9.8 ± 1.0 ind/km^2 (CV 0.108) in 2011. The owl density estimate for NL was 8.8 ± 1.0 ind/km^2 (CV 0.114) in 2010 and 7.3 ± 0.9 ind/km^2 (CV 0.123) in 2011. For SLP, the owl population density was 26.7 ± 6.2 ind/km^2 (CV 0.236) in 2010 and 47 ± 8.4 ind/km^2 (CV 0.180) in 2011 (Table 1). No significant differences were found among western burrowing owl densities (Wald test, $p = 0.431$) and the paired states of NL ($p = 0.967$) and SLP ($p = 0.635$).

Table 1 Western burrowing owl population density between 2010 and 2011 in Mexican prairie dog colonies in NL and SLP, Mexico

Model[a]	D[b]	N[c]	Estimated density (owl/ha)			CV[f]	No. colony[g]	Area (ha)
			Average	95 % IC[d]	95 % IC[e]			
Global								
2010								
	0.1	119	949	788	1039	0.107	9	9620*
			2026	1765	2326		50	20,529**
2011								
	0.09	116	944	783	1035	0.108	9	9620*
			2015	1753	2317		50	20,529**
NL								
2010								
	0.08	100	809	698	937	0.114	6	9170*
			1747	1508	2024		50	20,529**
2011								
	0.074	82	678	578	794	0.124	6	9170*
			1464	1248	1716		50	20,529**
SLP								
2010								
	0.26	19	118	37	87	0.236	3	450*
			190	60	141		12	727**
2011								
	0.47	34	211	48	167	0.180	3	450*
			341	170	397		12	727**

* Total area of sampled colonies

** Total area of colonies in SLP and NL

[a] Model base done AIC criteria: half-normal + cosine

[b] Western burrowing owl density (owl/ha)

[c] Total number of detections in both years

[d] Upper confidence intervals

[e] Lower confidence intervals

[f] Variation coefficient for the estimated density

[g] Number of Mexican prairie dog colonies

Applying the overall yearly estimates of western burrowing owl density to the entire area of the 50-colony complex of prairie dogs in NL and SLP resulted in a population size of 2026 (CV 0.173) in 2010 and 2015 (CV 0.213) in 2011. For colonies in NL, an average population size of 1747 (CV 0.178) was obtained in 2010 and 1464 (CV 0.218) for 2011, while in SLP, population estimates were between 190 (CV 0.312) and 341 (CV 0.322) for each year.

Discussion

To date, many density estimates have been made for the western burrowing owl in Canada and the USA, with quite variable results [1, 6, 7, 9, 11–15, 20–23, 25, 26]. The resulting variation in the population sizes can be attributed to the size of sample area, methodology, analytical precision, timing, observer skill, and so on; these have contributed to an inexact picture of the density of the western burrowing owl populations [6, 15]. Therefore, a comparison of our results with those of the USA and Canada could be difficult.

During the last 30 years, the North American Breeding Bird Survey has estimated a negative trend for the western burrowing owl population for Canada and the USA. Similarly, the United States Geological Survey (2014) has reported the same negative trend in the Chihuahuan Desert region [16].

Even though Mexico has not established systematic surveys that allow the establishment of a population trend, some studies (the present one included) can form the basis to achieve this goal of documenting population trends in the future.

In NL and SLP, the average density of breeding pairs (9.8 ind/km^2) in 2010 and 2011 is greater than that reported by Macias-Duarte in Sonora (6.4 ind/km^2) and similar to the Sinaloa average (9.2 ind/km^2) [29]. However, in Baja California, Itubarria-Rojas reported an average of 14.1 ind/km^2, which is a value higher than that determined by the present study [28]. This difference could be caused by the habitat quality among sites as reported in NL and SLP where burrow competition is related to the abundance of prairie dogs per colony or Baja California where the owls use irrigation canals to create burrows.

Our overall estimates of population size for western burrowing owls reveal the relative importance of Mexican prairie dog colonies to owl population viability. No previous data related to population size estimates in owls is available for the study area. However, the precision of these estimates must be taken with caution because of the variability between sites. However, we believe the extrapolation is correct because we sampled over 55 % of the current area with the active prairie dog colonies in both states. The range of the western burrowing owl in northeastern (NL, SLP, and Coahuila) Mexico includes viable colonies of Mexican prairie dogs. These areas provide an optimal habitat for the prairie dogs to feed on grasses and this contributes to a low height of herbaceous plants and allows the owls greater visual access to the foraging area. This species uses prairie dog colonies as a place for nesting, protection against climatic factors (extreme temperatures, flooding by rain, and strong winds). The owls also respond to alarm calls by prairie dogs, alerting them to the presence of predators. The western burrowing owl colonies in Mexico have declined from 88 colonies to 53, equivalent to a loss of 37 % in 10 years (1992–2003) [37, 38].

Many of the problems in northeastern Mexico that involve the western burrowing owl are directly related to loss of habitat from agriculture, but some direct mortality has been caused by collisions with vehicles. However, another possible cause of morbidity and mortality could be the direct or incidental (by bioaccumulation) exposure to pesticides used in neighboring areas.

Conclusions
These results represent the first systematic effort to address the conservation status of the western burrowing owl populations in Mexican prairie dog colonies located in northeastern Mexico. This geographic area is considered to contain the largest preserved Mexican prairie dog colonies in the country and deserves attention from the scientific and conservation communities. Furthermore, these results contribute new information to our understanding of the population dynamics of this

kind of species across North America, and highlight the urgent need to preserve grasslands, particularly those in the southern part of the Chihuahuan Desert, which harbor many bird species cataloged as threatened or endangered.

Abbreviations
Regions
USA: United States of America; NL: Nuevo Leon; SLP: San Luis Potosi.

Units
km: kilometers; ind/km^2: individual per square kilometer; ha: hectare; n: number of line transects; m: meters; mm: millimeter; °C: celsius; hr: hours.

Statistical
gx: probability of detection; x: given distance; $\pm$ SE: standard error; CV: coefficient variation; p: probability; AIC: akaike information criterion; IC: confidence intervals.

Orientation
N: north; W: western; GPS: global positioning system.

Authors' contributions
GRA conceived of and designed the study, collected the data and performed data analysis. AOK, AMD, AGV and JIGR contributed to study design and provided advice for data collection and analysis. All authors participated in drafting the manuscript. All authors read and approved the final manuscript.

Author details
[1] Facultad de Ciencias Biologicas, Universidad Autonoma de Nuevo Leon, Ave. Universidad s/n. Cd. Universitaria, 66455 San Nicolas de los Garza, Nuevo Leon, Mexico. [2] Ley Federal del Trabajo S/N, Universidad Estatal de Sonora, Col. Apolo, 83100 Hermosillo, Sonora, Mexico.

Acknowledgements
We are grateful to Pronatura Noreste Asociacion Civil for *Cynomys ludovicianus* photo (Fig. 2) and their facilities for hospitality throughout the sampling period.

Competing interests
The authors declare that they have no competing interests.

Funding
This research was funded with resources from the Universidad Autonoma de Nuevo Leon through the support program for Scientific and Technological Research (PAICyT).

References

1. Macias-Duarte A, Conway CC. Distributional changes in the western burrowing owl (*Athene cunicularia hypugaea*) in North America from 1967 to 2008. J Raptor Res. 2015;49:75–83.
2. ACA. Commission for Environmental Cooperation. In: North America conservation action plan (*Athene cunicularia hypugaea*). Commission for Environmental Cooperation. Printed in Canada; 2005. p. 1–55. http://www.cec.org Accessed 15 Apr 2009.
3. Clark RJ. A review of the taxonomy and distribution of burrowing owl (*Speotyto cunicularia*). J Raptor Res. 1997;9:14–23.
4. Coulombe HN. Behavior and population ecology of the burrowing owl, *Speotyto cunicularia*, in the Imperial Valley of California. Condor. 1971;73:162–76.
5. Howell GR, Webb S. A Guide to the birds of Mexico and Central America. Oxford University Press; 1995. pp. 364.
6. James PC, Espie RHM. Current status of the burrowing owl in North America: an agency survey. J Raptor Res. 1997;9:3–5.
7. COSEWIC. Assessment and update status report on burrowing owl *Athene cunicularia* in Canada. Committee on the status of endangered wildlife in Canada. Ottawa; 2006. pp. 31.
8. Desante DF, Ruhlen ED, Adamany SL, Butron KM, Amin S. A census of burrowing owls in central California in 1991. In: Lincer J, Steenhof K, editors. The burrowing owl, its biology and management including the proceedings of the first international burrowing owl symposium; 1997.
9. Desmond MJ, Savidge JA. Factors influencing burrowing owl (*Speotyto cunicularia*) nest densities and numbers in western Nebraska. Am Midl Nat. 1996;136:143–8.
10. Clayton KM, Schmutz JK. Is the decline of burrowing owls *Speotyto cunicularia* in prairie Canada linked to changes in great plains ecosystems? Bird Conserv Int. 1999;9(2):163–85.
11. Arrowood PC, Finley CA, Thompson C. Analyses of burrowing owl populations in New Mexico. J Raptor Res. 2001;35(4):362–70.
12. Korfanta NM, Ayers LW, Anderson SH, McDonald DB. A preliminary assessment of burrowing owl status in Wyoming. J Raptor Res. 2001;35:337–43.
13. Sheffield SR, Howery M. Current status, distribution, and conservation of the burrowing owl in Oklahoma. J Raptor Res. 2001;35:351–6.
14. Murphy RK, Hasselbland DW, Grondahl CD, Sidle JG, Martin RE, Feed DW. Status of the burrowing owl in North Dakota. J Raptor Res. 2001;35:322–30.
15. Klute DS, Green TM, Howe WH, Jones ST, Shaffer JL, Sheffield SR, Zimmerman TS. Status assessment and conservation plan for the western burrowing owl in the United States. US Department of Interior, Fish & Wildlife Service, Biological Technical Publication FWS/BTP-R6001-2003, Washington, D.C.; 2003.
16. Sauer JR, Hines J E, Fallon JE, Pardieck KL, Ziolkowski DJ. Link the North American breeding bird survey, results and analysis 1966–2013. Version 01.30.2015 USGS Patuxent Wildlife Research Center, Laurel. USA; 2014.
17. Butts KO, Lewis JC. The importance of prairie dog towns to burrowing owls in Oklahoma. Proc Okla Acad Sci. 1982;62:46–52.
18. Enriquez-Rocha P, Rangel-Salazar DW. Presence and distribution of Mexican owls: a review. J Raptor Res. 1993;27:154–60.
19. Trulio L. Burrowing owl demography and habitat use at two urban sites in Santa Clara County, California. J Raptor Res. 1997;9:84–9.
20. Conway CJ, Simon JC. Comparison of detection probability associated burrowing owl survey methods. J Wildl Manag. 2003;67(3):501–11.
21. Desante DF, Ruhlen ED, Rosenberg DK. Density and abundance of burrowing owls in the agricultural matrix in the Imperial Valley. Stud Avian Biol. 2004;27:116–9.
22. Manning JA. Burrowing owl population size in the Imperial Valley, California: survey and sampling methodologies for estimation. Final report to the Imperial irrigation district, Imperial, California, USA; 2009. http://www.iid.com/Modules/ShowDocument.aspx?documentid=8172. Accessed 15 Apr 2009.
23. Tipton HC, Doherty PF, Dreitz VJ. Abundance and density of mountain plover (*Charadrius montanus*) and burrowing owl (*Athene cunicularia*) in eastern Colorado. Auk. 2009;126:493–9.
24. Berardelli D, Desmond JM, Murray L. Reproductive success of burrowing owls in urban and grassland habitats in southern New Mexico. Wilson J Ornithol. 2010;122(1):51–9.

25. Crowe D, Longshore K. Population status and reproduction ecology of the western burrowing owl (*Athene cunicularia hypugaea*) in Clark Contry, Nevada. Report final 2005.USGS-582-P. United States Geological Survey; 2010. pp. 31.
26. Wilkerson RL, Sigel RB. Distribution an abundance of western burrowing owls (*Athene cunicularia hypugaea*) in southeastern California. Southwest Nat. 2011;56(3):378–84.
27. NOM-059-SEMARNAT-2010. Protección ambiental-Especies nativas de Mexico de flora y fauna silvestres-categorias de riesgo y especificaciones para su inclusion, exclusion o cambio-lista de especies en riesgo. DIARIO OFICIAL DE LA FEDERACION; 2010. http://www.profepa.gob.mx/inno-vaportal/file/435/1/NOM_059_SEMARNAT_2010.pdf. Accessed 30 Dec 2010.
28. Itubarria-Rojas H. Estimacion de abundacia y afinidad de habitat del tecolote llanero (*Athene cunicularia*) en el Valle de Mexicali California y Sonora, Mexico. Universidad Autonoma de Guadalajara. Facultad de Ciencias Quimicas y Biologicas. Tesis de licenciatura. Guadalajara, Jalisco, Mexico; 2002. pp. 36.
29. Macias-Duarte A. Change in migratory behavior as possible explanation for burrowing owl population declines in northern latitudes. The University of Arizona. School of Natural Resources and the Environmental. PhD Thesis. USA; 2011. pp. 145.
30. Valdez-Gomez HE, Holroyd GL. The burrowing owl, habits and distribution center in western Mexico. Boletin de la Sociedad de Ciencias Naturales de Jalisco. 2000;1:57–63.
31. Cruz-Nieto MA. Ecologia invernal de la lechuza llanera (*Athene cunicularia*), en los pastizales ocupados por los perritos llanero Mexicano (*Cynomys mexicanus*), Nuevo Leon, Mexico. Universidad Autonoma de Nuevo Leon. Facultad de Ciencias Biologicas. Laboratorio de Ornitologia. San Nicolas de los Garza. Tesis Doctorado en Ciencias; 2006. pp. 118.
32. Desmond MJ, Savidge JA, Eskridge KM. Correlations between burrowing owl and black-tailed prairie dog declines: a 7-year analysis. J Wild Manag. 2000;64(4):1067–75.
33. Griebel RL, Savidge JA. Factors related to body condition of nestling burrowing owls in Buffalo Gap National Grassland, South Dakota. Wilson J Ornithol. 2003;115:477–80.
34. McNicolle JL. Burrowing owl (*Athene cunicularia*) nest site selection in relation to prairie dog colony characteristics and surrounding land-use practices in Janos, Chihuahua, Mexico. Las Cruces, New Mexico, New Mexico State University. Ms Thesis. USA; 2005. pp. 54.
35. Ruiz-Ayma G. Exito reproductive, entrega de presas y dieta del tecolote (*Athene cunicularia hypugaea*) en el complejo de la colonias de perrito de la pradera Mexicano (*Cynomys mexicanus*) en Galeana, Nuevo Leon, Mexico. Universidad Autonoma de Nuevo Leon. Facultad de Ciencias Biologicas. Tesis Maestria en Ciencias. Mexico; 2009. pp. 85.
36. Ceballos GOG. Los mamiferos silvestres de Mexico. Ed. Fondo de Cultura Economica de España; 2009. pp. 986.
37. Carrera MMA. Situacion actual, estrategias de conservacion y bases para recuperacion del perrito llanero mexicano (*Cynomys mexicanus*). Universidad Autonoma de Mexico. Tesis Maestria en Ciencias. Mexico; 2008. pp. 72.
38. Scott-Morales L, Estrada E, Chavez-Ramirez M, Cotera M. Continued decline in geographic distribution of Mexican prairie dog (*Cynomys mexicanus*). J Mammal. 2004;85(6):1095.
39. Green GA, Anthony RG. Nesting success and habitat relationships of burrowing owl in the Columbia Basin. Oregon Condor. 1989;91:347–54.
40. Haug EA, Millsap BA, Martell MS. Burrowing owl (*Speotyto cunicularia*). In: Poole A, Gill F, editors, The birds of North America, No. 61. Academy of Natural Sciences, Philadelphia, and American Ornithologists' Union, Washington, DC; 1993. pp. 20.
41. Sheffield SR. Current status, distribution, and conservation of the burrowing owl (*Speotyto cunicularia*) in Midwestern North America. 1997. In: Duncan JR, Johnson DH, Nicholls TH, editors. Biology and conservation of owls of the Northern Hemisphere, USDA Forest Service, General Technical Report NC-190. North Central Forest Experiment Station, St. Paul, Minnesota;1997. pp. 399–407.
42. Wellicome TI. Effects of food on reproduction in burrowing owl (*Athene cunicularia*) during three stages of the breeding season. University of Alberta. Edmonton. PhD Thesis. Canada; 2000. pp. 113.

43. Holroyd GR, Rodriguez RE, Sheffield S. Conservation of the burrowing owl in western North America, challenges and recommendations. J Raptor Res. 2001;35(3):399–407.
44. McDonald D, Korfanta M, Lantz SJ. The burrowing owl (*Athene cunicularia*): a technical conservation assessment Wyoming, USDA Forest Service, Rocky Mountain Region; 2004.
45. CONABIO. Comision Nacional para el Conocimiento y Uso de la Biodiversidad. Mexico; 2008. http://www.conabio.gob.mx/informacion/metadata/gis/ecort08gw.xml?_xsl=/db/metadata/xsl/fgdc_html.xsl&_indent=no. Accessed 26 Sept 2009.
46. CONAGUA. Comisión Nacional del Agua. Consulta base de datos. Distrito Federal, Mexico. http://www.smn.cna.gob.mx/es/emas. Accessed 15 Sept 2009.
47. Instituto Nacional De Estadística Geografia e Informatica. Conjunto de datos vectoriales de la carta de uso del suelo y vegetacion, escala 1:250,000, Serie III. INEGI. Mexico; 2005.
48. Periodico Oficial. Monterrey, N. L., Gobierno Constitucional del Estado Libre y Soberano de Nuevo Leon, Mexico. Tomo CXXXIX; 2002.
49. WHSRN. Designación de sitio en categoria de importancia internacional para la conservacion de aves playeras de la red hemisferica de reservas para aves playeras. 2005. http://www.whsrn.org/site-profile/llano-de-la-soledad. Accessed 03 Dec 2014.
50. Macias-Duarte A, Panjabi AO, Pool D, Youngberg E, Levandoski G. Wintering grassland bird density in Chihuahuan Desert grassland priority conservation areas, 2007–2011. Rocky Mountain Bird Observatory, Brighton, CO, RMBO Technical Report INEOTROP- MXPLAT-10-2; 2011. pp. 164.
51. Del Coro-Arizmendi, Marquez VL. Areas de importancia para la conservacion de las aves, CONABIO & Fondo Mexicano para la Conservacion de la Naturaleza; 2000. http://www.conabioweb.conabio.gob.mx/aicas/doctos/NE-36.html. Accessed 03 Dec 2010.
52. Johnston MC. Past and present grassland of southern Texas and northeastern Mexico. Ecology. 1963;44:456–66.
53. Rojas MP. Generalidades sobre la vegetacion del estado de Nuevo Leon y datos acerca de su flora. Facultad de Ciencias. Mexico D.F, Universidad Nacional Autonoma de Mexico; 1965. pp. 124.
54. Rivera RE. Caracterizacion y productividad invernal de tres areas de pastizal habitat para la lechuza llanera (*Athene cunicularia*) en el Municipio de Galeana, Nuevo Leon; Mexico. Universidad Autonoma de Nuevo Leon. Facultad de Ciencias Biologicas. Tesis licenciatura. México; 2006. pp. 72.
55. Garcia RAM. Habitat reproductivo del gorrion de Worthen (*Spizella wortheni*) en cuatro localidades del noreste de Mexico. Universidad Autonoma de Nuevo Leon. Facultad de Ciencias Biologicas. Tesis licenciatura. México; 2008. pp. 71.
56. Martinez RLM. Caracterizacion de los sitios de anidación del gorrion de Worthen (*Spizella wortheni*) en los estados de Nuevo Leon y Coahuila de Zaragoza, Mexico. Universidad Autonoma de Nuevo Leon. Facultad de Ciencias Biologicas. Tesis licenciatura. México; 2009. pp. 47.
57. Ralph CJ, Geupel GR, PYLE P, Martin TE, Desante DF, Milá B. Manual de metodos de campo para el monitoreo de aves terrestres. General Technical Report PSW–GTR–159, USDA Forest Service, Albany; 1996.
58. Buckland ST, Anderson DR, Burnham KP, Laake JL, Borchers DL, Thomas L. Introduction to distance sampling. 3rd ed. Oxford: Oxford University Press; 2001. p. 466.
59. Buckland ST, Anderson DR, Burnham KP, Laake JL, Borchers DL, Thomas L. Introduction to distance sampling. 4th ed. Oxford: Oxford University Press; 2004. p. 435.
60. Manning JA. Factors affecting detection probability of burrowing owls in southwest agroecosystem environments. J Wild Manag. 2011;75:1558–67.
61. Manning JA, Kaler RSA. Effects of survey methods on burrowing owl behaviors. J Wild Manag. 2011;75:525–30.
62. Thomas L, Buckland ST, Rexstad EA, Laake JL, Strindberg S, Hedley SL, Bishop JRB, Marques TA, Burnham KP. Distance software: design and analysis of distance sampling surveys for estimating population size. J Appl Ecol. 2010;47:5–14.
63. Burnham KP, Anderson DR. Model selection and multimodel inference: a practical information-theoretic approach. 2nd ed. New York: Springer; 2002. p. 448.
64. MCulloch CE, Searle SR, Neuhaus JM. Generalized, linear and mixed models. 2nd ed. Wiley: New York; 2008.
65. Ruizayma G, Olallakerstupp A, Maciasduarte A, Antonio G, Gonzalezrojas JI. Data from: population density of the burrowing owl (*Athene cunicularia hypugaea*) in Mexican prairie dog (*Cynomys mexicanus*) colonies at northeastern México. BMC Ecol. http://dx.doi.org/10.5061/dryad.pm362.

7

Broad and flexible stable isotope niches in invasive non-native *Rattus* spp. in anthropogenic and natural habitats of central eastern Madagascar

Melanie Dammhahn[1*], Toky M. Randriamoria[2,3] and Steven M. Goodman[2,4]

Abstract

Background: Rodents of the genus *Rattus* are among the most pervasive and successful invasive species, causing major vicissitudes in native ecological communities. A broad and flexible generalist diet has been suggested as key to the invasion success of *Rattus* spp. Here, we use an indirect approach to better understand foraging niche width, plasticity, and overlap within and between introduced *Rattus* spp. in anthropogenic habitats and natural humid forests of Madagascar.

Results: Based on stable carbon and nitrogen isotope values measured in hair samples of 589 individual rodents, we found that *Rattus rattus* had an extremely wide foraging niche, encompassing the isotopic space covered by a complete endemic forest-dwelling Malagasy small mammal community. Comparisons of Bayesian standard ellipses, as well as (multivariate) mixed-modeling analyses, revealed that the stable isotope niche of *R. rattus* tended to change seasonally and differed between natural forests and anthropogenic habitats, indicating plasticity in feeding niches. In co-occurrence, *R. rattus* and *Rattus norvegicus* partitioned feeding niches. Isotopic mismatch of signatures of individual *R. rattus* and the habitat in which they were captured, indicate frequent dispersal movements for this species between natural forest and anthropogenic habitats.

Conclusions: Since *R. rattus* are known to transmit a number of zoonoses, potentially affecting communities of endemic small mammals, as well as humans, these movements presumably increase transmission potential. Our results suggest that due to their generalist diet and potential movement between natural forest and anthropogenic habitats, *Rattus* spp. might affect native forest-dependent Malagasy rodents as competitors, predators, and disease vectors. The combination of these effects helps explain the invasion success of *Rattus* spp. and the detrimental effects of this genus on the endemic Malagasy rodent fauna.

Keywords: Bayesian standard ellipse, Coexistence, Habitat use, Humid forest, Invasion ecology, Invasive species, *Rattus rattus*, *Rattus norvegicus*, Rodents, Fur, Stable carbon isotope, Stable nitrogen isotope

Background

Invasion of ecosystems by non-native species is a critical contemporary conservation threat because these species change ecological interactions, modify ecosystem functionality, and even cause extinctions of indigenous species [1]. Hence, the study of patterns, mechanisms, and consequences of the introduction of non-indigenous organisms is a major field of interest in ecology (e.g. [1, 2]). Important questions include whether invasive species are characterized by a specific set of traits (e.g. [3]) and, if so, whether certain traits make invasive species particularly devastating for indigenous ecological communities. In mammals, for example, species with large

---

*Correspondence: melanie.dammhahn@uni-potsdam.de
[1] Animal Ecology, Institute for Biochemistry and Biology, Faculty of Natural Sciences, University of Potsdam, Maulbeerallee 1, 14469 Potsdam, Germany
Full list of author information is available at the end of the article

litter size, frequent breeding, and long reproductive life-span are more likely to establish populations and spread after introduction [4]. The presence of invasive species can change selection pressures and, hence, the evolution of native species by direct genetic interaction such as hybridization or introgression [5] or by altering ecological interactions in communities, such as direct and apparent competition, predation, and parasitism [1]. In extreme cases, invasive species drive indigenous species to extinction (e.g. [6, 7]) or lead to the disassembly of communities [8]. Invasive species that are ecological generalists most likely have the strongest impact on native communities.

Murid rodents of the genus *Rattus* are a pervasive example of an invasive group of animals that is widespread and acts as the jack of all ecological trades. The ability of these animals to live successfully with humans has facilitated their transport [9], and today the black or ship rat (*Rattus rattus*), the Norway rat (*Rattus norvegicus*), and the Pacific rat (*Rattus exulans*) are established in many ecosystems worldwide and are among the most widespread and problematic invasive animals [10, 11]. The introduction of these rats to different island communities has severely impacted ground-nesting bird colonies in many areas of the world (reviewed in [12]). Moreover, being alternative prey for apex predators, *Rattus* spp. may also indirectly affect other prey species via apparent competition [13, 14]. For example, on the western Indian Ocean islands of Europa and Juan de Nova, introduced rats prey on ground-nesting birds but also constitute a food source for both native and introduced predators, which also prey on birds, leading to hyperpredation processes [15]. Furthermore, new parasites and pathogens carried by introduced *Rattus* spp. negatively impact native communities (reviewed in [16]). In Madagascar, dietary overlap [17], as well as external and craniodental morphological characteristics [18], suggest interference and exploitation competition for the same resources between rats and indigenous rodents. Hence, invasive rats may affect endemic species by the combined effects of predation, competition, and disease transmission. Indigenous organisms with independent evolutionary histories are therefore the most vulnerable to these altered ecological interactions [1, 19].

Separated from mainland Africa and India for approximately 90 million years [20], Madagascar has several independent terrestrial mammal radiations represented by four taxonomic orders: Afrosoricida, Carnivora, Primates, and Rodentia [21]. Two species of invasive rats, *R. rattus* and *R. norvegicus*, have been introduced to the island [22]. *Rattus rattus*, considered in the top 100 of the world's worst invasive organisms [23], is more pervasive than *R. norvegicus*, particularly outside of urban

areas, and has been able to colonize different forest habitats across the island; *R. norvegicus* tends to be more synanthropic [22]. Malagasy populations of *R. rattus* are from two successful introduction events from the same source population [24]. Seafarers from the Arabian Peninsula with *entrepôts* in eastern Africa and on Indian Ocean islands brought the species to Madagascar during the Middle Ages, when there was an increase in the movements of humans and trade goods across the Indian Ocean world [24, 25].

Today, *R. rattus* has overrun Madagascar [26] and occurs in virtually all natural and synanthropic habitats [27], representing >95% of rodent captures in certain natural forests, agricultural fields, and village settings [28, 29]. In areas where *R. rattus* have colonized natural habitats, which range from degraded to intact forests, there is an apparent concordant decline in endemic rodents of the subfamily Nesomyinae, which are strictly forest dwelling. However, direct impact of *R. rattus* on the endemic Malagasy mammalian fauna has been difficult to assess to date [18, 19], partly due to a lack of information on diet composition and habitat use of *R. rattus*. The main aim of this study was to investigate the trophic ecology of invasive *R. rattus* and *R. norvegicus* in natural humid forests, where they coexist with native mammals, and in anthropogenic village settings by using stable carbon and nitrogen isotope values of hair to assess indirectly trophic niche variation.

Stable isotope analyses of tissues, such as hair, have a considerable advantage over traditionally used methods of measuring diet such as fecal and stomach content analyses, because such samples provide information about diet integrated over several weeks (e.g. [30]). Metabolically inert hair reflects the diet of an individual for the period when the animal is replacing a large proportion of its fur [31, 32], which in laboratory *R. rattus* is ca. 40 days [33]. Other laboratory experiments with controlled diets revealed that in the fur of *R. norvegicus*, isotopic half-lives and retention times (the time taken to replace isotopes of different sources in tissues) for stable carbon and nitrogen ranges from 65 to 100 days [34]. If food sources available to a consumer differ in stable isotope signatures, the stable isotope niche (i.e. the area covered by individual consumer signatures in stable isotope bi-plots) can be used as a proxy for the trophic niche [35–37]. For example, animals feeding at several trophic levels such as omnivores have a larger range of stable nitrogen values; whereas variation in stable carbon values of the consumer mainly reflects variation in basal resources, such as different carbon cycles of plants. This approach is not without caveats because capturing all sources of variation in stable isotopes is difficult (e.g. [38, 39]). However, several empirical and experimental validations suggest that stable isotope niches can reflect both quantitative and qualitative aspects of trophic niches [37, 40]. For example, comparing stable isotope and stomach content analyses, Rodriguez and Herrera [41] showed that isotopic niches were an accurate measure of trophic niche width of *R. rattus* living on islands in the Gulf of California, USA.

In this study, we analyzed stable carbon and nitrogen isotope values in hair samples collected from individual *Rattus* spp. in different habitat types in central eastern Madagascar, which included agricultural fields, anthropogenic steppe, and natural humid forest formations (ranging from degraded to relatively intact). Specifically, we addressed the following hypotheses:

1. Invasive *R. rattus* has a generalist diet. Both in its native and invasive range, *R. rattus* is known to be an omnivorous generalist feeding on invertebrates, vertebrates, and plants (e.g. [42–49]), which has been shown by fecal and stomach content analyses and stable isotope analyses [41, 50, 51]. Therefore, we expect that *R. rattus* has a large stable isotope space ranging over several trophic levels and including various basal source pools.
2. Invasive *R. rattus* has a flexible diet. We predict that these omnivorous and generalist animals adjust their feeding niche to food availability and, therefore, occupy different stable isotope niches in natural and anthropogenic habitats. Since food sources are more diverse in forest habitats, we expect *R. rattus* to occupy wider stable isotope niches in natural forest settings as compared to anthropogenic steppe and agricultural fields. Moreover, we predict narrower stable isotope niches in the more food-rich wet season in comparison to the food-limited dry season [52].
3. Individual *R. rattus* disperse, traversing different habitat types. We expect a certain proportion of individuals to be isotopically mismatched to the habitat in which they were captured. Forest habitats, as the sampled evergreen mid-elevation humid forest, are dominated by C_3 plants. Agricultural fields of a variety of crops (e.g., banana, sugar cane, maize, cassava, and rice) and anthropogenic steppe (secondary vegetation regrown after slash-and-burn agriculture) also have C_4 plants (more details in Additional file 1: Text S1). C_3 plants generally have lower $\delta^{13}C$ values than C_4 plants (global averages: −28 vs −14‰: [53, 54]). Therefore, we assume that natural forests differ from anthropogenic steppe and agricultural fields in $\delta^{13}C$ baselines. Consequently, we predict *R. rattus* sampled in non-forested areas to have higher $\delta^{13}C$ values than those from natural forests. If individuals

move between habitat types, they should be isotopically mismatched particularly in $\delta^{13}C$. Since males are often cited as the dispersing sex [55], we expect a disproportional number of young adult males to be isotopically mismatched.
4. In sympatry, invasive *R. norvegicus* and *R. rattus* partition their feeding niches. Feeding niche partitioning is one of the principal mechanisms maintaining coexistence of congeneric species [56]. Previous studies revealed that in co-occurrence, morphologically smaller *Rattus* spp. have wider feeding niches than larger congeners [42]. Therefore, we predict that *R. rattus* (Madagascar adult animals, mean total body length 365.0 mm, mean body mass 116.2 g, [22]) has a wider stable isotope niche than the larger *R. norvegicus* (Madagascar adult animals, mean total body length 405.0 mm, mean body mass 259.0 g, [22]) [18].

Methods
Field sites and fieldwork
Non-primate terrestrial small mammal surveys were carried out at 12 sites from July 2013 to March 2015 in the Moramanga District (Alaotra Mangoro Region) of central eastern Madagascar (Fig. 1; Additional file 1: Text S1, Additional file 2: Table S1). The work was conducted at three different types of sites:

1. Natural humid forest sites (n = 4): Antavibe, Avondrona, Lakato, and Sahandambo.
2. Villages sites (n = 5): Ambalafary, Antanambao, Antsahatsaka, Antsirinala, and Maridaza.
3. Sites with a combination of natural humid forest and villages (n = 3): Besakay, Mahatsara, and Sahavarina.

Habitat classification
In accordance with the aims of this project, the habitats at the different study sites were classified as:

(1) Evergreen mid-elevation humid forest from 800 to 1800 m [43]. (2) Anthropogenic steppe or shrub thickets. This habitat included a formation known as *savoka*, referring to secondary vegetation, generally of non-native and invasive plants, that grow after removal of natural forest and at least at the initial stages associated with slash-and-burn agriculture (*tavy*). At such sites, within a few seasons of planting, the soils are depleted, rendering such areas poor for agriculture production. In turn, these sites are often abandoned, allowing the colonization of often dense secondary vegetation. (3) Village settings, which include habitats around houses, agricultural fields, grain storage buildings, and associated immediate areas. The agricultural zones include those with hill rice, paddy rice, and a variety of other crops (e.g. banana, sugar cane, maize, cassava, and rice).

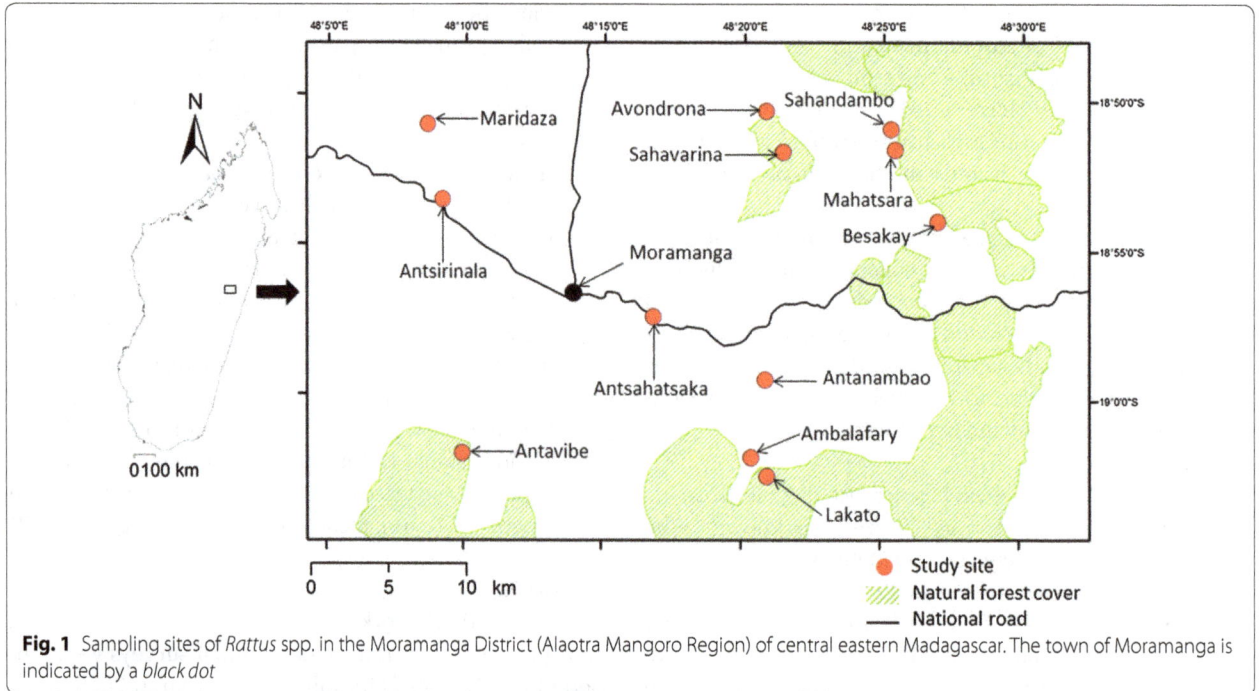

Fig. 1 Sampling sites of *Rattus* spp. in the Moramanga District (Alaotra Mangoro Region) of central eastern Madagascar. The town of Moramanga is indicated by a *black dot*

Trapping and sampling

Based on satellite images, each site was delimited into an area of 700 × 700 m or 49 ha. The number of trap lines installed in each habitat type per site was proportional to the representation of the habitat in the 49 ha parcel. Trap lines were installed a minimum distance of 60 m from each other.

Three different trap types were employed: BTS traps, 30 × 10 × 10 cm (BTS Company, Besançon, France); National traps, 39.2 × 12.3 × 12.3 cm (Tomahawk Trap Company, Hazelhurst, Wisconsin, USA); and Sherman traps, 22.5 × 8.6 × 7.4 cm (H.B. Sherman Traps Inc., Tallahassee, Florida, USA). Within each 49 ha site, 120 traps were distributed into 10 trap lines, each composed of 12 traps alternated as BTS-National-Sherman and the series repeated four times. Adjacent traps within a single line were separated from one another by a distance of 8–10 m. Traps were baited daily with a mixture of peanut butter and fresh cassava, visited in the morning and in the late afternoon, and during the latter period fresh bait was added and the previous bait removed.

A few milligrams of hair were collected from the back of each captured *Rattus* spp. (for sample sizes see Table 1). Each sample was placed in an Eppendorf tube, bearing the unique specimen number of the individual, and then exposed to indirect sunlight for about 15 min for thorough drying. To measure the isotopic variability in plants and soils of the habitat, i.e. an isotope habitat baseline, plant (n = 57) and soil (n = 49) samples were collected in close vicinity to the trap lines within each

49 ha site. At each study site, soil samples without leaf litter were collected in a defined position at 27 cm depth; plant samples were obtained in a more random manner and included leaves of the dominant floristic species, including C_3 and C_4 plants, during the season of animal capture at a particular site.

Ethics statement

The study has been conducted in accordance with the Institut Pasteur (Paris) guidelines (http://www.pasteur.fr/ip/easysite/pasteur/en/institut-pasteur/ethics-charter) for animal husbandry and experiments. As no national committee for animal welfare existed on Madagascar during the period of this study, the protocol was approved and validated by an Ad hoc committee of the Institut Pasteur de Madagascar, which included representatives from the Veterinary School of the University of Antananarivo and the Madagascar Ministry of Livestock.

Stable isotope analyses

Samples were analyzed at the Centre for Stable Isotopes Research and Analyses (KOSI) in Göttingen, Germany. Prior to analyses, each hair sample was cleaned with ethanol and all samples were oven dried at 60 °C until weight was constant. Lipid extraction was not performed for hair samples because C:N ratio was 3.3 ± 0.17 (mean ± SD) which was below that suggested for extraction (3.5; [57]). Plant samples were ground and homogenized with a ball mill. Soil samples were homogenized.

Table 1 Summary of trapping results and sampling for stable isotope analyses of *R. rattus* and *R. norvegicus* in the three habitat types and by site

Type of site	Site name	Season	R. rattus			R. norvegicus		
			Forest	Anthropogenic steppe	Agricultural field	Forest	Anthropogenic steppe	Agricultural field
Natural forest	Antavibe	Dry	60 (26)	–	–	–	–	–
		Wet	15 (11)	–	–	–	–	–
	Avondrona	Dry	11 (11)	–	–	–	–	–
		Wet	70 (40)	–	–	–	–	–
	Lakato	Dry	130 (1)	–	–	1 (1)	–	–
		Wet	28 (26)	–	–	–	–	–
	Sahandambo	Wet	47 (15)	–	–	–	–	–
Village	Ambalafary	Dry	–	156 (37)	16 (4)	–	2 (2)	–
		Wet	–	66 (64)	7 (6)	–	–	–
	Antanambao	Wet	–	62 (29)	21 (4)	–	–	1 (1)
	Antsahatsaka	Dry	–	68 (3)	95 (1)	–	–	7 (7)
		Wet	–	30 (13)	22 (8)	–	–	3 (3)
	Antsirinala	Dry	–	22 (12)	15 (10)	–	–	1 (1)
	Maridaza	Wet	–	82 (40)	8 (7)	–	2 (2)	4 (4)
Natural forest-village	Besakay	Dry	90 (0)	104 (2)	35 (0)	–	–	–
		Wet	8 (1)	60 (16)	21 (7)	–	1 (1)	–
		Dry	34 (24)	94 (38)	28 (13)	–	–	–
	Mahatsara	Wet	3 (3)	33 (3)	31 (3)	–	–	–
	Sahavarina	Dry	17 (10)	20 (11)	60 (43)	–	–	2 (2)
		Wet	8 (8)	2 (2)	17 (13)	–	–	–

As not all captured animals were sampled for stable isotope analyses, these different values are presented. The first figure under the habitat types is the number of captured individuals based on 600 standard live trap-nights per site and the second figure, in parentheses, is the number of individuals included in the stable isotope analyses. In the "natural forest" sites, traps were limited to forest habitat; in "village" sites, traps were in anthropogenic steppe and agricultural fields; and in "natural forest-village" sites, traps were in all three habitats. -: not present

For determination of $\delta^{13}C$ and $\delta^{15}N$, approximately 1 mg of total hair, homogenized soil, and plants was enclosed into tin capsules, and subsequently processed through an isotope ratio mass spectrometer (Delta Plus, Finnigan MAT, Bremen, Germany) in an online-system after passage through an element analyzer (NA 1110, Carlo Erba, Milan, Italy). The isotope data are presented in ‰ as $\delta^{15}N$ relative to nitrogen in air or $\delta^{13}C$ relative to Pee Dee Belemnite calculated as follows:

$$\delta X(‰) = \left[(R_{sample}/R_{standard}) - 1 \right] \times 1000$$

where δX is $\delta^{15}N$ or $\delta^{13}C$, and R is the respective $^{15}N/^{14}N$ or $^{13}C/^{12}C$ ratio. Analytical precision (±SD) was calculated based on 300 acetanilide laboratory working standard (KOSI2, Göttingen, Germany) replicates (two after each 10 unknowns) and was 0.12‰ for $\delta^{15}N$ and 0.12‰ for $\delta^{13}C$.

Statistical analyses

To compare stable isotope niches among *R. rattus* sampled in different habitats, we used two approaches. A description of our habitat classification is presented above. First, based on the R package *SIAR* [58, 59], we calculated standard ellipses and convex hulls for *R. rattus* within each habitat and tested for differences in size of stable isotope niches among habitats. Thereafter, we calculated overlap between standard ellipses and convex hulls among *R. rattus* sampled in different habitats. Using Bayesian estimates of standard ellipses (based on 10,000 posterior draws and default settings of *SIAR*), we compared the size of stable isotope niche widths among *R. rattus* of different habitats. This technique has two advantages: (1) it is relatively robust against unequal sample sizes and (2) it includes measures of uncertainty around the estimates [59]. We followed the same procedure to estimate size and overlap of stable isotope niches between *R. rattus* and *R. norvegicus*; these comparisons were based only on samples obtained from locations where the two species co-occurred (Table 1).

Second, to test for differences in stable isotope niches among *R. rattus* sampled in different habitats, we used multivariate mixed models with Gaussian error distributions run with the R package *MCMCglmm* (Markov-chain Monte-Carlo generalized linear mixed models, [60]). These models use a Bayesian approach and provide a control for heterogeneity in variances and sample sizes between sampling sites by adding the random effect sampling site (specified as a random intercept) and test for differences between habitats specified as fixed effects. In addition, we entered season (defined as dry and wet, see below), as well as sex and age classes (defined as non-reproductive and reproductive, see below) to test whether these factors explain variation in stable isotope values of *R. rattus*. Aspects of season were derived from climatological data of the general study area and two periods were recognized: (1) dry season from July to October and (2) wet season from December to April. We classified individuals into two age classes based on breeding status: (1) non-reproductive—males with small abdominal testes and undeveloped epididymides and females with imperforated vagina and small mammae, and (2) reproductive—males with scrotal testes and developed epididymides and females with embryos, large mammae, and/or lactating. Since males are documented to be the dispersing sex in *R. rattus* [55], and, in general, rodents tend to disperse before breeding [61], we further entered an interaction between sex and age class, specified as a fixed effect, in this multivariate mixed model.

We set slightly informative priors by dividing the total phenotypic variance of δ^{13}C values and δ^{15}N values by the number of random effects in the model and set a low degree of belief (nu = 1) [60]. We used 200,500 iterations, a thinning interval of 200, and a burn-in of 500, which resulted in low temporal autocorrelation between estimates of subsequent models. Subsequently, we used univariate linear mixed models and the procedures described in Zuur et al. [62] to test for differences in δ^{13}C and δ^{15}N among *R. rattus* sampled in different habitats. We obtained the explained variances (pseudo-R^2) using the approach suggested by Nakagawa and Schielzeth [63]. We followed the same method to test for differences between *R. rattus* and *R. norvegicus* stable isotope niches, but we included only habitat type as a fixed effect.

To explore potential stable isotope mismatch between individual signatures and the habitat a given animal was trapped, we first tested our assumption of habitat-specific stable isotope baselines. Using soil and plant samples collected along trapping transects (see above), we tested whether habitat types, specified as fixed effects, differ in δ^{13}C and δ^{15}N, respectively, using restricted maximum likelihood linear mixed modeling (LMM) with normal errors and the R-package *lme4* [62]. The sampling site was entered as random effect (specified as random intercept) in this model. Assuming that (1) habitats differ in stable isotope baselines and (2) *R. rattus* sampled

in different habitats differ isotopically, we first classified all individuals outside of the statistical non-outlier range of the population of a specific habitat as stable isotopically mismatched. Here, statistical outliers were δ^{13}C values (x) with either $x < Q1 - 1.5 * (Q3 - Q1)$ or $x > Q3 + 1.5 * (Q3 - Q1)$, with Q1 = first quartile and Q3 = third quartile of the distribution of individual δ^{13}C distributions. Forested habitats are dominated by C_3 plants, while agricultural fields and anthropogenic steppe have C_4 plants. Therefore, we based this classification only on the stable carbon isotope.

Further, we used linear discriminant function analysis with leave one out cross (jackknifed) validation, run with the *lda* function of the R-package *MASS*, to (1) quantify overall and habitat-specific percentages of correct classifications of individual *R. rattus* to their habitat of origin, and (2) to identify misclassified individuals. Since sampling sites were isotopically heterogeneous, we corrected individual *R. rattus* isotope signatures by the isotope baseline of each sampling site ($\delta^{15}N_{cor} = \delta^{15}N_{rat} - \delta^{15}N_{plants}$ and $\delta^{13}C_{cor} = \delta^{13}C_{rat} - \delta^{13}C_{plants}$), with $\delta^{15}N_{plants}$ and $\delta^{13}C_{plants}$ being the median of all plant samples collected in forested areas at each site. We used only one habitat type for this baseline correction to retain habitat-type specific variation and selected forest because we were mainly interested in detecting signs of movement between forested and human-modified habitats.

Finally, using generalized mixed effect models (GLMM) with binomial error distribution and sampling site as a random effect (specified as random intercept), we tested whether the probability of being a statistical outlier for δ^{13}C values and of being misclassified, respectively, was higher for males than for females and higher for non-reproductive than for reproductive individuals. Since males are assumed to be the dispersing sex, particularly from natal areas, and such movements occur before breeding (see above), we further entered an interaction between sex and age class specified as a fixed effect in these models. Moreover, we entered habitat type as a fixed effect to control for potential habitat type differences. For each model, we checked whether residuals of the model followed a Gaussian distribution. All data were analyzed using R 2.15 (The R Foundation for Statistical Computing, Vienna, Austria, http://www.r-project. org) and the specified packages. Values of *P* were two tailed throughout and the accepted significance level was $P < 0.05$.

Results

Habitats and sites in which *Rattus* spp. were captured

In total, 565 individuals of *R. rattus* and 24 individuals of *R. norvegicus* were sampled across the different sites and habitat types [Table 1, raw data at Dryad Digital Repository: http://dx.doi.org/10.5061/dryad.j04ff)]. *Rattus rattus* was present in all habitats, whereas *R. norvegicus* was mainly found in agricultural fields and less frequently in anthropogenic steppe. Lakato was the only forested site in which *R. norvegicus* was captured.

Within-species niche variation in *R. rattus*

The stable isotope niche width of *R. rattus* was large in all sampled habitat types (Fig. 2; Additional file 3: Table S2). This species had a wider stable isotope niche in natural forests as compared to anthropogenic steppe (comparison of standard ellipses: $P = 0.032$) and tended to cover a wider niche in natural forests as compared to agricultural fields ($P = 0.072$) (Fig. 2; Additional file 3: Table S2). The stable isotope niches of *R. rattus* in agricultural fields, natural forests, and anthropogenic steppe showed notable overlap (45–72%; Fig. 2; Additional file 4: Table S3).

Stable isotope niches in *R. rattus* differed between natural forests and agricultural fields (multivariate mixed model: $P = 0.038$, Table 2) but not between natural forests and anthropogenic steppe ($P = 0.14$). *Rattus rattus* sampled in the various habitats differed in overall $\delta^{13}C$ (Table 3, pseudo-$R^2 = 0.12$): those in agricultural fields tended to have higher $\delta^{13}C$ than those sampled in natural forests, and those in anthropogenic steppe had higher $\delta^{13}C$ than those sampled in natural forests (Table 3). The various individuals of this species showed no differences between habitats in $\delta^{15}N$ (Table 3).

When controlling for habitat in *R. rattus*, no difference was found in stable isotope niches between males

Table 2 Results of multivariate Bayesian mixed models on within-species niche variation in *R. rattus*

Parameter	Posterior mean	l-95% CI	u-95% CI	P
Intercept $\delta^{13}C$	−22.44	−23.17	−21.71	<0.001
Intercept $\delta^{15}N$	7.20	5.97	8.40	<0.001
Anthropogenic steppe[a]	0.24	−0.05	0.59	0.138
Agricultural field[a]	0.40	0.03	0.75	0.038

Shown are posterior means, lower, and upper 95% credibility intervals and P values, which are based on 10,000 simulations

Significant results are marked in italic

Reference level is [a]natural forest

and females ($P = 0.17$), and between non-reproductive and reproductive individuals ($P = 0.92$) (Additional file 5: Table S4). The interaction between sex and age class did not improve model fit ($\Delta DIC < 2$) and was removed from the model. Stable isotope niches of *R. rattus* tended to be smaller in the wet season as compared to the dry season (posterior mean = -0.23; CI: -0.49, 0.04, $P = 0.088$) (Additional file 5: Table S4). Univariate models for $\delta^{13}C$ and $\delta^{15}N$ revealed no effect of sex, age class, and season for both stable isotopes (Additional file 6: Table S5). Moreover, including the interaction between sex and age class did not improve model fit. *Rattus rattus* sampled in different habitats differed only in $\delta^{13}C$ values, more specifically, those sampled in anthropogenic steppe and agricultural fields had higher $\delta^{13}C$ values than those sampled in natural forests (Additional file 6: Table S5).

Habitat mismatching

First, we tested our assumption of isotopic difference between habitat types. Controlling for sampling site, we found that *R. rattus* sampled in natural forest had lower $\delta^{13}C$ values than those from anthropogenic steppe and

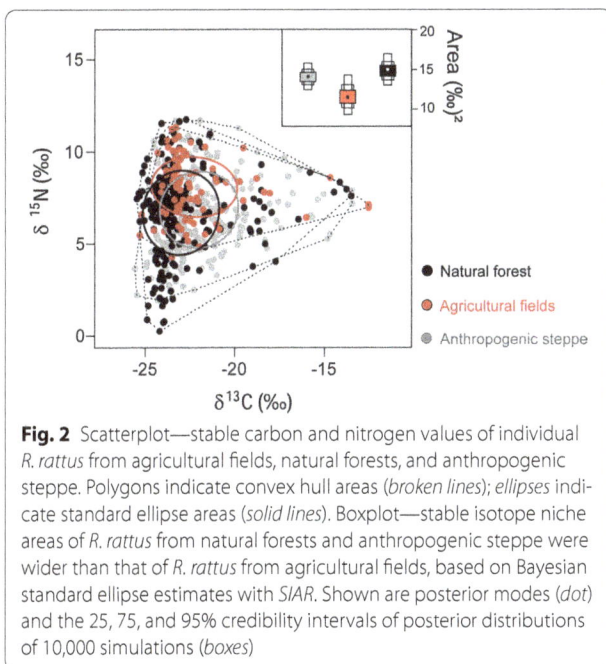

Fig. 2 Scatterplot—stable carbon and nitrogen values of individual *R. rattus* from agricultural fields, natural forests, and anthropogenic steppe. Polygons indicate convex hull areas (*broken lines*); *ellipses* indicate standard ellipse areas (*solid lines*). Boxplot—stable isotope niche areas of *R. rattus* from natural forests and anthropogenic steppe were wider than that of *R. rattus* from agricultural fields, based on Bayesian standard ellipse estimates with *SIAR*. Shown are posterior modes (*dot*) and the 25, 75, and 95% credibility intervals of posterior distributions of 10,000 simulations (*boxes*)

Table 3 Results of univariate LMMs for $\delta^{13}C$ and $\delta^{15}N$ on within-species niche variation in *R. rattus*

Parameter	β ± SE	t	P	X^2*	P*
$\delta^{13}C$					
Intercept	−22.88 ± 0.30	75.96	<0.001		
Anthropogenic steppe[a]	0.71 ± 0.26	2.75	0.006		
Agricultural field[a]	0.58 ± 0.32	1.80	0.072	7.60	0.022
$\delta^{15}N$					
Intercept	7.02 ± 0.54	13.11	<0.001		
Anthropogenic steppe[a]	0.07 ± 0.18	0.39	0.696		
Agricultural field[a]	0.34 ± 0.22	1.58	0.114	2.80	0.246

Significant results are marked in italic

Reference level is [a]natural forest

*X^2 and P values are based on log-likelihood-ratio tests (LRT) comparing models with and without the main effect of habitat type with $df = 2$

Table 4 Results of multivariate Bayesian mixed models on niche differentiation between *R. rattus* and *R. norvegicus*

Parameter	Posterior mean	l-95% CI	u-95% CI	P
Intercept $\delta^{13}C$	−21.70	−22.68	−20.72	<0.001
Intercept $\delta^{15}N$	7.99	6.77	9.27	<0.001
Species[a]	−1.03	−1.56	−0.45	<0.001
Anthropogenic steppe[b]	0.25	−0.06	0.55	0.116
Agricultural field[b]	0.29	−0.05	0.66	0.126

Shown are posterior means, lower and upper 95% credibility intervals, and *P* values, which are based on 10,000 simulations

Significant results are marked in italic

Reference levels are [a] *R. norvegicus* and [b] natural forest

agricultural fields (see above). Taking only trapping sites into account where all three habitat types were represented and with sufficient sample sizes (Ambalafary, Besakay, and Sahavarina), *R. rattus* from natural forest had lower $\delta^{13}C$ values than those sampled in anthropogenic steppe and agricultural fields (Additional file 7: Table S6). Moreover, habitat baseline measurements differed between habitat types (soil: $X^2 = 6.37$, $df = 2$, $P < 0.041$; plants: $X^2 = 15.64$, $df = 2$, $P < 0.001$, Additional file 8: Table S7, Additional file 9: Table S8). Soil samples tended to have lower $\delta^{13}C$ values in natural forest as compared to anthropogenic steppe ($\beta = -1.18 \pm 0.63$, $t = 1.85$, $P = 0.064$) and were lower in $\delta^{13}C$ values in forests than in agricultural fields ($\beta = -1.41 \pm 0.62$, $t = 2.29$, $P = 0.022$). In addition, forest plants had lower $\delta^{13}C$ values than plants collected in or near agricultural fields ($\beta = -6.95 \pm 1.70$, $t = 4.09$, $P < 0.001$), but did not differ from plants collected in anthropogenic steppe ($\beta = -1.53 \pm 1.01$, $t = 1.52$, $P = 0.13$).

Based on stable isotope values, we found indirect evidence of dispersal movements of *R. rattus* between habitat types. This species varied widely in $\delta^{13}C$ values with many individuals having values outside the non-outlier range of the habitat in which they were trapped (Fig. 2). These were not predominantly young males (interaction sex × age class: $X^2 = 1.67$, $df = 1$, $P = 0.20$), but included females, as well as non-reproductive and reproductive animals.

Stable isotope signatures of individual *R. rattus* allowed moderate prediction inference of the habitat type in which the animal was captured. Discriminant function 1 explained 90.4% of the total variance and habitat type was correctly classified in 54.6% of the cases. Classification was similar for *R. rattus* sampled in natural forest (59%) and anthropogenic steppe (67%), but individuals sampled from agricultural fields could not be classified (4% correctly classified). The majority of misclassified anthropogenic steppe samples were classified as from natural forest or vice versa. Misclassification of natural forest and anthropogenic steppe samples to agricultural fields was rare (natural forest: 1/207; anthropogenic steppe: 11/246). Using only samples from natural forest and agricultural fields, correct classification of forest samples was high (90%), but those from agricultural fields remained moderate (41%). Neither sex ($X^2 = 0.61$, $df = 1$, $P = 0.44$), age class ($X^2 = 0.70$, $df = 1$, $P = 0.40$) or the interaction between sex and age class ($X^2 = 1.29$, $df = 1$, $P = 0.26$) affected the probability of an individual being misclassified in the DFA.

Niche differentiation between *R. rattus* and *R. norvegicus*
In co-occurrence, stable isotope niches differed between *Rattus* spp. (multivariate mixed model: $P < 0.001$) but not between habitats (Table 4; Fig. 3). This difference is due to variation between species in $\delta^{15}N$ values ($X^2 = 21.16$, $df = 1$, $P < 0.001$, pseudo-$R^2 = 0.58$), but not in $\delta^{13}C$ values ($X^2 = 0.71$, $df = 1$, $P = 0.40$, pseudo-$R^2 = 0.08$) (Table 5). Combining all habitat types, in which the two species were captured in sympatry, *R. rattus* covered a larger stable isotope niche width than *R. norvegicus* (Fig. 3); the areas of standard ellipses and convex hulls were 14.3 and 90.4‰² for *R. rattus* and 9.4 and 26.7‰² for *R. norvegicus*. The core stable isotope niche (estimated as a standard ellipse) of *R. rattus* was larger than that of *R. norvegicus* combining all habitat types ($P = 0.027$). *Rattus rattus* overlapped with 35% of the core stable isotope niche of *R. norvegicus*, whereas only 24% of core isotope niche of *R. rattus* was shared with *R. norvegicus*.

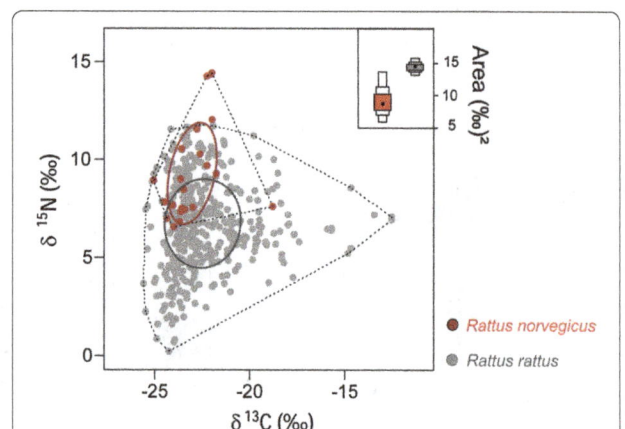

Fig. 3 Scatterplot—stable carbon and nitrogen values of individual co-occurring *R. rattus* and *R. norvegicus*. Polygons indicate convex hull areas (*broken lines*) and ellipses indicate standard ellipse areas (*solid lines*). Boxplot—stable isotope niches of *R. rattus* and *R. norvegicus* differ markedly in width, based on Bayesian standard ellipse estimates with *SIAR*. Shown are posterior modes (*dot*) and the 25, 75, and 95% credibility intervals of posterior distributions of 10,000 simulations (*boxes*)

Table 5 Results of univariate LMMs for $\delta^{13}C$ and $\delta^{15}N$ on niche differentiation between *R. rattus* and *R. norvegicus*

Parameter	β ± SE	t	P	X^{2*}	P*
$\delta^{13}C$					
Intercept	−23.42 ± 0.49	47.70	<0.001		
Species[a]	0.34 ± 0.43	0.78	0.433	0.71	0.39
Anthropogenic steppe[b]	*0.62 ± 0.24*	*2.64*	*0.008*		
Agricultural field[b]	*0.71 ± 0.29*	*2.43*	*0.015*	*7.92[c]*	*0.019[c]*
$\delta^{15}N$					
Intercept	8.58 ± 0.83	10.25	<0.001		
Species[a]	*−1.60 ± 0.36*	*4.45*	*<0.001*	*21.16*	*<0.001*
Anthropogenic steppe[b]	0.05 ± 0.20	0.26	0.796		
Agricultural field[b]	0.13 ± 0.24	0.54	0.585	0.31	0.85

Significant results are marked in italic

Reference levels are [a] *R. norvegicus* and [b] natural forest

*X^2 and P values are based on log-likelihood-ratio tests (LRT) comparing models with and without the respective term with $df = 1$; [c] for the main effect of habitat type with $df = 2$

Discussion

Species in the genus *Rattus* are among the most pervasive and successful invasive mammals in the world and responsible for major changes in native ecological communities in areas where they have been introduced. Besides dispersal and life-history characteristics of these animals [4], a flexible generalist diet appears to be another important aspect that explains their successful invasion of non-native ecosystems [41]. Here, we used stable isotope analyses to study feeding niches of introduced *Rattus* on Madagascar and found that *R. rattus* had an extremely broad feeding niche (see next section). In co-occurrence, *R. rattus* and *R. norvegicus* appear to partition their feeding niches, enlarging the trophic niche space covered by the genus. Moreover, we found indirect evidence that individual *R. rattus* disperse between anthropogenic habitats and natural forest. These results suggest that due to their generalist diet, rats presumably affect endemic small mammal species as competitors and predators, and, as shown by other studies, as vectors of parasites and zoonotic diseases [64, 65]. The combination of these effects might explain both the invasion success of rats, as well as their detrimental effects on the native fauna. Below we discuss our results in more detail.

Introduced *R. rattus* have an exceedingly wide feeding niche

Feeding niches of *R. rattus* on Madagascar were notably wide. As indicated by variation in $\delta^{15}N$ values, the diet of this species might include food sources from several trophic levels, with trophic levels being generally separated by 3–5‰ in $\delta^{15}N$ [66–68]. Variation in $\delta^{13}C$ values

further suggests that *R. rattus* incorporated C_3- and C_4/CAM-based source pools in their diets. Forest ecosystems are dominated by C_3 plants, which have lower $\delta^{13}C$ values than C_4 plants (global average −28 vs −14‰; [54]; −29.8 ± 1.6‰ (mean ± SD) in our study) and drought-adapted CAM plants [53]. Overall, the feeding niche of this introduced rodent was notably larger than that of any native Malagasy mammal species or genus studied so far (e.g. [69–75]), and about three-times larger than the complete native small mammal community of Nesomyinae rodents and Tenrecidae tenrecs at a site in the Malagasy Central Highlands [71].

Many other dietary studies of *R. rattus* outside of Madagascar, both in its native and invasive range, described the species as an omnivorous generalist at the population level. For example, it is known to feed on earthworms [43–45], terrestrial mollusks [46], crabs [47], snails [49], arthropods [42], and birds [49]. In the Hawaiian Islands, *R. rattus* contributes to the dispersal of some native and non-native seeds, including the highly invasive shrub *Clidemia hirta* [42]. Moreover, stable isotope analyses have shown that invasive *Rattus* spp. are characterized by a generalist diet (*R. rattus*: [41, 50]; *R. norvegicus*: [51]).

Our stable isotope samples were derived from specimens originating from different habitat types, including natural forest and various forms of anthropogenic steppe, ranging from grasslands to shrublands, as well as agricultural fields with cassava, rice, legumes, maize, sugar cane, and banana. Nevertheless, the trophic space covered by *R. rattus* sampled in these different habitats overlapped largely. Only animals from agricultural fields had a narrower feeding niche than those from natural forest or anthropogenic steppe. These differences were driven by variation in $\delta^{13}C$ values, but not in $\delta^{15}N$ values, reflecting the C_3 and C_4 plant dominated source pools (see above). During the wet season, when food availability is presumed to be higher [52], *R. rattus* tended to have narrower feeding niches as compared to the more food-limited dry season. Hence, a generalist diet with high among-individual variation appears to be a common and habitat-independent feature of *R. rattus* trophic ecology in the habitats we studied.

Rattus rattus is similar in external and craniodental morphological traits to some endemic forest-dwelling rodents of the subfamily Nesomyinae, namely terrestrial *Nesomys* and *Gymnuromys* and scansorial *Eliurus* spp. [18], and they might compete with these species for resources. Indeed, inference of diet overlap has been found based on seeds excavated from burrows of nesomyines and *R. rattus* in natural forest of the Parc National d'Andringitra, in central southern Madagascar [17]. Future studies should focus more closely on food resource and trophic niche overlap between *R. rattus* and nesomyines to assess interference and exploitation competition between these animals. In summary, an extraordinary large trophic niche space of *R. rattus* might explain its invasion success of different habitats and ecosystems, as this allows (1) establishing new populations under novel environmental conditions and (2) population maintenance due to reduced intra-specific competition.

Introduced *Rattus* spp. partition generalist feeding niches

Feeding niche partitioning is one of the principal mechanisms maintaining stable coexistence of species in communities. Particularly, the coexistence of closely related congeneric taxa has attracted much attention from ecologists because these species should show closer similarities in their feeding niches than more distantly related species [76]. However, studying coexistence mechanisms in extant communities is not always tractable and is shadowed by the "ghost of competition past" [77]. In the case of Madagascar, the presence of two species of introduced congeneric rodents, *R. norvegicus* and *R. rattus*, offers an excellent opportunity to study coexistence mechanisms at work, at least on a short-time scale. Here, we indirectly analyzed trophic niches of these two widespread species and found that they differ in feeding niche width. The larger and presumably more competitive *R. norvegicus* covers a smaller feeding niche than the smaller and presumably less competitive *R. rattus*. These results should be interpreted with caution because sample sizes for *R. norvegicus* was only moderate and not evenly spread over the habitat types. Nevertheless, we think that the results are not simply an artifact of differences in sample size between species, as the employed Bayesian approach and the standard ellipse comparisons are not particularly sensitive to heterogeneous sample sizes. Moreover, the stable isotope niches of both species only overlap moderately. Similarly, in the Hawaiian Islands, as revealed by stomach content analyses, the smaller *R. exulans* had a wider feeding niche than its larger congener *R. rattus* [42]. In the context of our Madagascar study, higher $\delta^{15}N$ values in *R. norvegicus* might indicate a greater percentage of animal source food in their diet as compared to *R. rattus*. Besides this feeding niche partitioning, both species appear to segregate spatially on Madagascar, with *R. norvegicus* being mainly around villages and in anthropogenic habitats, rarely captured in natural forest, in contrast to *R. rattus* [22]. Stable isotope niche differentiation between invasive *Rattus* spp. enlarges the trophic space covered by the genus and, in turn, might increase the ecological effects of these species on native mammal communities.

Stable isotope mismatch suggests movement between habitats

Within *R. rattus*, several non-reproductive and reproductive individuals of both sexes and at all sampling sites showed mismatching between their individual stable isotopic signature and that of the habitat where they were trapped. Differential resources use exploitation of C_3 and C_4 plants by individuals might theoretically lead to such a pattern for anthropogenic steppe and agricultural fields, where both types of plants exist, but cannot explain why certain individuals have higher $\delta^{13}C$ values in natural forests, where only C_3 plants occur. Therefore, we suggest that given general differences between habitat types in stable isotope baselines, as well as habitat-specific signatures of individuals of this species, these mismatches indicate that regardless of age and sex, animals move between habitats on a spatial scale of hundreds of meters to a few kilometers. In studies of *R. rattus* outside of Madagascar, males, particularly subadults, have been cited as the predominantly dispersing sex [55], and dispersal occurred mainly before the reproductive season [61]. Therefore, we anticipated males in particular to have a greater proportion of stable isotopic signatures outside the habitat-specific range in which they were captured. However, our results suggest that these mismatched signals occur in the different age and sex classes of *R. rattus* and with no strong seasonal component.

Hitherto, no study has tracked the exact movements of individual *R. rattus* on Madagascar. In geographical areas outside of Madagascar, where this species has been introduced, its home range size shows considerable variation between habitats. For example, Whisson et al. [78] found average home range sizes of 0.45 ha for females and 0.78 ha for males based on radio-telemetry over 2 months in a riparian habitat of California, which was similar to the findings of Hooker and Innes [79] of 0.49 ha for females and 1.1 ha for males in New Zealand. However, in a short-term 4-day radio-telemetry study in a New Zealand forest, males ranged up to 11 ha [80]. During one night, a given individual can travel up to 900 m [79] and traverse areas without vegetational cover approaching 500 m [81].

On Madagascar, movements of *R. rattus* between different types of anthropogenic habitats have been shown. For example, based on trapping studies at two east coast sites, this species shifted seasonally to villages, particularly around houses, during the period of rice harvest or food scarcity in secondary forested habitat [82]. Using Rhodamine B as a marker, Rahelinirina et al. [65] demonstrated that in eastern Madagascar individuals of this species move regularly between houses, adjacent sisal plantations, and irrigated rice fields, and up to 350 m over a 3-month period. Moreover, *R. rattus* is well known to occupy rice fields, to ravage crops [83], and then displace to higher ground when food resources decrease or burrows are flooded during the wet season [84]. Our indirect results confirm these observations, but we could not detect seasonal patterns in dispersal, presumably associated with our stable isotope measurements of hair integrating information over at least a period of several weeks.

We present evidence for *R. rattus* ranging between anthropogenic and natural forests; this aspect has several important implications. First, tracking resource availability by switching seasonally between habitats might provide rats another competitive advantage as compared to the endemic nesomyine rodents, which are largely forest-dependent. Second, *Rattus* spp. and their associated parasites are known to transmit a number of zoonotic diseases, which might negatively affect native communities and increase the invasion success of *Rattus* spp. (see [16] for a summary of parasite-related effects and consequences of invasion by *Rattus*). Such mobility increases the sphere of transmission potential of *Rattus*. For example, the fleas of *R. rattus* in some areas of Madagascar are a bubonic plague reservoir [85], and at a forested site in the Central Highlands, *Yersinia pestis* (plague bacteria) was isolated from endemic mammals, as well as *R. rattus* [27]. Investigations on the occurrence and morphology of trypomastigotes, a flea-transmitted parasite, suggest that introduced *R. rattus* might contribute to the decline of native endemic rodents by the transmission of this parasite [86]. In order to understand in better detail the role and associated probability of *R. rattus* to transmit endo- and ectoparasites between anthropogenic and natural habitats on the island, future studies should focus on tracking the movements of individual animals. Moreover, phylogeographical studies of parasites at a local scale [87] or capture-mark-recapture of ectoparasites [88] might provide further insight into the potential of *R. rattus* to introduce parasites and diseases into native mammalian communities.

Conclusions

Although the analysis of stable isotopes allows only indirect inferences, our results suggest that on Madagascar introduced *Rattus* spp. might affect endemic forest-dependent rodents as competitors, predators, and disease vectors. *Rattus rattus* had an extremely broad and flexible feeding niche. Moreover, in co-occurrence *R. rattus* and *R. norvegicus* appear to partition their feeding niches, enlarging the trophic niche space covered by the genus. Movements between anthropogenic habitats and natural forests might increase the transmission potential of *R. rattus*, a well known vector of a number of zoonoses. The combination of these effects helps explain the invasion success of *Rattus* spp. and the detrimental effects of members of this genus on the endemic Malagasy rodent fauna.

Additional files

Additional file 1: Text S1. Characterization of sampling sites.

Additional file 2: Table S1. Summary of the sampling period and geographic position for each sampling site.

Additional file 3: Table S2. Areas of standard ellipses and convex hulls for *Rattus rattus* in different habitats.

Additional file 4: Table S3. Pairwise overlap between stable isotope niches of *Rattus rattus* in different habitats.

Additional file 5: Table S4. Results of extended multivariate Bayesian mixed model for $\delta^{13}C$ and $\delta^{15}N$ of *Rattus rattus*.

Additional file 6: Table S5. Results of extended univariate LMMs for $\delta^{13}C$ and $\delta^{15}N$ of *Rattus rattus*.

Additional file 7: Table S6. Results of univariate LMMs for $\delta^{13}C$ and $\delta^{15}N$ of *Rattus rattus* sampled in Ambalafary, Antsahatsaka, Besakay, and Sahavarina.

Additional file 8: Table S7. Raw stable carbon and nitrogen data of soil samples.

Additional file 9: Table S8. Raw stable carbon and nitrogen data of plant samples.

Authors' contributions
Conceived and designed the experiments: TMR, SMG. Performed the formal analyses: MD. Analyzed the data: MD, SMG. Contributed reagents/materials/analysis tools: TMR, SMG. Wrote the paper: MD, SMG, TMR. All authors read and approved the final manuscript.

Author details
[1] Animal Ecology, Institute for Biochemistry and Biology, Faculty of Natural Sciences, University of Potsdam, Maulbeerallee 1, 14469 Potsdam, Germany. [2] Association Vahatra, BP 3972, 101 Antananarivo, Madagascar. [3] Mention Zoologie et Biodiversité Animale, Université d'Antananarivo, BP 906, 101 Antananarivo, Madagascar. [4] Field Museum of Natural History, 1400 South Lake Shore Drive, Chicago, IL 60605, USA.

Acknowledgements
We would like to thank the Mention Zoologie et Biodiversité Animale, Université d'Antananarivo (formerly known as Département de Biologie Animale), and the Direction du Système des Aires Protégées de Madagascar for their help in obtaining and issuing research permits to conduct associated fieldwork. We are grateful to two anonymous reviewers for their helpful comments on an earlier version of this paper.

Competing interests
The authors declare that they have no competing interests.

Funding
The field portion of this study was funded by a Grant from the Wellcome Trust (095171) to The University of Aberdeen (Sandra Telfer) and the laboratory portion by "StopRats" (European Union, European Development Fund FED 2013330-223).

References

1. Mack RN, Simberloff D, Lonsdale WM, Evans H, Clout M, Bazzaz FA. Biotic invasions: causes, epidemiology, global consequences, and control. Ecol Appl. 2000;10:689–710.
2. Elton C. The ecology of invasions by animals and plants. London: Methuen; 1958.
3. van Kleunen M, Dawson W, Schlaepfer D, Jeschke JM, Fischer M. Are invaders different? A conceptual framework of comparative approaches for assessing determinants of invasiveness. Ecol Lett. 2010;13:947–58.
4. Capellini I, Baker J, Allen WL, Street SE, Venditti C. The role of life history traits in mammalian invasion success. Ecol Lett. 2015;18:1099–107.
5. Mooney HA, Cleland EE. The evolutionary impact of invasive species. Proc Nat Acad Sci. 2001;98:5446–51.
6. Fritts TH, Rodda GH. The role of introduced species in the degradation of islands ecosystems: a case history of Guam. Ann Rev Ecol Syst. 1998;29:113–40.
7. Davis MA. Biotic globalization: does competition from introduced species threaten biodiversity? Bioscience. 2003;53(5):481–9.
8. Sanders NJ, Gotelli NJ, Heller NE, Gordon DM. Community disassembly by an invasive species. Proc Nat Acad Sci. 2003;100:2474–7.
9. Alderton D. Rodents of the world. New York: Facts on File; 1996.
10. Towns DR, Atkinson IAE, Daugherty CH. Have the harmful effects of introduced rats on islands been exaggerated? Biol Invasions. 2006;8:863–91.
11. Angel A, Wanless RM, Cooper J. Review of impacts of the introduced house mouse on islands in the Southern Ocean: are mice equivalent to rats? Biol Invasions. 2009;11:1743–54.
12. Jones HP, Tershy BR, Zavaleta ES, Croll DA, Keitt BS, Finkelstein ME, Howald GR. Severity of the effects of invasive rats on seabirds: a global review. Conserv Biol. 2008;22:16–26.
13. Holt RD. Apparent competition and the structure of prey communities. Theor Popul Biol. 1977;12:197–229.
14. Holt RD, Grover J, Tilman D. Simple rules for interspecific dominance in systems with exploitative and apparent competition. Am Nat. 1994;144(5):741–71.
15. Ringler D, Russell JC, Le Corre M. Trophic roles of black rats and seabird impacts on tropical islands: mesopredator release or hyperpredation? Biol Conserv. 2015;185:75–84.
16. Morand S, Bordes F, Chen H-W, Claude J, Cosson J-F, Galan M, CzirjÁK GÁ, Greenwood AD, Latinne A, Michaux J, et al. Global parasite and Rattus rodent invasions: the consequences for rodent-borne diseases. Integr Zool. 2015;10(5):409–23.
17. Goodman SM, Sterling EJ. The utilization of Canarium (Burseraceae) seeds by vertebrates in the Réserve Naturelle Intégrale d'Andringitra, Madagascar. Fieldiana: Zool. 1996;85:83–9.
18. Miljutin A, Lehntonen JT. Probability of competition between introduced and native rodents in Madagascar: an estimation based on morphological traits Estonian. J Ecol. 2008;57:133–12.
19. Ganzhorn JU. Effects of introduced Rattus rattus on endemic small mammals in dry deciduous forest fragments of western Madagascar. Anim Conserv. 2003;6:147–57.
20. Rabinowitz PD, Coffin MF, Falvey D. The separation of Madagascar and Africa. Science. 1983;220:67–9.
21. Poux C, Madsen O, Marquard E, Vietites DR, De Jong WW, Vences M. Asynchronous colonization of Madagascar by four endemic clades of primates, tenrecs, carnivores, and rodents as inferred from nuclear genes. Syst Biol. 2005;54:719–30.
22. Soarimalala V, Goodman SM. Les petits mammifères de Madagascar. Antananarivo: Association Vahatra; 2011.
23. Lowe S, Browne M, Boudjelas S, De Poorter M. 100 of the World's worst invasive alien species. A selection from the Global Inasive Species Database. In: The Invasive Invasive Species Specialist Group (ISSG) a specialist group of the Species Survival Comminssion (SSC) of the World Conservation Union (IUCN); 2004.
24. Brouat C, Tollemaere C, Estoup A, Loiseau A, Sommer S, Soanandrasana R, Rahalison L, Rajerison M, Piry S, Goodman SM. Invasion genetics of a human commensal rodent: the black rat Rattus rattus in Madagascar. Mol Ecol. 2014;23:4153–67.
25. Tollemaere C, Brouat C, Duplantier JM, Rahalison L, Rahelinirina S, Pascal M, Mone H, Mouahid G, Leirs H, Cosson J-F. Phylogeography of the introduced species Rattus rattus in the western Indian Ocean, with special emphasis on the colonization history of Madagascar. J Biogeogr. 2010;37:398–410.
26. Goodman SM. Rattus on Madagascar and the dilemma of protecting the endemic rodent fauna. Conserv Biol. 1995;9:450–3.
27. Duplantier J-M, Duchemin J-B, Chanteau S, Carniel E. From the recent lessons of the Malagasy foci towards a global understanding of the factors involved in plague re-emergence. Vet Res. 2005;36:437–53.
28. Goodman SM, Ganzhorn JU, Olson LE, Pidgeon M, Soarimalala V. Annual variation in species diversity and relative density of rodents and insectivores in the Parc National de la Montagne d'Ambre, Madagascar. Ecotropica. 1997;3:109–18.
29. Duplantier J-M, Rakotondravony D. The rodent problem in Madagascar: agricultural pest and threat to human health. In: Singleton GR, Hinds LA, Leirs HZZ, editors. Ecologically-based management of rodent pests. Camberra: Australian Centre for International Agricultural Research; 1999. p. 441–59.
30. Cerling TE, Wittemyer G, Ehleringer JR, Remien CH, Douglas-Hamilton I. History of animals using isotope records (HAIR): a 6-year dietary history of one family of African elephants. Proc Nat Acad Sci. 2009;106(20):8093–100.
31. Fraser EE, Longstaffe FJ, Fenton MB. Moulting matters: the importance of understanding moulting cycles in bats when using fur for endogenous marker analysis. Can J Zool. 2013;91:533–44.
32. Tieszen LL, Boutton TW, Tesdahl KG, Slade NA. Fractionation and turnover of stable carbon isotopes in animal tissues: implications for $\delta^{13}C$ analysis of diet. Oecologia. 1983;57:32–7.
33. Caut S, Angulo E, Courchamp F. Discrimination factors ($\delta^{15}N$ and $\delta^{13}C$) in an omnivorous consumer: effect of diet isotopic ratio. Funct Ecol. 2008;22:255–63.
34. Kurle CM. Interpreting temporal variation in omnivore foraging ecology via stable isotope modelling. Funct Ecol. 2009;23(4):733–44.
35. Bearhop S, Adams CE, Waldron S, Fuller RA, Macleod H. Determining trophic niche width: a novel approach using stable isotope analysis. J Anim Ecol. 2004;73:1007–12.
36. Newsome SD, Tinker MT, Monson DH, Oftedal OT, Ralls K, Staedler MM, Fogel ML, Estes JA. Using stable isotopes to investigate individual diet specialization in California sea otters (Enhydra lutris nereis). Ecology. 2009;90(4):961–74.
37. Layman CA, Arrington DA, Montana CG, Post DM. Can stable isotope ratios provide for community-wide measures of trophic structure? Ecology. 2007;88(1):42–8.
38. Flaherty EA, Ben-David M. Overlap and partitioning of the ecological and isotopic niches. Oikos. 2010;119(9):1409–16.
39. Boecklen WJ, Yarnes CT, Cook BA, James AC. On the use of stable isotopes in trophic ecology. Annu Rev Ecol Syst. 2011;42(1):411–40.
40. Perkins MJ, McDonald RA, van Veen FJF, Kelly SD, Rees G, Bearhop S. Application of nitrogen and carbon stable isotopes ($\delta^{15}N$ and $\delta^{13}C$) to quantify food chain length and trophic structure. PLoS ONE. 2014;9(3):e93281.
41. Rodriguez MA, Herrera MLG. Isotopic niche mirrors trophic niche in a vertebrate island invader. Oecologia. 2013;171:537–44.
42. Shiels AB, Flores CA, Khamsing A, Krushelnycky PD, Mosher SM, Drake DR. Dietary niche differentiation among three species of invasive rodents (Rattus rattus, R. exulans, Mus musculus). Biol Invasions. 2013;15:1037–48.
43. Norman FI. Food preferences of an insular population of Rattus rattus. J Zool Lond. 1970;162:493–503.
44. Clark DB. Age- and sex-dependent foraging strategies of a small mammalian omnivore. J Anim Ecol. 1980;49:549–63.
45. Copson GR. The diet of introduced rodents Mus musculus L. and Rattus rattus L. on subantarctic Macquarie Island. Aust Wildl Res. 1986;13:441–5.
46. Stclair JJH. The impacts of invasive rodents on island invertebrates. Biol Conserv. 2011;144:68–81.
47. Fall MW, Medina AB, Jackson WB. Feeding patterns of Rattus rattus and Rattus exulans on Eniwetok Atoll, Marshall Islands. J Mammal. 1971;51:69–76.
48. Chiba S. Invasive rats alter assemblage characteristics of land snails in the Ogasawara islands. Biol Conserv. 2000;143:1558–63.
49. VanderWerf EA. Rodent control decreases predation on artificial nests in O'ahu 'elepaio habitat. J Field Ornithol. 2001;72:448–57.

50. Riofrio-Lazo M, Paez-Rosas D. Feeding habits of introduced black rats, *Rattus rattus*, in nesting colonies of Galapagos Petrel on San Cristóbal Island, Galapagos. PLoS ONE. 2015;10(5):e0127901.
51. Major HL, Jones IL, Charette MR, Diamond AW. Variations in the diet of introduced Norway rats (*Rattus norvegicus*) inferred using stable isotope analysis. J Zool. 2007;271:463–8.
52. Wright PC. Lemur traits and Madagascar ecology: coping with an island environment. Yearb Phys Anthropol. 1999;42:31–2.
53. Marshall JD, Brooks R, Lajtha K. Sources of variation in the stable isotopic composition of plants. In: Michener R, Lajtha K, editors. Stable isotopes in ecology and environmental science, vol. 2. Malden: Blackwell; 2007. p. 22–60.
54. O'Leary MH. Carbon isotopes in photosynthesis. Bioscience. 1988;38:328–36.
55. King CM, Innes JG, Gleeson D, Fitzgerald N, Winstanley T, O'Brien B, Bridgman L, Cox N. Reinvasion by ship rats (*Rattus rattus*) of forest fragments after eradication. Biol Invasions. 2011;13:2391–408.
56. Chase JM, Leibold MA. Ecological niches: linking classical and contemporary approaches. Chicago: University of Chicago Press; 2003.
57. Post DG, Layman CA, Arrington DA, Takimoto G, Quattrochi JP, Montana CG. Getting to the fat of the matter: models, methods and assumptions for dealing with lipids in stable isotope analyses. Oecologia. 2007;152:179–89.
58. Parnell AC, Inger R, Bearhop S, Jackson AL. Source partitioning using stable isotopes: coping with too much variation. PLoS ONE. 2010;5(3):e9672.
59. Jackson AL, Inger R, Parnell AC, Bearhop S. Comparing isotopic niche widths among and within communities: SIBER—Stable Isotope Bayesian Ellipses in R. J Anim Ecol. 2011;80(3):595–602.
60. Hadfield JD. MCMC methods for multi-response generalized linear mixed models: the MCMCglmm R package. J Stat Softw. 2010;33(2):1–36.
61. Wolff JO. Social biology of rodents. Integr Zool. 2007;2:193–204.
62. Zuur AF, Ieno EN, Walker NJ, Saveliev AA, Smith GM. Mixed effects models and extensions in ecology with R. New York: Springer; 2009.
63. Nakagawa S, Schielzeth H. A general and simple method for obtaining R2 from generalized linear mixed-effects models. Methods Ecol Evol. 2013;4:133–42.
64. Andrianaivoarimanana V, Kreppel K, Elissa N, Duplantier J-M, Carniel E, Rajerison M, Jambou R. Understanding the persistence of plague foci in Madagascar. PLoS Negl Trop Dis. 2013;7(11):e2382.
65. Rahelinirina S, Duplantier JM, Ratovonjato J, Ramilijaona O, Ratsimba M, Rahalison L. Study on the movement of *Rattus rattus* and evaluation of the plague dispersion in Madagascar. Vector Borne Zoonotic Dis. 2010;10:77–84.
66. DeNiro MJ, Epstein S. Influence of diet on the distribution of nitrogen isotopes in animals. Geochim Cosmochim Acta. 1981;45:341–51.
67. McCutchan JH Jr, Lewis WM Jr, Kendall C, McGrath CC. Variation in trophic shift for stable isotope ratios of carbon, nitrogen, and sulfur. Oikos. 2003;102:378–90.
68. Vanderklift MA, Ponsard S. Sources of variation in consumer-diet δ^{15}N enrichment: a meta-analysis. Oecologia. 2003;136:169–82.
69. Crowley BE, Thorén S, Rasoazanabary E, Vogel ER, Barrett MA, Zohdy S, Blanco MB, McGoogan KC, Arrigo-Nelson SJ, Irwin MT, et al. Explaining geographical variation in the isotope composition of mouse lemurs (*Microcebus*). J Biogeogr. 2011;38:2106–21.
70. Crowley BE, Blanco MB, Arrigo-Nelson SJ, Irwin MT. Stable isotopes document resource partitioning and effects of forest disturbance on sympatric cheirogaleid lemurs. Naturwissenschaften. 2013;100(10):943–56.
71. Dammhahn M, Soarimalala V, Goodman SM. Trophic niche differentiation and microhabitat utilization in a species-rich montane forest small mammal community of eastern Madagascar. Biotropica. 2013;45:111–8.
72. Dammhahn M, Rakotondramanana CF, Goodman SM. Coexistence of morphologically similar bats (Vespertilionidae) on Madagascar: stable isotopes reveal fine-grained niche differentiation among cryptic species. J Trop Ecol. 2015;31:153–64.
73. Dammhahn M, Goodman SM. Trophic niche differentiation and microhabitat utilization revealed by stable isotope analyses in a dry-forest bat assemblage at Ankarana, northern Madagascar. J Trop Ecol. 2014;30:97–109.
74. Dammhahn M, Kappeler PM. Stable isotope analyses reveal dense trophic species packing and clear niche differentiation in a Malagasy primate community. Am J Physical Anthropol. 2014;153:249–59.
75. Lührs M-L, Dammhahn M, Kappeler P. Strength in numbers: males in a carnivore grow bigger when they associate and hunt cooperatively. Behav Ecol. 2013;24(1):21–8.
76. Violle C, Nemergut DR, Pu Z, Jiang L. Phylogenetic limiting similarity and competitive exclusion. Ecol Lett. 2011;14(8):782–7.
77. Connell JH. Diversity and the coevolution of competitors, or the ghost of competition past. Oikos. 1980;35:131–8.
78. Whisson DA, Quinn JH, Collins KC. Home range and movements of roof rats (*Rattus rattus*) in an old-grown riparian forest, California. J Mammal. 2007;88:589–94.
79. Hooker S, Innes J. Ranging behaviour of forest-dwelling ship rats, *Rattus rattus*, and effects of poisoning with brodifacoum. N Z J Zool. 1995;22:291–304.
80. Pryde M, Dilks P, Fraser I. The home range of ship rats (*Rattus rattus*) in beech forest in the Eglinton Valley, Fiordland, New Zealand: a pilot study. N Z J Zool. 2005;32:139–42.
81. Taylor KD. Range of movement and activity of common rats (*Rattus norvegicus*) on agricultural land. J Appl Ecol. 1978;15:663–77.
82. Rasolozaka IN: Biologie et migration des rats dans les polycultures de la côte Est. In Rongeurs et Lutte Antimurine à Madagascar. In: Zehrer W, Rafanomezana S, editors. *Promotion de la Protection Intégrée des Cultures et des Denrées Stockées à Madagascar.* Antananarivo: Projet DPV/GTZ; 1998. p. 59–67.
83. Rasamoelina G, Rasamoel M, Rakotovao J-M, Rafanomezana S. Résultats d'une évaluation des dégâts de rats sur le riz irrigué à Madagascar pendant la saison 1996–1997. In: Zehrer W, Rafanomezana S, editors. Rongeurs et Lutte Antimurine à Madagascar Promotion de la Protection Intégrée des Cultures et des Denrées Stockées à Madagascar. Antananarivo: Projet DPV/GTZ; 1998. p. 219–26.
84. Ratsimanosika L. Biologie et migration des rats dans la riziculture de Manaratsandry/Marovoay. In: Zehrer W, Rafanomezana S, editors. Rongeurs et Lutte Antimurine à Madagascar Promotion de la Protection Intégrée des Cultures et des Denrées Stockées à Madagascar. Antananarivo: Project DPV/GTZ; 1998. p. 69–90.
85. Duchemin J-B: Biogéographie des puces de Madagascar. Thèse de Doctorat (Parasitologie). *Thèse de Doctorat (Parasitologie).* Créteil Val de Marne, France: Université de Paris XII; 2003.
86. Laakkonen J, Goodman SM, Duchemin JB, Duplantier JM. Trypomastigotes and potential flea vectors of the endemic rodents and the introduced *Rattus rattus* in the rainforests of Madagascar. Biodivers Conserv. 2003;12:1775–83.
87. van der Mescht L, Matthee S, Matthee CA. Comparative phylogeography between two generalist flea species reveal a complex interaction between parasite life history and host vicariance: parasite-host association matters. BMC Evol Biol. 2015;15(1):1–15.
88. Zohdy S, Kemp A, Durden L, Wright P, Jernvall J. Mapping the social network: tracking lice in a wild primate (*Microcebus rufus*) population to infer social contacts and vector potential. BMC Ecol. 2012;12(1):4.

8

Palaeoenvironmental drivers of vertebrate community composition in the Belly River Group (Campanian) of Alberta, Canada, with implications for dinosaur biogeography

Thomas M. Cullen[1,2]* and David C. Evans[1,2]

Abstract

Background: The Belly River Group of southern Alberta is one of the best-sampled Late Cretaceous terrestrial faunal assemblages in the world. This system provides a high-resolution biostratigraphic record of terrestrial verte-brate diversity and faunal turnover, and it has considerable potential to be a model system for testing hypotheses of dinosaur palaeoecological dynamics, including important aspects of palaeoecommunity structure, trophic interactions, and responses to environmental change. Vertebrate fossil microsites (assemblages of small bones and teeth concentrated together over a relatively short time and thought to be representative of community composition) offer an unparalleled dataset to better test these hypotheses by ameliorating problems of sample size, geography, and chronostratigraphic control that hamper other palaeoecological analyses. Here, we assembled a comprehensive relative abundance dataset of microsites sampled from the entire Belly River Group and performed a series of analyses to test the influence of environmental factors on site and taxon clustering, and assess the stability of faunal community composition both temporally and spatially. We also test the idea that populations of large dinosaur taxa were particularly sensitive to small-scale environmental gradients, such as the paralic (coastal) to alluvial (inland) regimes present within the time-equivalent depositional basin of the upper Oldman and lower Dinosaur Park Formations.

Results: Palaeoenvironment (i.e. reconstructed environmental conditions, related to relative amount of alluvial, fluvial, and coastal influence in associated sedimentary strata) was found to be strongly associated with clustering of sites by relative-abundance faunal assemblages, particularly in relation to changes in faunal assemblage composition and marine-terrestrial environmental transitions. Palaeogeography/palaeolandscape were moderately associated to site relative abundance assemblage clustering, with depositional setting and time (i.e. vertical position within stratigraphic unit) more weakly associated. Interestingly, while vertebrate relative abundance assemblages as a whole were strongly correlated with these marine-terrestrial transitions, the dinosaur communities do not appear to be particularly sensitive to them.

Conclusions: This analysis confirms that depositional setting (i.e. the sediment type/sorting and associated characteristics) has little effect on faunal assemblage composition, in contrast to the effect of changes in the broader palaeoenvironment (e.g. upper vs. lower coastal plain, etc.), with marine-terrestrial transitions driving temporal faunal dynamics within the Belly River Group. The similarity of the dinosaur faunal assemblages between the time-equivalent

*Correspondence: thomas.cullen@mail.utoronto.ca
[1] Department of Ecology and Evolutionary Biology, University of Toronto, Toronto, ON, Canada
Full list of author information is available at the end of the article

portions of the Dinosaur Park Formation and Oldman Formation suggests that either these palaeoenvironments are more similar than characterized in the literature, or that the dinosaurs are less sensitive to variation in palaeoenvironment than has often been suggested. A lack of sensitivity to subtle environmental gradients casts doubt on these forces acting as a driver of altitudinal zonation of dinosaur communities in the Late Cretaceous of North America.

Keywords: Palaeoenvironments, Vertebrate microfossil sites, Palaeoecology, Altitudinal sensitivity, Biogeography, Dinosaurs, Faunal turnover, Latitudinal climate gradients, Belly River Group

Background

Differences in faunal composition in the Late Cretaceous of Western North America have been hypothesized to reflect adaptation to latitudinal and altitudinal climatic gradients [1–7]. Environmental changes caused by transgression-regression cycles of the Western Interior Sea have been suggested to drive the high diversity and high faunal turnover rates of non-avian dinosaurs [4, 5, 8–11], along with changes in the vertebrate community structure more generally [12–16]. However, global-scale analyses of dinosaurs, as a whole [17] and at the family level [18], indicate that large-scale changes in sea level may not have had a significant influence on broad patterns of diversity, evolution, or migration. This suggests that putative patterns in dinosaur ecology and evolution related to sea level, such as those described from Western North America, may be either the result of other factors, such as sampling biases, or may be occurring on a scale that is too small to be readily detected in such coarse-scale analyses [4]. Previous studies have suggested that the composition and diversity of taxa recovered from specific fossil localities across Western North America varies depending on their distance from the palaeoshoreline of the Western Interior Sea [1–3, 5, 7, 19, 20]. This has led to considerable debate regarding the degree of provinciality/endemism in dinosaur populations [1–7, 9, 10, 14, 15, 19–28], the putatively high diversity and restricted range of dinosaur taxa when compared to modern large mammals [2, 5, 7, 19, 20, 22, 29–37], as well as discussions of niche-partitioning in dinosaurs across environmental gradients in a single depositional basin [2, 5, 10, 19, 20, 23, 24, 29, 30, 38–45]. These noted variations in palaeocommunities over sub-million year timeframes and over relatively small palaeogeographic areas suggests that many taxa, but particularly large-bodied dinosaurs, may have been sensitive to palaeoclimatic and palaeoenvironmental change [1–3, 5, 7, 9, 43, 44, 46]. However, this model has been challenged for reliance on data derived from disjunct geographic areas that are poorly constrained chronostratigraphically [28]. the ability to test hypotheses about dinosaur biogeography, endemism, and environmental sensitivity has historically been difficult, as many specimens were collected with only limited geological data or stratigraphic information, and known by very low sample sizes [24, 43, 47], though ongoing work collecting new specimens, relocating these sites, and incorporating them into the broader stratigraphy is ameliorating some of these issues [23, 24, 43, 44, 48].

The Belly River Group of southern Alberta is one of the best-sampled Late Cretaceous vertebrate fossil deposits in the world [44], providing a high-resolution biostratigraphic record of terrestrial vertebrate diversity and faunal turnover, and has considerable potential to be as a model system for testing hypotheses of dinosaur palaeoecological dynamics, including important aspects of palaeocommunity structure, trophic interactions, and responses to environmental change [48]. The Belly River Group is composed of the Foremost, Oldman, and Dinosaur Park formations, and spans a large portion of the Campanian from a period of time from approximately 79–74 Ma [1, 49, 50]. The full extent of the Belly River Group records two major regional sea level changes in the Western Interior Seaway (the relatively shallow, inland seaway that at its greatest extent stretched from the Arctic Ocean to the Gulf of Mexico), the first of which is a regressive event in the Foremost and lower Oldman formations, and the second of which is a major transgressive event recorded in the uppermost Oldman and Dinosaur Park formations that marks the transition between the Belly River Group and the overlying Bearpaw Formation [49, 50]. The Foremost Formation is the stratigraphically lowest unit within the Belly River Group, and conformably overlies the marine shales of the Pakowki Formation. The earliest Foremost sediments show considerable marine influence, and the formation, going from lowest to highest outcrops, is generally composed of paralic to non-marine sediments, following a coarsening upwards succession [50, 51]. Conformably overlying the Foremost Formation is the Oldman Formation. The Oldman Formation is broadly considered to represent more fluvial, inland conditions, and is made up of series of upward fining palaeochannel sandstone successions, with a variety of channel top, channel margin, and overbank facies [12, 14, 50, 52]. The amount of exposure of upper Oldman sediments is dependent on sampling location, with southern sites showing a thick succession of strata, and northern sites truncated above the middle Oldman by a regional disconformity [14,

50]. To the north, upper Oldman Formation strata are replaced disconformably by those of the Dinosaur Park Formation as a result of clastic-wedge replacement [50]. The Dinosaur Park Formation (DPF) is considered to have a greater coastal influence than the underlying Oldman Formation (OM), and is characterized by sandy to muddy, alluvial, estuarine, and paralic facies [12, 49, 50, 52]. The transition to more marine environment in the upper Dinosaur Park Formation is marked by the deposition of the Lethbridge Coal Zone (LCZ), which interfingers with and is overlain by marine shales of the Bearpaw Formation [13, 49, 53]. The pre-LCZ section of the Dinosaur Park Formation in Dinosaur Provincial Park (DPP) is thought to be broadly time-equivalent to the upper Oldman Formation in the Milk River/Manyberries (MRM) regions of southern Alberta [14, 50]. The muddy strata of the upper Oldman Formation in Milk River/Manyberries are typically thought to be more environmentally similar to that of the middle Oldman 'Comrey sandstone' (representing a more seasonally arid, inland, fully non-marine fluvial landscape) than to the time-equivalent pre-LCZ Dinosaur Park Formation in DPP (representing a wetter, more marine-influenced coastal plain) [14, 49, 50]. This view has been questioned by other studies suggesting the time-equivalent Oldman and Dinosaur Park formations share a generally wet, coastal environment and that variations in recorded wet-dry signal may represent seasonality and/or spatial variation in local habitat [22, 54]. Interpreting these units is complicated by the existence of heterogeneity in their geographic extent and palaeoenvironment through time, particularly in their relation to the shore of the Western Interior Seaway [5, 8, 19, 20, 24]. Meaningful comparison between these units requires detailed chronostratigraphic control to mitigate the confounding effects of temporal changes in community structure. This is particularly important because the distribution of dinosaurs has been hypothesized to be sensitive to even small-scale palaeoenvironment differences, such as those between the lower and upper coastal plain settings of the time-equivalent Oldman and Dinosaur Park formations [5, 14, 24, 50, 52].

Vertebrate microfossil sites from the Belly River Group have provided a wealth of knowledge on vertebrate palaeoecology [12, 14, 15, 52, 55, 56]. Vertebrate microfossil sites, sensu [57], are useful in overcoming issues of low sample size that commonly hinder palaeontological investigations, as they are both abundant and each site can preserve large numbers of small teeth, bones, and scales of numerous taxa thought to represent much of the vertebrate community composition of a given area [12–16, 53, 55, 58]. These sites are concentrated in a number of ways, with most representing in-channel deposits, crevasse splays, low energy ponds, or shoreface lag deposits

[14, 16, 52, 59, 60]. While vertebrate microfossil sites in a given area or formation may represent the same broader palaeoenvironment (e.g. upland fluvial system, lowland coastal plain, etc.), their method of deposition may differ (e.g. flow rates, depositional energy, sediment size/sorting, etc.), and this has led to concerns that microsites with differing depositional setting may not preserve comparable vertebrate material or faunal assemblages [14, 59]. These concerns regarding the effect of depositional setting on the presence/absence and ranked abundances of taxa at a given microvertebrate site have been partially lessened by a new taphonomic model for microsite formation suggesting that depositional setting may not play a large role when comparing sites of different type, as channel deposits may represent the short-term erosion and local re-deposition of lower-energy pond deposits [59]. Microsite studies in the Belly River Group of Alberta have so far focused on identifying the taxa present across the region, and assessing any broad trends or associations that may be present through specific stratigraphic intervals [8, 12–14, 16, 53, 58, 61]. These studies have identified aquatic and terrestrial communities, along with some evidence of potential endemism. Additionally, they have demonstrated that terrestrial-marine environmental transitions drive at least some of the changes in vertebrate faunal assemblages, such as an increase in ceratopsid dinosaur abundance correlated with increasing marine influence in the Dinosaur Park Formation of DPP [8], and an inverse relationship between sharks and lissamphibian abundances in the Foremost Formation [16]. However, major quantitative studies have focused on either the transgressive or regressive sequences, and not the entirety of the Belly River Group, resulting in our understanding of the environmental drivers behind microsite faunal assemblage structure remaining incomplete despite the abundance and quality of the available data. In addition, new data suggests that dinosaur species found in the Dinosaur Park Formation of DPP are also found in the time-equivalent sections of the Oldman Formation in MRM [22, 24]. These units have been described as representing different palaeoenvironmental regimes [14, 22, 24, 50]. However, the Oldman Formation is palaeoenvironmentally dynamic throughout its history, shifting from lowland coastal plain to more inland braided rivers, and back to lowland coastal plain [14, 16, 22, 50]. As a result, comparisons of vertebrate microfossil sites can provide an important test of habitat sensitivity in dinosaurs.

Given the ongoing debate regarding the putatively narrow associations of dinosaurs with particular environments, locations, and/or geological formations, this study uses the largest Cretaceous vertebrate microsite dataset yet assembled to first test the previously suggested associations between faunal assemblages and differing environments, and then use those as a proxy to test for differences in dinosaur assemblages in the time-equivalent sections of the Dinosaur Park and Oldman formations. We hypothesize that altitudinal (inland vs. coastal) effects will act as the largest driver of faunal assemblage change, following previous results on more limited datasets, with other taphonomic or temporal effects acting minimally on the preserved ecological signal. We also hypothesize that dinosaurs, particularly those of large body size, will not be sensitive to altitudinal change as recorded in the Belly River Group. Although it is frequently hypothesized that large dinosaurs are sensitive to environmental variation [1–9, 11, 14, 15, 19–24, 26, 27, 42, 43, 62], many groups of large mammals today are resilient to environmental variation and have broad latitudinal distributions [2, 7, 32].

Methods

Dataset integration and taxonomic consistency

A dataset (Table 1) of vertebrate microfossil taxon abundance for Belly River Group sites (N = 48) was created through the merger of multiple literature sources [12–14, 16, 53, 58, 61]. These sites span the duration of the Belly River Group (BRG), and were sampled spatially from Dinosaur Provincial Park (DPP) and from the area around the Milk River south of Manyberries (MRM) (Fig. 1). Some revision to the taxonomic categories was required in order to successfully merge these datasets, with specific changes related to differences in the inclusivity of taxonomic categories in each source dataset (e.g. 'mammals' vs. genus-level assignment of respective taxa, due to a lack of genus resolution in included DPP data; inclusion of specimens of Allocaudata within Caudata in early studies and separation in later studies, Brinkman pers. comm. 2016), updating terminology to reflect more recent taxonomic revisions (e.g. *Atractosteus* to *Lepisosteus* for BRG gars, 'teleost D' to *Coriops*), noting areas of potential future taxonomic revision (e.g. *Myledaphus* to *Myledaphus* + *Pseudomyledaphus*, as the latter may exist in multiple sites under the former name), and revisions to identifications based on discussions with the collectors of the source microsite data (e.g. Pachycephalosaur material being now referred to hypsilophodont, Brinkman pers. comm. 2016) [14, 16, 63, 64].

Dataset standardization and R- vs. Q-mode cluster analysis

The combined data matrix was rarefied in R using functions contained within the 'vegan' package [65]. This was done to compare sampling intensity between sites, and found that few sample issues exist between sites, though most sites are likely somewhat undersampled compared

to their theoretical optimum. Rarefied data were further Wisconsin double standardized after conversion to relative abundances, again through use of the functions contained in the 'vegan' package [65]. The standardized relative abundance dataset was converted to the percentage-difference dissimilarity index (also referred to as Bray-Curtis) [66]. R- vs. Q-mode cluster analyses were generated for the resultant dissimilarity dataset, with UPGMA linkage, using the 'tabasco' function of the 'vegan' R package [65]. DPP and Milk River sites were identified on the resulting plots, and major site/taxon clusters were highlighted.

The R- vs. Q-mode cluster analysis was repeated for three sub-sample analyses focused on the time-equivalent upper Oldman and Dinosaur Park formations, one including only dinosaur proportions, another including only theropod proportions, and a third including proportions of all non-dinosaurs shared between these sites. The first two subsamples contained different source data (Additional file 1) for Dinosaur Provincial Park sites than the comparisons of total vertebrate assemblages, being derived from the re-sampled dinosaur material of Brinkman et al. [8] instead of Brinkman [12]. The dataset of Brinkman et al. [8] allows for more detailed taxonomic comparisons when restricted to dinosaur data (due to the inclusion of several additional small theropod genera), but is not considered appropriate for the broader vertebrate faunal comparisons due to the lack of methodological consistency (associated with the targeted nature of this additional dinosaur sampling) relative to the faunal data for other vertebrate material in Dinosaur Provincial Park sites. The non-dinosaur subsample used the same source data as the primary analysis of all vertebrates, the only difference being the removal of dinosaurs from the dataset. This last analysis was performed to confirm the effect of dinosaurs on overall trends in the data, and avoid possible issues of circularity in interpreting the dinosaur data in isolation.

Environmental factors and redundancy analysis

Data relating to the palaeoenvironmental setting (marine, transitional, terrestrial—paralic/lower coastal plain, terrestrial—alluvial/upper coastal plain), palaeogeography (Dinosaur Provincial Park localities, Milk River localities), stratigraphic interval (Foremost Formation, lower Oldman Formation, middle Oldman Formation or 'Comrey sandstone', time-equivalent upper Oldman and pre-LCZ Dinosaur Park formations, Lethbridge Coal Zone of Dinosaur Park Formation), and depositional setting (shoreface deposit, crevasse splay deposit, in-channel deposit) were taken from the literature [12–14, 16, 52, 53, 58, 61] and assembled into an environmental data matrix (Additional file 2). Though the palaeogeographical

Table 1 Taxon abundance table for each site analyzed in this study, with information derived and standardized from literature [12–14, 16, 53, 58, 61]

Taxon / Site	Myledaphus + Pseudomyledaphus	Protoplatyrhina	Hybodus	Centrophoroides	Odontaspididae	Cretolamna	Archaeolamna	Orectolobidae	Synechodus	Rhinobatos	Ischyrhiza	Chiloscyllium	Squatina	Elasmobranchii indet.
BB96 (DPF/LCZ; DPP)	496	79	49	0	5	0	46	3	0	0	12	0	0	2
L2377 (DPF/LCZ; DPP)	1019	356	88	0	91	12	37	38	0	3	16	0	27	4
BB115 (DPF/pre-LCZ; DPP)	243	0	2	0	0	0	0	0	0	0	0	0	0	0
BB119 (DPF/pre-LCZ; DPP)	129	0	0	0	0	0	0	0	0	0	0	0	0	0
BB108 (DPF/pre-LCZ; DPP)	354	0	0	0	0	0	0	0	0	0	0	0	0	0
BB102 (DPF/pre-LCZ; DPP)	948	0	2	0	0	0	0	0	0	0	0	0	0	0
BB94 (DPF/pre-LCZ; DPP)	107	0	0	0	0	0	0	0	0	0	0	0	0	0
BB75 (DPF/pre-LCZ; DPP)	55	0	0	0	0	0	0	0	0	0	0	0	0	0

Palaeoenvironmental drivers of vertebrate community composition in the Belly River Group (Campanian)... 71

Table 1 continued

Taxon	Myleda-phus + Pseudo-myledaphus	Protoplat-yrhina	Hybodus	Centro-phoroides	Odontas-pididae	Cretol-amna	Archae-olamna	Orectolo-bidae	Synecho-dus	Rhinoba-tos	Ischyrhiza	Chiloscyl-lium	Squatina	Elasmo-branchii indet.
BB54 (DPF/pre-LCZ; DPP)	218	0	0	0	0	0	0	0	0	0	0	0	0	0
BB120 (DPF/pre-LCZ; DPP)	228	0	0	0	0	0	0	0	0	0	0	0	0	0
BB106 (DPF/pre-LCZ; DPP)	13	0	0	0	0	0	0	0	0	0	0	0	0	0
BB25 (DPF/pre-LCZ; DPP)	54	0	0	0	0	0	0	0	0	0	0	0	0	0
BB117 (DPF/pre-LCZ; DPP)	61	0	0	0	0	0	0	0	0	0	0	0	0	0
BB78 (DPF/pre-LCZ; DPP)	6	0	0	0	0	0	0	0	0	0	0	0	0	0
BB104 (DPF/pre-LCZ; DPP)	110	0	0	0	0	0	0	0	0	0	0	0	0	0
BB97 (DPF/pre-LCZ; DPP)	92	0	0	0	0	0	0	0	0	0	0	0	0	0
BB86 (DPF/pre-LCZ; DPP)	657	0	0	0	0	0	0	0	0	0	0	0	0	0

Table 1 continued

Taxon	Myleda-phus + Pseudo-myledaphus	Protoplat-yrhina	Hybodus	Centro-phoroides	Odontas-pididae	Cretol-amna	Archae-olamna	Orectolo-bidae	Synecho-dus	Rhinoba-tos	Ischyrhiza	Chiloscyl-lium	Squatina	Elasmo-branchii indet.
BB51 (DPF/ pre-LCZ; DPP)	163	0	0	0	0	0	0	0	0	0	0	0	0	0
BB31 (DPF/ pre-LCZ; DPP)	54	0	0	0	0	0	0	0	0	0	0	0	0	0
BB98 (DPF/ pre-LCZ; DPP)	34	0	0	0	0	0	0	0	0	0	0	0	0	0
BB107 (OM/ Com-rey; DPP)	68	0	0	0	0	0	0	0	0	0	0	0	0	0
BB100 (OM/ Com-rey; DPP)	93	0	0	0	0	0	0	0	0	0	0	0	0	0
BB103 (OM/ Com-rey; DPP)	9	0	0	0	0	0	0	0	0	0	0	0	0	0
BB121 (OM/ Com-rey; DPP)	52	0	0	0	0	0	0	0	0	0	0	0	0	0
BB71 (OM/ Com-rey; DPP)	9	0	0	0	0	0	0	0	0	0	0	0	0	0
BB118 (OM/ Com-rey; DPP)	2	0	0	0	0	0	0	0	0	0	0	0	0	0

Table 1 continued

Taxon	Myledaphus + Pseudomyledaphus	Protoplatyrhina	Hybodus	Centrophoroides	Odontaspididae	Cretolamna	Archaeolamna	Orectolobidae	Synechodus	Rhinobatos	Ischyrhiza	Chiloscyllium	Squatina	Elasmobranchii indet.
BB105 (OM/Comrey; DPP)	20	0	0	0	0	0	0	0	0	0	0	0	0	0
PLS (OM/upper; MRM)	14	0	0	0	0	0	0	0	0	0	0	0	0	0
HAS (OM/upper; MRM)	2	0	0	0	0	0	0	0	0	0	0	0	0	0
BMC (OM/upper; MRM)	0	0	0	0	0	0	0	0	0	0	0	0	0	0
SalS (OM/upper; MRM)	1	0	0	0	0	0	0	0	0	0	0	0	0	0
CS (OM/upper; MRM)	2	0	0	0	0	0	0	0	0	0	0	1	0	0
HS (OM/upper; MRM)	6	0	0	0	0	0	0	0	0	0	0	0	0	0
RDS (OM/upper; MRM)	102	0	0	0	0	0	0	0	0	0	0	0	0	0
CN-1 (OM/upper; MRM)	2	0	0	0	0	0	0	0	0	0	0	1	0	0
CN-2 (OM/upper; MRM)	0	0	0	0	0	0	0	0	0	0	0	0	0	0
CBC (OM/upper; MRM)	1	0	0	0	0	0	0	0	0	0	0	0	0	0
ORS (OM/Comrey; MRM)	2	0	0	0	0	0	0	0	0	0	0	0	0	0

Table 1 continued

Taxon	Myledaphus + Pseudomyledaphus	Protoplatyrhina	Hybodus	Centrophoroides	Odontaspididae	Cretolamna	Archaeolamna	Orectolobidae	Synechodus	Rhinobatos	Ischyrhiza	Chiloscyllium	Squatina	Elasmobranchii indet.
PHS (OM/lower; MRM)	19	0	0	0	0	0	0	0	0	0	0	0	0	0
EZ (OM/lower; MRM)	8	0	0	0	0	0	0	0	0	0	0	0	0	0
PHR93-2 (OM/lower; MRM)	24	0	0	0	0	0	0	0	0	0	0	1	0	0
WS (OM/lower; MRM)	9	0	0	0	0	0	0	0	0	0	0	1	0	0
HoS (OM/lower; MRM)	15	0	0	0	0	0	0	0	0	0	0	0	0	0
SPS (Foremost; MRM)	235	0	0	0	0	0	0	0	0	0	0	5	0	0
PK (Foremost; MRM)	29	0	1	0	2	0	3	0	0	0	0	0	0	2
PHR-1 (Foremost; MRM)	1952	0	10	0	23	0	0	0	0	0	8	0	1	0
PHR-2 (Foremost; MRM)	2164	0	15	0	79	0	10	1	1	9	16	6	8	0
PHRN (Foremost; MRM)	3780	382	18	15	50	4	27	5	10	3	20	0	34	690

Taxon	Elasmodus sp.	'Holostean A'	'Holostean B'	Acipenseriformes	Belonostomus	Lepisosteus	Amiidae	Phyllodontidae	Paratarpon	Esocidae	Enchodus	Coriops	Teleostei	Anura
Site
BB96 (DPF/LCZ; DPP)	5	0	0	14	0	0	0	215	0	0	4	0	0	0
L2377 (DPF/LCZ; DPP)	40	0	0	1	0	0	0	679	0	0	0	0	3	0

Table 1 continued

Taxon	Elasmodus sp.	'Holostean A'	'Holostean B'	Acipenseriformes	Belonostomus	Lepisosteus	Amiidae	Phyllodontidae	Paratarpon	Esocidae	Enchodus	Coriops	Teleostei	Anura
BB115 (DPF/pre-LCZ; DPP)	0	0	3	2	2	39	0	16	1	0	0	1	3	0
BB119 (DPF/pre-LCZ; DPP)	0	26	2	1	2	204	1	6	0	1	0	0	1	1
BB108 (DPF/pre-LCZ; DPP)	0	13	0	0	5	90	1	6	0	5	0	1	4	7
BB102 (DPF/pre-LCZ; DPP)	0	668	11	39	22	2714	18	36	0	14	0	17	56	23
BB94 (DPF/pre-LCZ; DPP)	0	44	1	10	13	161	5	4	0	2	0	2	15	3
BB75 (DPF/pre-LCZ; DPP)	0	116	1	5	10	157	2	4	1	3	0	4	11	2
BB54 (DPF/pre-LCZ; DPP)	0	1582	196	15	5	742	149	13	0	35	0	155	331	8
BB120 (DPF/pre-LCZ; DPP)	0	105	1	8	16	155	9	12	0	6	0	47	32	14
BB106 (DPF/pre-LCZ; DPP)	0	66	0	4	1	91	7	6	0	3	0	8	49	29
BB25 (DPF/pre-LCZ; DPP)	0	80	1	1	6	29	12	1	0	4	0	6	28	13
BB117 (DPF/pre-LCZ; DPP)	0	17	0	1	6	41	3	4	0	4	0	3	19	8
BB78 (DPF/pre-LCZ; DPP)	0	175	0	0	0	130	5	35	0	3	0	33	57	24
BB104 (DPF/pre-LCZ; DPP)	0	213	3	7	3	16	42	22	0	14	0	56	163	112
BB97 (DPF/pre-LCZ; DPP)	0	86	1	0	3	62	7	6	0	8	0	16	40	35

Table 1 continued

Taxon	Elasmodus sp.	'Holostean A'	'Holostean B'	Acipenseriformes	Belonostomus	Lepisosteus	Amiidae	Phyllodontidae	Paratarpon	Esocidae	Enchodus	Coriops	Teleostei	Anura
BB86 (DPF/pre-LCZ; DPP)	0	223	6	9	12	25	58	11	0	16	0	65	190	177
BB51 (DPF/pre-LCZ; DPP)	0	218	0	6	8	10	19	4	0	2	0	23	119	20
BB31 (DPF/pre-LCZ; DPP)	0	255	0	1	1	19	11	3	0	12	0	33	79	64
BB98 (DPF/pre-LCZ; DPP)	0	124	0	0	0	3	28	10	0	6	0	25	63	65
BB107 (OM/Comrey; DPP)	0	334	0	3	8	48	18	14	0	10	0	30	136	24
BB100 (OM/Comrey; DPP)	0	268	0	5	2	46	17	11	0	15	0	34	109	77
BB103 (OM/Comrey; DPP)	0	359	0	0	0	379	16	14	0	17	0	73	213	85
BB121 (OM/Comrey; DPP)	0	62	0	9	10	152	9	6	0	2	0	20	60	6
BB71 (OM/Comrey; DPP)	0	36	0	3	1	12	6	3	0	0	0	6	15	5
BB118 (OM/Comrey; DPP)	0	105	0	0	0	4	3	2	0	5	0	7	26	20
BB105 (OM/Comrey; DPP)	0	212	0	0	0	46	11	9	0	16	0	123	51	89
PLS (OM/upper; MRM)	0	9	0	0	0	30	3	2	0	5	0	0	1	15
HAS (OM/upper; MRM)	0	32	0	0	0	126	17	8	0	10	0	41	141	92
BMC (OM/upper; MRM)	0	0	0	0	0	6	0	0	0	0	0	2	10	35

Table 1 continued

Taxon	Elasmodus sp.	'Holostean A'	'Holostean B'	Acipenseriformes	Belonostomus	Lepisosteus	Amiidae	Phyllodontidae	Paratarpon	Esocidae	Enchodus	Coriops	Teleostei	Anura
SalS (OM/upper; MRM)	0	32	0	0	0	10	44	0	0	7	0	20	48	40
CS (OM/upper; MRM)	0	49	0	0	0	169	0	4	0	18	0	18	6	66
HS (OM/upper; MRM)	0	36	0	0	0	107	2	9	0	13	0	22	197	90
RDS (OM/upper; MRM)	0	73	0	0	0	0	1	9	0	15	0	16	34	44
CN-1 (OM/upper; MRM)	0	57	0	0	0	30	0	16	0	22	0	48	230	102
CN-2 (OM/upper; MRM)	0	5	0	0	0	14	0	0	0	2	0	8	36	33
CBC (OM/upper; MRM)	0	19	0	0	6	123	7	12	0	10	0	32	73	147
ORS (OM/Comrey; MRM)	0	55	0	0	0	10	0	2	0	10	0	13	144	33
PHS (OM/lower; MRM)	0	1	0	6	0	339	0	45	0	1	0	35	148	29
EZ (OM/lower; MRM)	0	1	4	0	0	147	18	8	0	4	0	42	150	43
PHR93-2 (OM/lower; MRM)	0	31	5	4	0	126	14	19	0	6	0	43	101	70
WS (OM/lower; MRM)	0	138	25	0	11	404	11	16	0	15	0	63	225	75
HoS (OM/lower; MRM)	0	54	0	6	0	1001	10	12	0	7	0	67	100	47
SPS (Foremost; MRM)	0	229	45	14	0	204	5	139	0	29	0	22	46	24

Table 1 continued

Taxon	Elasmodus sp.	'Holostean A'	'Holostean B'	Acipenseriformes	Belonostomus	Lepisosteus	Amiidae	Phyllodontidae	Adocus	Paratarpon	Esocidae	Enchodus	Coriops	Teleostei	Anura
PK (Foremost; MRM)	4	0	6	0	0	43	0	0	0	0	0	0	0	1	1
PHR-1 (Foremost; MRM)	0	94	355	90	185	2463	9	1156	0	0	4	0	10	89	33
PHR-2 (Foremost; MRM)	0	22	166	25	160	834	17	2016	0	0	7	0	30	265	31
PHRN (Foremost; MRM)	0	8	56	2	0	143	0	2281	0	0	0	0	0	305	10

Taxon	Caudata + Allocaudata	Mosasauridae	Squamata	Plesiosauria	Testudines indet.	Solemydidae	Basilemys	Trionychidae	Adocus	Chelydridae	Baenidae	Champsosaurus	Eusuchia	Ceratopsidae
Site														
BB96 (DPF/LCZ; DPP)	0	1	0	2	0	0	0	0	0	0	0	0	1	1
L2377 (DPF/LCZ; DPP)	0	0	0	1	0	0	0	0	0	0	0	1	1	0
BB115 (DPF/pre-LCZ; DPP)	13	0	3	0	0	0	1	8	0	0	3	14	15	0
BB119 (DPF/pre-LCZ; DPP)	13	0	0	0	0	0	0	15	0	1	1	15	29	8
BB108 (DPF/pre-LCZ; DPP)	25	0	11	0	0	0	2	10	0	3	4	7	96	6
BB102 (DPF/pre-LCZ; DPP)	244	0	69	0	0	0	2	79	0	13	47	93	414	37

Table 1 continued

Taxon	Caudata + Allocaudata	Mosasauridae	Squamata	Plesiosauria	Testudines indet.	Solemydidae	Basilemys	Trionychidae	Adocus	Chelydridae	Baenidae	Champsosaurus	Eusuchia	Ceratopsidae
BB94 (DPF/pre-LCZ; DPP)	45	0	7	0	0	0	0	25	0	3	16	7	27	7
BB75 (DPF/pre-LCZ; DPP)	47	0	11	0	0	0	2	15	0	2	5	15	28	4
BB54 (DPF/pre-LCZ; DPP)	347	0	12	0	0	0	0	12	0	5	6	109	53	8
BB120 (DPF/pre-LCZ; DPP)	283	0	17	0	0	0	0	9	0	4	10	41	36	6
BB106 (DPF/pre-LCZ; DPP)	125	0	9	0	0	0	0	38	0	3	14	14	54	7
BB25 (DPF/pre-LCZ; DPP)	120	0	5	0	0	0	0	4	0	3	0	7	7	11
BB117 (DPF/pre-LCZ; DPP)	64	0	3	0	0	0	0	5	0	0	8	4	24	4
BB78 (DPF/pre-LCZ; DPP)	194	0	22	0	0	0	0	31	0	5	3	25	59	13
BB104 (DPF/pre-LCZ; DPP)	340	0	27	0	0	0	0	10	0	11	12	26	30	21
BB97 (DPF/pre-LCZ; DPP)	266	0	12	0	0	0	0	12	0	4	7	7	25	7
BB86 (DPF/pre-LCZ; DPP)	550	0	61	0	0	0	0	28	0	21	30	91	123	9
BB51 (DPF/pre-LCZ; DPP)	235	0	25	0	0	0	0	1	0	3	5	68	33	0
BB31 (DPF/pre-LCZ; DPP)	178	0	13	0	0	0	0	29	0	1	1	55	13	7

Table 1 continued

Taxon	Caudata + Allocaudata	Mosasauridae	Squamata	Plesiosauria	Testudines indet.	Solemydidae	Basilemys	Trionychidae	Adocus	Chelydridae	Baenidae	Champsosaurus	Eusuchia	Ceratopsidae
BB98 (DPF/pre-LCZ; DPP)	148	0	5	0	0	0	0	15	0	1	0	24	5	3
BB107 (OM/Comrey; DPP)	166	0	14	0	0	0	0	7	0	0	24	24	21	0
BB100 (OM/Comrey; DPP)	296	0	18	0	0	0	0	12	0	2	4	33	29	9
BB103 (OM/Comrey; DPP)	322	0	46	0	0	0	0	15	0	3	0	16	32	0
BB121 (OM/Comrey; DPP)	93	0	7	0	0	0	0	19	0	0	3	8	61	0
BB71 (OM/Comrey; DPP)	33	0	2	0	0	0	0	5	0	0	4	2	7	1
BB118 (OM/Comrey; DPP)	68	0	5	0	0	0	0	12	0	0	1	6	15	4
BB105 (OM/Comrey; DPP)	592	0	35	0	0	0	0	14	0	12	2	13	31	0
PLS (OM/upper; MRM)	39	0	8	0	0	0	0	3	0	2	2	10	30	6
HAS (OM/upper; MRM)	474	0	9	0	0	0	0	3	0	0	5	6	21	10
BMC (OM/upper; MRM)	59	0	9	0	0	0	0	6	0	12	2	0	1	5
SalS (OM/upper; MRM)	105	0	5	0	0	0	0	1	0	3	4	7	20	15

Table 1 continued

Taxon	Caudata + Allocaudata	Mosasauridae	Squamata	Plesiosauria	Testudines indet.	Solemydidae	Basilemys	Trionychidae	Adocus	Chelydridae	Baenidae	Champsosaurus	Eusuchia	Ceratopsidae
CS (OM/ upper; MRM)	378	0	18	0	0	0	0	18	7	15	16	37	93	10
HS (OM/ upper; MRM)	393	0	27	0	0	0	0	14	0	1	6	14	82	12
RDS (OM/ upper; MRM)	80	0	7	0	0	0	0	14	0	10	14	9	118	19
CN-1 (OM/ upper; MRM)	425	0	11	0	0	0	0	6	0	9	5	1	11	18
CN-2 (OM/ upper; MRM)	84	0	8	0	0	0	0	4	0	15	8	2	5	2
CBC (OM/ upper; MRM)	158	0	11	0	0	0	0	17	0	6	6	2	21	17
ORS (OM/ Comrey; MRM)	284	0	9	0	0	0	0	12	0	23	10	4	4	5
PHS (OM/ lower; MRM)	73	0	25	0	0	0	0	24	15	0	13	27	51	2
EZ (OM/ lower; MRM)	82	0	60	0	0	0	0	6	4	2	1	6	13	11
PHR93-2 (OM/ lower; MRM)	223	0	11	0	0	0	0	2	0	0	4	18	20	10
WS (OM/ lower; MRM)	317	0	45	0	0	0	0	14	8	5	11	214	82	8
HoS (OM/ lower; MRM)	250	0	8	0	0	0	0	2	13	14	2	17	31	3
SPS (Fore- most; MRM)	409	0	7	0	0	0	0	1	0	4	18	50	38	2
PK (Fore- most; MRM)	1	1	5	0	3	35	2	24	9	1	8	28	51	2

Table 1 continued

Taxon	Caudata + Allocaudata	Mosasauridae	Squamata	Plesiosauria	Testudines indet.	Solemydidae	Basilemys	Trionychidae	Adocus	Chelydridae	Baenidae	Champsosaurus	Eusuchia	Ceratopsidae
PHR-1 (Foremost; MRM)	41	0	12	0	0	1	0	106	129	21	93	97	122	52
PHR-2 (Foremost; MRM)	103	0	11	0	0	0	0	7	3	1	2	213	205	12
PHRN (Foremost; MRM)	24	0	0	0	0	272	0	21	15	0	2	64	151	4

Taxon	Ankylosauria	Hypsilophodont	Hadrosauridae	Theropoda indet.	Dromaeosauridae	Saurornitholestinae	Richardoestesia	Troodon	Paronychodon	Tyrannosauridae	cf. Aves	Mammalia
Site												
BB96 (DPF/LCZ; DPP)	0	0	0	0	0	0	0	0	0	0	0	0
L2377 (DPF/LCZ; DPP)	0	0	0	0	0	0	0	0	0	0	1	1
BB115 (DPF/pre-LCZ; DPP)	1	0	21	0	1	0	0	0	0	1	0	1
BB119 (DPF/pre-LCZ; DPP)	0	0	56	0	0	2	0	0	0	2	0	0
BB108 (DPF/pre-LCZ; DPP)	38	0	110	0	0	35	0	2	0	4	0	3
BB102 (DPF/pre-LCZ; DPP)	126	0	276	0	1	47	0	3	0	2	0	36
BB94 (DPF/pre-LCZ; DPP)	3	0	93	0	1	7	0	0	0	1	0	11
BB75 (DPF/pre-LCZ; DPP)	3	0	81	0	1	0	0	0	0	1	0	10
BB54 (DPF/pre-LCZ; DPP)	1	0	13	0	1	8	0	0	0	0	0	8

Table 1 continued

Taxon	Ankylosauria	Hypsilopho-dont	Hadrosauri-dae	Theropoda indet.	Dromaeosau-ridae	Saurorni-tholestinae	Richardoes-tesia	Troodon	Paronycho-don	Tyrannosau-ridae	cf. Aves	Mammalia
BB120 (DPF/ pre-LCZ; DPP)	3	0	113	0	0	1	0	0	0	3	0	24
BB106 (DPF/ pre-LCZ; DPP)	25	0	100	0	2	8	0	1	0	1	0	1
BB25 (DPF/ pre-LCZ; DPP)	6	0	73	0	1	4	0	0	0	3	0	3
BB117 (DPF/ pre-LCZ; DPP)	13	0	67	0	0	6	0	0	0	1	0	2
BB78 (DPF/ pre-LCZ; DPP)	8	0	174	0	1	8	0	1	0	1	0	0
BB104 (DPF/ pre-LCZ; DPP)	14	0	186	0	1	17	0	1	0	1	0	15
BB97 (DPF/ pre-LCZ; DPP)	5	2	121	0	1	7	0	0	0	3	0	2
BB86 (DPF/ pre-LCZ; DPP)	76	4	424	0	2	43	0	4	0	1	0	9
BB51 (DPF/ pre-LCZ; DPP)	12	3	116	0	2	9	0	0	0	0	0	7
BB31 (DPF/ pre-LCZ; DPP)	11	3	104	0	1	21	0	0	0	5	0	8
BB98 (DPF/ pre-LCZ; DPP)	6	0	106	0	0	8	0	0	0	1	0	7
BB107 (OM/ Comrey; DPP)	75	1	65	0	0	1	0	1	0	2	0	7
BB100 (OM/ Comrey; DPP)	69	1	161	0	0	17	0	0	0	3	0	4
BB103 (OM/ Comrey; DPP)	342	2	263	0	3	4	0	3	0	5	0	16

Table 1 continued

Taxon	Ankylosauria	Hypsilophodont	Hadrosauridae	Theropoda indet.	Dromaeosauridae	Saurornitholestinae	Richardoestesia	Troodon	Paronychodon	Tyrannosauridae	cf. Aves	Mammalia
BB121 (OM/Comrey; DPP)	13	0	92	0	0	6	0	1	0	1	0	2
BB71 (OM/Comrey; DPP)	3	0	26	0	0	2	0	2	0	0	0	1
BB118 (OM/Comrey; DPP)	8	0	62	0	0	7	0	1	0	0	0	1
BB105 (OM/Comrey; DPP)	11	4	418	0	1	28	0	3	0	3	0	26
PLS (OM/upper; MRM)	10	1	128	0	0	25	1	2	2	3	1	6
HAS (OM/upper; MRM)	5	0	268	0	1	8	2	0	2	3	2	13
BMC (OM/upper; MRM)	1	3	54	0	0	4	0	0	1	4	0	19
SalS (OM/upper; MRM)	1	0	132	0	0	5	0	0	1	2	0	6
CS (OM/upper; MRM)	5	2	219	0	0	6	2	1	1	4	2	9
HS (OM/upper; MRM)	2	1	108	0	0	4	2	0	3	1	4	7
RDS (OM/upper; MRM)	6	2	295	0	2	41	7	12	0	4	2	6
CN-1 (OM/upper; MRM)	0	0	136	0	0	10	0	2	0	3	2	8
CN-2 (OM/upper; MRM)	1	0	50	0	0	1	2	1	1	1	1	5
CBC (OM/upper; MRM)	1	0	87	0	0	7	3	4	0	3	0	12

Table 1 continued

Taxon	Ankylosauria	Hypsilopho-dont	Hadrosauri-dae	Theropoda indet.	Dromaeosau-ridae	Saurorni-tholestinae	Richardoes-tesia	Troodon	Paronycho-don	Tyrannosau-ridae	cf. Aves	Mammalia
ORS (OM/Comrey; MRM)	5	0	99	0	1	1	2	0	1	1	3	8
PHS (OM/lower; MRM)	2	0	93	0	0	7	0	0	1	2	0	9
EZ (OM/lower; MRM)	1	1	101	0	0	1	0	0	0	1	1	4
PHR93-2 (OM/lower; MRM)	0	0	65	0	0	11	0	0	1	0	1	3
WS (OM/lower; MRM)	0	1	61	0	0	7	0	0	2	3	3	8
HoS (OM/lower; MRM)	2	0	83	0	0	6	2	0	0	2	0	7
SPS (Fore-most; MRM)	2	2	162	0	0	2	0	0	2	1	3	38
PK (Foremost; MRM)	6	1	6	0	0	4	0	0	0	1	0	0
PHR-1 (Foremost; MRM)	8	2	178	0	0	7	3	0	0	7	3	11
PHR-2 (Foremost; MRM)	8	0	263	0	0	18	3	0	1	9	0	9
PHRN (Foremost; MRM)	8	0	6	5	0	0	1	0	0	1	1	2

separation between the sampling localities is relatively small in the context of latitudinal climate gradients (~150 km apart, with Dinosaur Provincial Park at ~50.75° latitude and Milk River at ~49.15° latitude), it is of a similar magnitude to previous analyses of vertebrate endemism in microsites [8, 14–16, 53, 58] and endemism/provinciality of large dinosaur macrofossils [1, 5, 19, 20, 37, 43, 44]. The environmental matrix and the dissimilarity matrix generated for the cluster analyses were then ordinated via redundancy analysis (RDA) in order to assess the relationship between each environmental variable and the clustering of site faunal assemblages.

Pair-wise site assemblage similarity

Relative abundance data of site faunal assemblages were split into two smaller datasets corresponding to sampling location (DPP vs. MRM), and ordered stratigraphically. The proportions of each taxonomic group were then plotted and compared using the R packages 'ggplot2' and 'reshape2' [67, 68]. Additionally, pair-wise Bray-Curtis similarity values were computed for sites from each sampling area using the 'fossil' package [69], and plotted as curves showing relative changes in site similarity through stratigraphy. Average faunal proportions for each stratigraphic interval were also produced and plotted.

This was repeated in three sub-sampling analyses focused on the time-equivalent upper Oldman and Dinosaur Park formations, one including only dinosaur proportions, another including only theropod proportions, and one including non-dinosaur proportions. As with the R- vs. Q-mode subsamples, the first two sub-sample analyses used the re-sampled dataset of Dinosaur Provincial Park sites (from Brinkman et al. [8]), for the same purpose of more specific dinosaur taxon comparability, while the third sub-sample used the primary dataset (with dinosaurs removed).

Results
R- vs. Q-mode cluster analysis

Cluster analyses of sites (Q-mode) and taxa (R-mode) were performed and compared to identify major clustering trends among Belly River Group microsites (Fig. 2). Two primary site clusters were identified, with an additional grade of sites between them. The largest site cluster (yellow highlighted component in Fig. 2) contains all Oldman Formation sites (N = 23), along with a majority of the pre-LCZ Dinosaur Park Formation sites (N = 14, out of a possible 18), and one Foremost Formation site ('SPS'). This cluster contains two large sub-clusters, which broadly group sites based on their sampling region (either DPP or MRM). The second primary site cluster (blue highlighted component in Fig. 2) contains the three stratigraphically lowest Foremost Formation sites

('PHR-1', 'PHR-2', 'PHRN') and both Lethbridge Coal Zone sites ('BB96', 'L2377'). The grade of sites (green highlighted component) situated between the two primary clusters contains one Foremost Formation site ('PK') and four of the stratigraphically highest pre-LCZ Dinosaur Park Formation sites ('BB102', 'BB119', 'BB108', 'BB115'). These stratigraphically high pre-LCZ Dinosaur Park Formation sites (along with 'BB94', 'BB75', 'BB54', and 'BB120', situated in the yellow highlighted component of Fig. 2) are positioned near the locally-variable conformable boundary between the informal lower 'sandy' and upper 'muddy' units within the pre-LCZ Dinosaur Park Formation, a transition thought to indicate the acceleration of the transgressive sequence leading into the LCZ and Bearpaw Formation [12, 50, 52]. The two primary site clusters correspond broadly to the clustering of taxa in the R-mode analysis, with the larger site cluster associated (yellow in Fig. 2) most strongly with lissamphibians (e.g. Caudata + Allocaudata), dinosaurs (e.g. Hadrosauridae, Ceratopsidae, *Troodon*, Dromaeosauridae, cf. Aves), and actinopterygians (e.g. 'Holostean A', *Coriops*, Teleostei, Esocidae), and the smaller primary site cluster (blue in Fig. 2) most strongly associated with batoids (e.g. *Myledaphus* + *Pseudomyledaphus*, *Ischyrhiza*, *Protoplatyrhina*, *Rhinobatos*), sharks (e.g. *Hybodus*, *Archaeolamna*, Odontaspidae), and actinopterygians (e.g. *Belonostomus*, *Enchodus*, Phyllodontidae). The sub-clusters within the larger (yellow in Fig. 2) of the two primary clusters are broadly similar in taxonomic composition, with the only major difference being the lack of several taxa (e.g. *Paronychodon*, *Richardoestesia*, cf. Aves) in Dinosaur Provincial Park sites. The grade of sites (green in Fig. 2) not included in the two primary clusters is associated with aquatic taxa present to varying degrees in both other clusters (e.g. *Myledaphus* + *Pseudomyledaphus*, *Lepisosteus*, Eusuchia, Baenidae, *Champsosaurus*).

Redundancy analysis

Redundancy analysis (RDA) was carried out on the percent difference dissimilarity matrix computed from the Belly River Group microsite relative abundance dataset, with stratigraphic interval (a proxy for temporal change), depositional setting (site-specific sedimentological characteristics), palaeogeographic sampling location (DPP or MRM), and palaeoenvironment (as reconstructed for the broader area or interval within the geological formation) as explanatory factors (Fig. 3). Broad overlap exists in sites preserved as crevasse splays or in-channel deposits, with these two representing the depositional setting of the vast majority of sites (Fig. 3a, red and green polygons), though sites preserved as shoreface deposits did cluster separately from other sites (Fig. 3a, blue polygon).

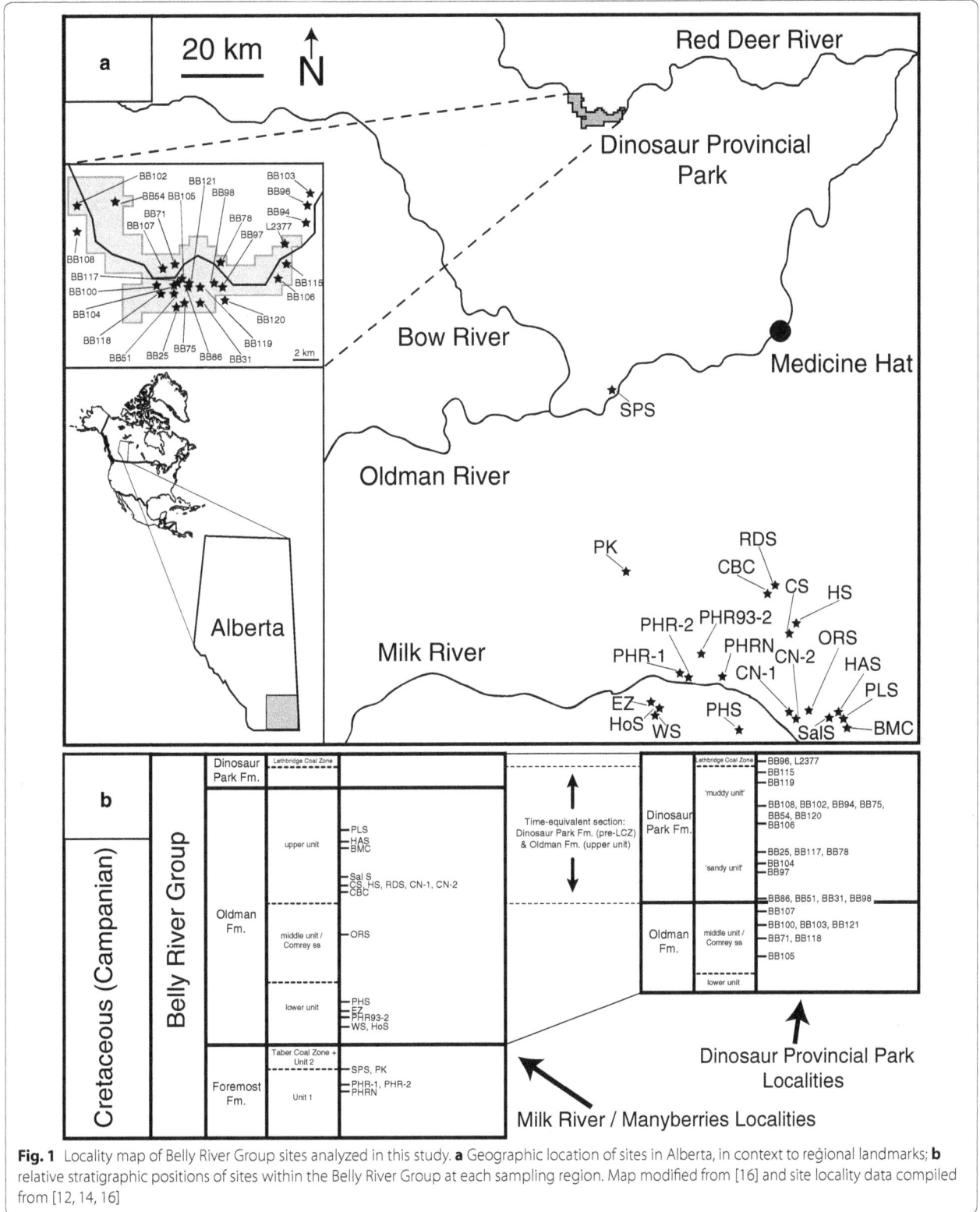

Fig. 1 Locality map of Belly River Group sites analyzed in this study. **a** Geographic location of sites in Alberta, in context to regional landmarks; **b** relative stratigraphic positions of sites within the Belly River Group at each sampling region. Map modified from [16] and site locality data compiled from [12, 14, 16]

Palaeogeographic sampling location was effective at separating site clusters in certain situations, such as for sites in the time-equivalent portion of the upper

Oldman and pre-LCZ Dinosaur Park formations (Fig. 3b, dark red and blue polygons). When expanded to all sites, sampling location did not produce distinct

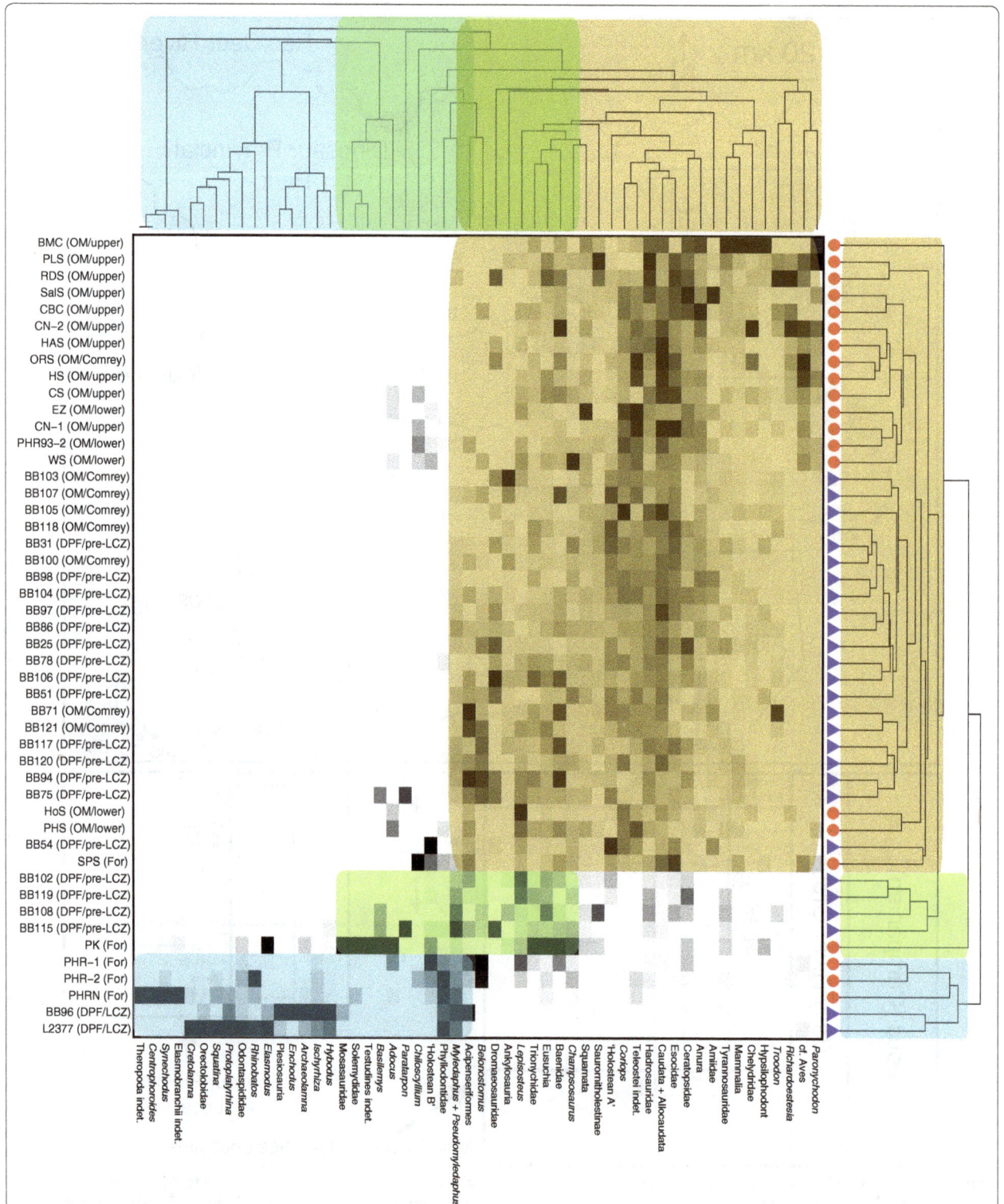

Fig. 2 R-mode vs. Q-mode cluster analysis of Belly River vertebrate microfossil sites. Dinosaur Provincial Park sites indicated with *blue text* and *triangles*. Milk River/Manyberries sites indicated with *red text* and *circles*. Coloured regions note environments associated with indicated sites and taxa: *yellow* terrestrial, *green* transitional, *blue* marine

clusters, and broad overlap was found relating to the position of lower Belly River Group sites from Milk River and upper Belly River Group sites from DPP (Fig. 3b, light red and blue polygons). When the stratigraphic interval of each site was analyzed as a clustering variable, considerable overlap was found and no directional organization could be found that would be consistent with a linear relationship through time (Fig. 3c). Sites from the lower and middle (Comrey sandstone) Oldman Formation (Fig. 3c, purple and yellow polygons) plotted adjacent to one another, with a broad overlapping distribution of sites from the upper Oldman and pre-LCZ Dinosaur Park formations (Fig. 3c, green polygon). Most Foremost Formation sites (Fig. 3c, blue polygon) plotted between pre-LCZ Dinosaur Park Formation sites from high in stratigraphic section (Fig. 3c, sites of green polygon with more negative positions on first RDA axis), near the boundary with the LCZ, and sites from within the Lethbridge Coal Zone itself (Fig. 3c, light blue polygon). The exception to this was the 'SPS' site, which plotted most closely to lower Oldman Formation sites. When using palaeoenvironment as a factor, sites were assigned to one of four settings, based on their lithology and predominant fauna: (1) marine, (2) transitional, (3) terrestrial (paralic; lower coastal plain), and (4) terrestrial (alluvial; upper coastal plain). The two terrestrial groupings correspond to the palaeoenvironmental conditions from which the vast majority of dinosaur fossils are known, and represent the two primary environmental regimes discussed in previous studies of dinosaur environmental sensitivity and/or provinciality/endemism [1–7, 9, 19, 20, 22–25, 36, 37, 42–45, 49, 70]. Three non-overlapping grouping were obtained: one for sites with palaeoenvironments reconstructed as marine (Fig. 3d, blue polygon), one for sites reconstructed as transitional and preserving a mix of marine and terrestrial sedimentological features and taxa (Fig. 3d, purple polygon), and a final grouping for sites reconstructed as being terrestrial (Fig. 3d, green and yellow polygons). Sites with terrestrial palaeoenvironments were further subdivided based on their prior associations with more paralic, lower coastal plains (Fig. 3d, green polygon) or more alluvial, upper coastal plains (Fig. 3d, yellow polygon). These further subdivisions followed a clustering pattern consistent with other sites, with more inland terrestrial sites plotting further from marine and transitional sites than more coastal plain terrestrial sites. Despite this trend, considerable overlap exists between more coastally influenced terrestrial sites and more alluvial terrestrial sites, indicating that the two cannot be considered truly distinct for the purpose of site clustering.

Pair-wise site assemblage similarity

Pair-wise Bray-Curtis (percentage-difference) similarity was computed for each consecutive pair of stratigraphically-ordered neighbouring sites from each sampling region (DPP and MRM), along with a visual representation of relative taxonomic group abundance at each site (Fig. 4). The Milk River/Manyberries sites (Fig. 4, red box at right) range stratigraphically from the Foremost Formation to the upper unit of the Oldman Formation (time-equivalent to the pre-LCZ Dinosaur Park Formation in DPP), and the Dinosaur Provincial Park sites (Fig. 4, blue box at left) stratigraphically range from the middle Oldman Formation ('Comrey sandstone') to the Lethbridge Coal Zone. Pair-wise similarity curves are relatively stable for much of the sampled intervals in both DPP (Fig. 4, blue curve) and MRM (Fig. 4, red curve), ranging from approximately 50–80% similarity. Two exceptions to this stability exist, one during the regressive phase recorded near the end of the Foremost Formation in MRM sites (Fig. 4, near base of red curve) and the other during the transgressive phase in the Lethbridge Coal Zone near the boundary between the Belly River Group and the overlying Bearpaw Formation in DPP sites (Fig. 4, near top of blue curve). In both of these cases, site similarity dropped to approximately 10–20%. Trends recorded across relative abundances of taxa in individual sites (Fig. 4, DPP sites in blue box at left and MRM sites in red box at right) and in formational average taxon abundance (Fig. 4, top right) both show that the site similarity drop in the Foremost Formation was associated with large reductions in relative abundances of chondrichthyans and large increases in relative abundances of lissamphibians, with the inverse seen in the similarity drop in the Lethbridge Coal Zone.

Sub-sample analyses of time-equivalent sites

Using three subsets of the broader relative abundance dataset, with re-sampled values of dinosaurs from Brinkman et al. [8] in place of Brinkman [12] for the first two, R- vs. Q-mode cluster analyses and pair-wise assemblage similarity analyses were performed for dinosaur-only (Fig. 5), theropod-only (Fig. 6), and non-dinosaur (Fig. 7) components of the assemblage. These subset comparisons were made only for sites in the overlapping stratigraphic intervals of each sampling area, namely the middle ('Comrey sandstone') Oldman Formation, and the upper Oldman Formation and pre-LCZ Dinosaur Park Formation.

In the dinosaur sub-sample (Fig. 5), hadrosaurs constituted the vast majority of assemblage relative abundance in almost all sites from both Dinosaur Provincial Park and Milk River. The only exception to this was the DPP 'BB54' site, which clustered away from all other sites (Fig. 5A).

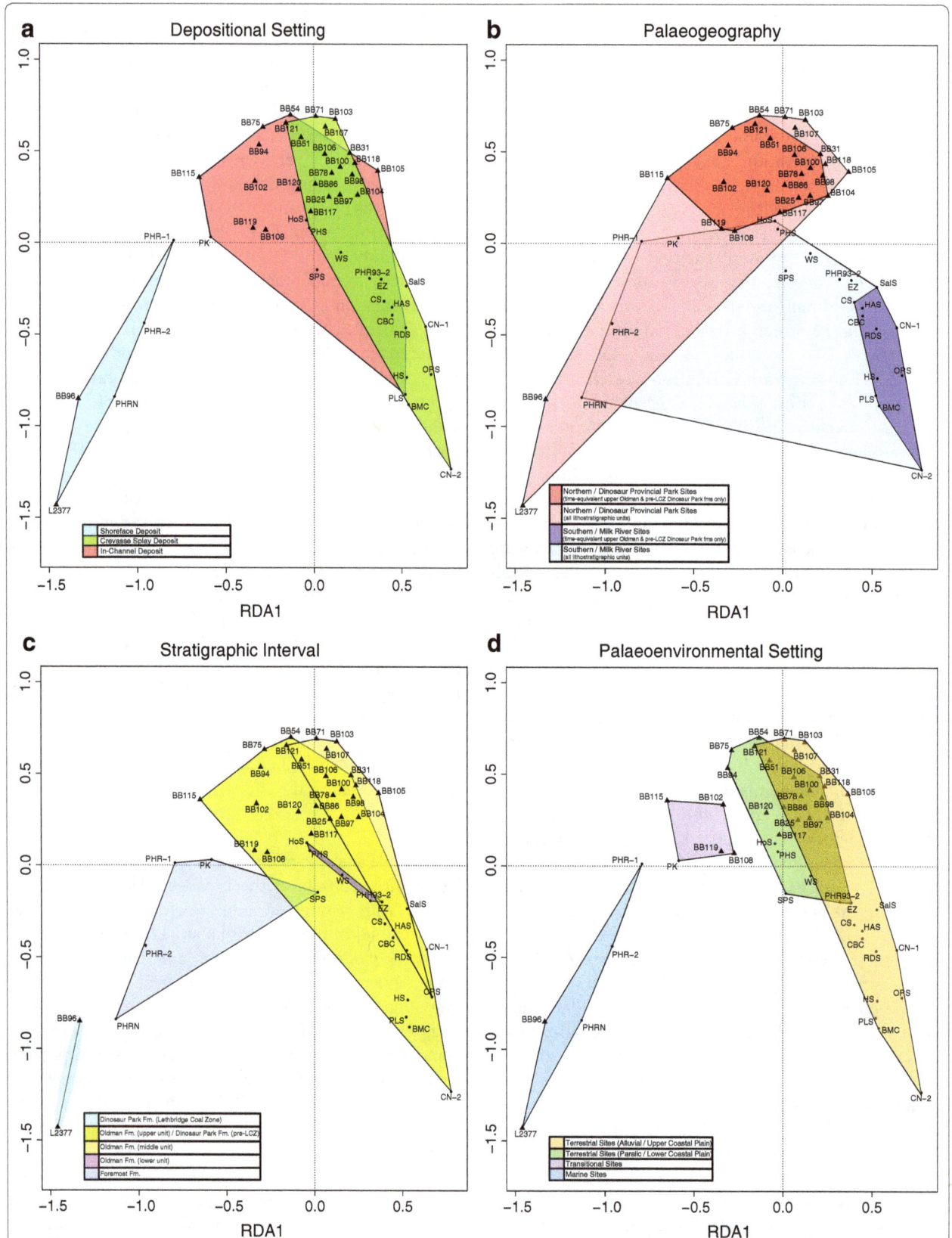

Fig. 3 Redundancy analysis of Belly River Group vertebrate microfossil site data and associated environmental factors. **a** Depositional setting; **b** palaeogeographic sampling location; **c** stratigraphic interval; **d** palaeoenvironmental setting. Subdivisions within each environmental factor noted in associated legends

Fig. 4 a Pair-wise Bray-Curtis similarity of Dinosaur Provincial Park (*blue square*) and Milk River/Manyberries (*red square*) vertebrate microfossil relative abundance assemblages through Belly River Group lithostratigraphic record. **b** Formational average proportions of each taxonomic group in sampled regions Taxonomic groups, site identifications, and other relevant information noted in legend

With the exception of hadrosaurs, no single dinosaur taxon was found to be driving large-scale site clustering, though increased *Troodon* relative abundance seems to drive the finer-scale clustering of two MRM sites ('RDS' and 'CBC') and a DPP site ('BB71'), and relative abundance of ankylosaurs drives the clustering of two DPP sites ('BB86' and 'BB100') and one MRM site ('PLS'). Unlike in the cluster analyses of the broader vertebrate assemblages, there is less distinct clustering of sites by sampling region (Fig. 5a, MRM sites indicated by red circles and DPP sites indicated by blue triangles). Pair-wise site similarity for dinosaur assemblages in the sub-sampled interval is relatively stable for both DPP and MRM, ranging approximately 40–80% similarity (Fig. 5b, DPP sites in blue box and represented by blue similarity curve and MRM sites in red box and

represented by red similarity curve). The only deviation from this trend is a drop to approximately 20% similarity for sites neighbouring 'BB54', which as noted above represents an apparent outlier due to a lower relative abundance of hadrosaurs, and much higher relative abundances of ceratopsians and *Richardoestesia* (Fig. 5b). Formational average abundances of dinosaurs (Fig. 5c) do not show any major shifts in assemblage between the middle Oldman Formation sites and the sites of the time-equivalent upper Oldman and pre-LCZ Dinosaur Park formations, nor is there considerable difference in assemblages between the DPP and Milk River localities. The only exceptions to this are a moderate increase in ceratopsians in DPP between the middle Oldman Formation (~1% relative abundance) and time-equivalent pre-LCZ Dinosaur Park

Formation (~7% relative abundance), a shift also found in the equivalent intervals of the Milk River sites (~4 to ~7% relative abundance, respectively). In the middle Oldman Formation and time-equivalent upper Oldman Formation of Milk River, there were also changes in small theropod relative abundances, with saurornitholestines increasing slightly (~1 to ~5%), and *Troodon* going from absent to present (with relative abundance in the upper Oldman Formation of ~1%).

The theropod-only sub-sample analyses (Fig. 6) produced similar results to the sub-sample of dinosaurs. Most sites did not show any strong signal from a particular taxon driving clustering patterns (Fig. 6a), though a few clustered together due to their greater association with tyrannosaurids and *Richardoestesia* ('BB120', 'BB115', 'BB75', BB119) or with cf. *Aves* ('BB61', 'CN-2', 'ORS', 'HS'). As with dinosaurs, the site similarity curves of the theropod sub-samples from DPP and MRM are very similar (Fig. 6b), both staying within a range of ~30 to ~80% similarity. The lower bound of that similarity range related to sites neighbouring those with very little theropod material (e.g. 'BB120', 'BB75', 'BB115', and 'BB119'). The formational average theropod relative abundances (Fig. 6c) in DPP show no appreciable differences, with slight increases in proportions of tyrannosaurids and *Richardoestesia*, and slight decrease in *Troodon*. In MRM, there are more considerable differences in formational average relative theropod abundances, with saurornitholestine proportions greatly increasing, *Troodon* appearing in the upper Oldman Formation while not being found in the middle Oldman Formation, and all other taxa proportionally decreasing slightly.

The non-dinosaur sub-sample analysis (Fig. 7) produced similar results to the non-marine components of the R- vs. Q-mode analysis of all microsite data (Fig. 2, yellow square). Clustering of sites based on their provenience was apparent, though sites did not cluster exclusively based on being from DPP or MRM (Fig. 7a). Sites from MRM clustered more closely to the majority of DPP sites than a number of DPP sites (e.g. 'BB106', 'BB117', 'BB121', 'BB94', 'BB75', 'BB119', 'BB102'), with those latter sites forming a cluster more similar to each other than to any other site. This cluster was associated the proportion of particular actinopterygians (e.g. *Lepisosteus*, *Paratarpon*, Acipenseriformes, *Belonostomus*), and turtles

(e.g. *Basilemys*, Baenidae, Trionychidae). Two sites, both stratigraphically high in the DPF and close to the LCZ, were associated with batoids (e.g. *Myledaphus* + *Pseudomyledaphus*), *Basilemys*, and in one case sharks (*Hybodus*, in 'BB115'). As in other analyses, 'BB54' grouped as something of an outlier, and was here associated strongly with 'Holostean A' and 'Holostean B'. Other DPP sites were broadly associated with many taxa, though in particular with actinopterygians (e.g. *Coriops*, 'Holostean A'), baenid turtles, and lissamphibians (e.g. Caudata + Allocaudata). Sites sampled from MRM as a whole were distinguished from DPP mainly due to an even stronger association with lissamphibians (e.g. Caudata + Allocaudata) and certain actinopterygians (e.g. esocids, teleosts), though particular MRM sites were also distinguished based on their association with taxa that were either absent or in low abundance at other sites, particularly those from DPP. For example, *Adocus* and *Chiloscyllium* are strongly associated with the 'CS' site, though absent or in very low abundance in most sites. One MRM site ('BMC') clustered as an outlier, and was distinguished through a suite of taxa (e.g. chelydrid turtles, mammals, anurans, squamates). Both of these outlier sites ('BB54' and 'BMC') have been noted in previous research to be sedimentologically distinct from other sites, possibly representing an exception to the general trend of depositional setting having a relatively small effect on microsite assemblage structure [14, 52]. Site similarity curves for DPP and MRM (Fig. 7b) were very similar, and very stable, both fluctuating around 60% similarity for much of the sampled interval. The only prominent exception to this came at the top of the time-equivalent interval in DPP, where similarity began to steadily drop, reaching approximately 40% similarity by the top of the sampled interval. Overall proportions of major taxonomic groups in DPP and MRM during this interval (Fig. 7c) were similar, with the notable exception being the higher proportion of batoids in DPP (<20%) when compared to MRM (<5%).

Discussion

Drivers of faunal assemblage clustering in the Belly River Group

Our results broadly support the conclusions of previous studies such as Brinkman et al. [14], expand on their

(See figure on next page.)

Fig. 5 **a** R-mode vs. Q-mode cluster analysis of dinosaur component of Belly River Group microsites from the time-equivalent interval of the Oldman and Dinosaur Park formations of Dinosaur Provincial Park and Milk River/Manyberries. Dinosaur Provincial Park sites indicated with *blue text* and *triangles*. Milk River/Manyberries sites indicated with *red text* and *circles*. **b** Pair-wise Bray-Curtis similarity of Dinosaur Provincial Park (*blue square*) and Milk River/Manyberries (*red square*) microsite dinosaur relative abundance assemblages through lithostratigraphic record of time-equivalent Oldman and Dinosaur Park formations. **c** Formational average proportions of each dinosaur group in sampled regions. Taxonomic groups, site identifications, and other relevant information noted in legend

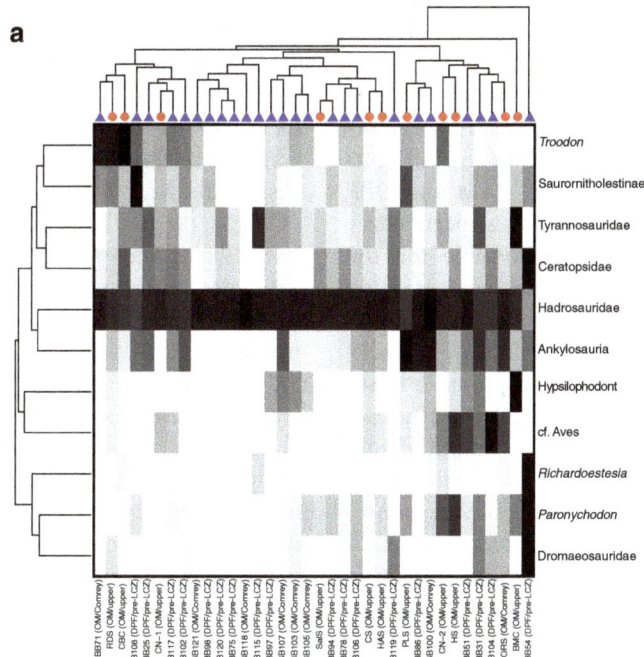

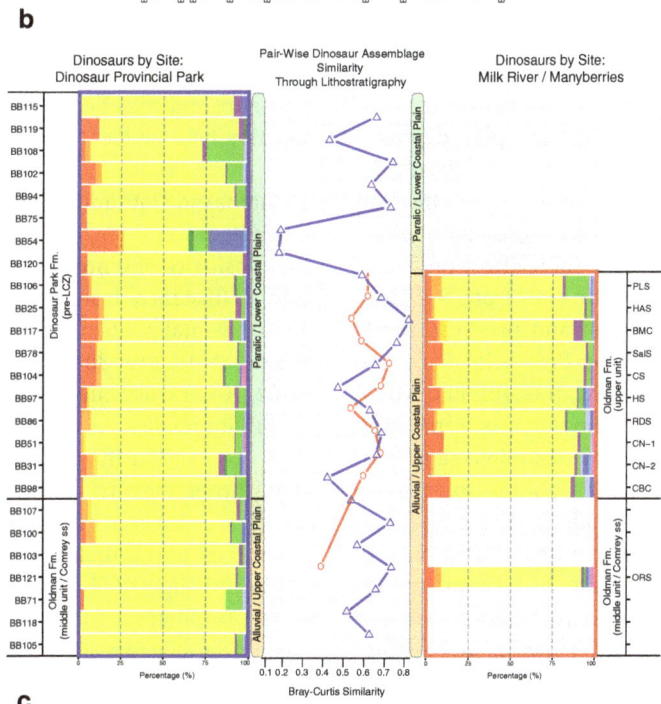

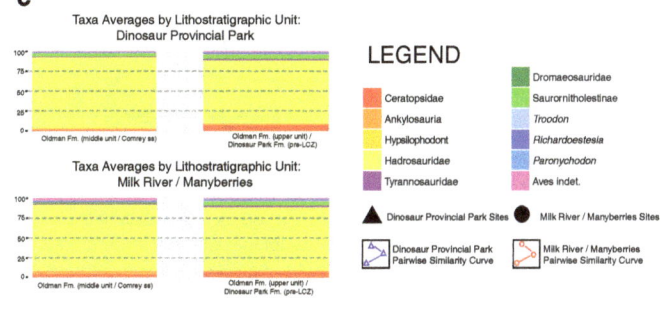

work to include sites from both MRM and DPP in a series of analyses, and more thoroughly test the role that various abiotic factors play in structuring the preserved faunal assemblages. The results of the R- vs. Q-mode cluster analyses (Fig. 2), RDA analyses (Fig. 3), and pair-wise site similarity comparisons (Fig. 4) for the full sample of microsite faunal assemblages of MRM and DPP indicate that palaeoenvironmental changes, particularly marine-terrestrial transitions, are responsible for the most significant changes in faunal assemblage structure, and that other factors like site depositional setting, relative stratigraphic position within the Belly River Group, and palaeogeographic sampling location played a lesser to negligible role. While depositional setting (Fig. 3a) appears superficially to explain site clustering, it should be noted that the only depositional setting that did not display considerable overlap with others were shoreface deposits, and each site characterized as a shoreface deposit was also characterized as preserving a primarily marine palaeoenvironment. Of the three depositional settings analyzed, there was broad overlap in sites characterized as crevasse splays or in-channel deposits (Fig. 3a), which is consistent with palaeoenvironment (Fig. 3d) driving site clustering, as there is also considerable overlap in terrestrial sites, all of which in this sample are preserved as either in-channel or crevasse splay deposits. The differing depositional characteristics of these terrestrial sites are therefore not having a strong effect on the preserved faunal assemblage. Palaeogeographic sampling location appears to also have some effect on faunal assemblage structure, at least within sites from the time-equivalent interval of the Oldman and pre-LCZ Dinosaur Park formations of Dinosaur Provincial Park and Milk River (Fig. 3b). In that context, the DPP and MRM sites formed distinct clusters, though this separation did not hold when expanded to the complete sample of Belly River Group microsites. The separate timeequivalent DPP and MRM clusters provide further support to the hypothesis that at least some of the differences in microsite faunal assemblage structure is the result of biogeographic differences related to environmental variation across the palaeolandscape [12, 14]. However, the distinct clustering of these sites may also be an artefact related to the absence of several taxa (e.g. *Paronychodon*, cf. Aves, *Richardoestesia*) in the DPP microsite data that are

moderately abundant in MRM sites, despite these taxa being reported in the Dinosaur Park Formation [71]. This effect is similarly seen in the PHRN site (Fig. 2) being associated with 'Theropoda indet' material (alongside the numerous marine chondrichthyan taxa that primarily characterize the site) due to this taxonomic category only existing for this site (based on the source data).

Overall, the results of these analyses for the entire Belly River Group microsite database build on prior research conducted on this subject [8, 12–14, 16, 53, 58, 61], and serve to more thoroughly and quantitatively establish the patterns that have been observed in this system. It is not particularly surprising that changes in sea level in the Belly River Group acted as a strong driver of environmental and faunal assemblage change, as the most significant change in faunal assemblage during these intervals is the inverse proportional change in chondrichthyans and lissamphibians, which is almost certainly related to the degree of marine preference (or lack-thereof in the latter case) in these taxa [16, 72, 73]. However, it is important to quantify and understand the exact nature of these trends, as these controvertial data form the baseline for future comparisons and facilitate the testing of more controversial questions, such as the environmental sensitivity of dinosaurs and the effects that more subtle environmental variation have on local palaeocommunity structure.

Altitudinal and latitudinal sensitivity of dinosaur assemblages

The sensitivity of dinosaur populations to changes in altitudinal (distance from palaeoshoreline) and latitudinal environmental gradients has been the subject of considerable debate for over 30 years [1–9, 11, 15, 23, 26, 27, 42–46, 55, 62, 70], and although it has been questioned [17, 18, 22, 28, 74], it remains one of the primary explanations for patterns observed in the evolution and distribution of dinosaurs throughout the Late Cretaceous of western North America. A focused sub-sampling of the time-equivalent interval of the Oldman and Dinosaur Park formations within the larger Belly River Group microsite abundance dataset facilitates a controlled and direct test of dinosaur assemblage changes across differing palaeoenvironments (Figs. 5, 6), while also allowing comparisons to the non-dinosaur component of the broader vertebrate assemblage (Fig. 7) albeit at a regional, rather than continental, scale.

(See figure on next page.)
Fig. 6 **a** R-mode vs. Q-mode cluster analysis of theropod component of Belly River Group microsites from the time-equivalent interval of the Oldman and Dinosaur Park formations of Dinosaur Provincial Park and Milk River/Manyberries. Dinosaur Provincial Park sites indicated with *blue text* and *triangles*. Milk River/Manyberries sites indicated with *red text* and *circles*. **b** Pair-wise Bray-Curtis similarity of Dinosaur Provincial Park (*blue square*) and Milk River/Manyberries (*red square*) microsite theropod relative abundance assemblages through lithostratigraphic record of time-equivalent Oldman and Dinosaur Park formations. **c** Formational average proportions of each theropod group in sampled regions. Taxonomic groups, site identifications, and other relevant information noted in legend

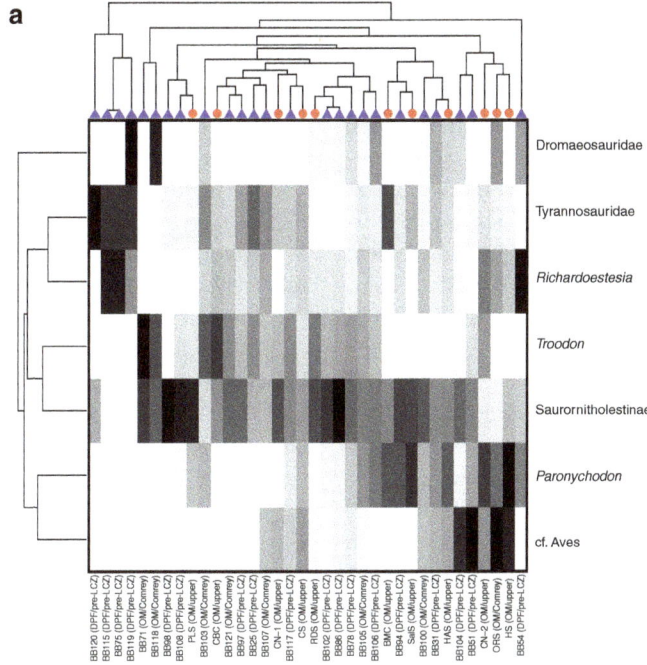

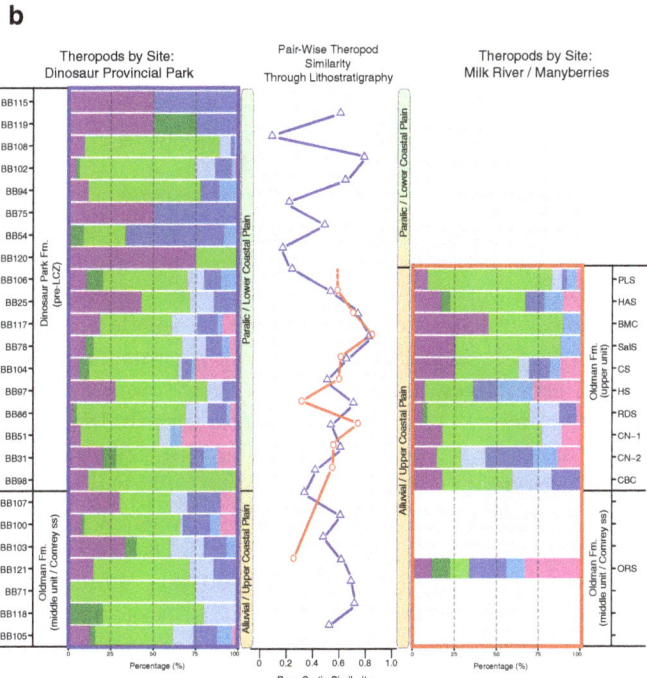

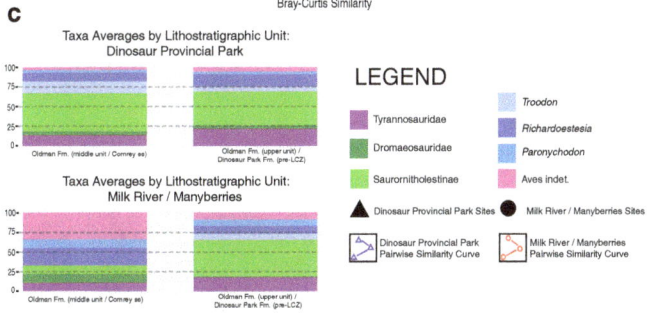

Within dinosaurs (Fig. 5) there is broad assemblage stability, despite the sampled regions representing differing terrestrial environments (lower coastal plain vs. upper coastal plain/inland alluvial fan). In both DPP and MRM, hadrosaurs dominate the preserved dinosaur assemblages, often representing over 80% of relative abundance. A moderate increase in ceratopsians is noted through time in DPP, though across time-equivalent intervals the proportional difference between ceratopsians in DPP and MRM is negligible. This stability between sampling areas is also generally seen in the theropods (Fig. 6), with *Troodon* representing the main exception, as it is not present in the earliest MRM site, appearing there later than in DPP. This was also noted by Brinkman et al. [14], and was attributed to southward migration/range-expansion. The later appearance of *Troodon* in MRM, along with an increase in saurornitholestine material from the middle to upper Oldman Formation in MRM, may be a genuine change in theropod assemblage that is not seen in DPP, or it may be a result of sampling issues, as the middle Oldman Formation of the MRM area is only represented by a single site ('ORS'). The non-dinosaur component of the vertebrate faunal assemblage was relatively stable across much of the sampled interval, although there were considerable differences in the relative abundances of taxa between DPP and MRM (Fig. 7), with batoids forming a much larger component of the overall fauna in DPP. Unlike the pattern shown in the broader vertebrate assemblages (Fig. 7), there does not appear to be strong clustering according to sampling region in dinosaurs (Figs. 5a, 6a), suggesting that the palaeogeographic signal in the broader analysis is due to abundance differences (e.g. batoids) or rare/endemic taxa of non-dinosaurian affiliation (e.g. lissamphibians, turtles, etc.). The similarity of faunal assemblages, and particularly dinosaur assemblages, within the two terrestrial palaeoenvironmental settings (coastal plain vs. alluvial/inland) indicates that these subtle variations in terrestrial palaeoenvironment may have less effect on faunal assemblage structure than previously suggested [1, 12, 14], a position supported by recent research on ceratopsids in the Oldman Formation of the Milk River/Manyberries region [22]. It is also possible that the palaeoenvironmental interpretation of these formations and sampling areas

is more dynamic than originally described [50], though, pending future geological investigations, there is currently no reason to think this is the case. The relative similarity of the dinosaur faunal assemblages of DPP and MRM, and how those contrast to the differences seen in the rest of the vertebrate faunal assemblage between these areas and throughout the extent of the Belly River Group, does not support the idea that dinosaurs, including large bodied taxa like hadrosaurs and ceratopsians, are sensitive to relatively small environmental changes across the regional palaeoenvironmental landscape [1–8, 11, 19–24, 27, 39, 43, 70]. How this plays out over continental scales and between basins is currently unclear, but at least in terms of community composition measured at the family level, large bodied herbivore communities seem to exhibit little variation over the altitudinal transects considered here.

Conclusions

The results of this study demonstrate that palaeoenvironmental setting is the primary driver of differences in vertebrate faunal assemblages throughout the Belly River Group, with palaeogeography/palaeolandscape acting as another factor in structuring these assemblages. Depositional setting and stratigraphic interval do not have particularly strong effects on the preserved faunal assemblage, confirming the results of other smaller-scale studies [14, 59].

The subsample analyses of dinosaur and theropod assemblages, and their comparisons to the broader vertebrate assemblages, suggest one of two possible conclusions: either (a) dinosaur community composition is not sensitive to subtle changes in altitudinal and latitudinal palaeoenvironmental gradients, and/or (b) the differences in environment between the pre-LCZ Dinosaur Park Formation of DPP and the upper Oldman Formation of MRM have been overstated. The higher proportion of batoids in DPP than MRM across this same interval suggests that the more coastally-influenced terrestrial environment of DPP is genuine, refuting the long-held idea that dinosaur communities were particularly sensitive to small-scale environmental gradients, such as paralic (coastal) to alluvial (inland) regimes within a single depositional basin. Further research is required to fully answer

(See figure on next page.)

Fig. 7 a R-mode vs. Q-mode cluster analysis of non-dinosaur component of Belly River Group microsites from the time-equivalent interval of the Oldman and Dinosaur Park formations of Dinosaur Provincial Park and Milk River/Manyberries. Dinosaur Provincial Park sites indicated with *blue text* and *triangles*. Milk River/Manyberries sites indicated with *red text* and *circles*. **b** Pair-wise Bray-Curtis similarity of Dinosaur Provincial Park (*blue square*) and Milk River/Manyberries (*red square*) microsite non-dinosaurian vertebrate relative abundance assemblages through lithostratigraphic record of time-equivalent Oldman and Dinosaur Park formations. **c** Formational average proportions of each non-dinosaurian vertebrate group in sampled regions. Taxonomic groups, site identifications, and other relevant information noted in legend

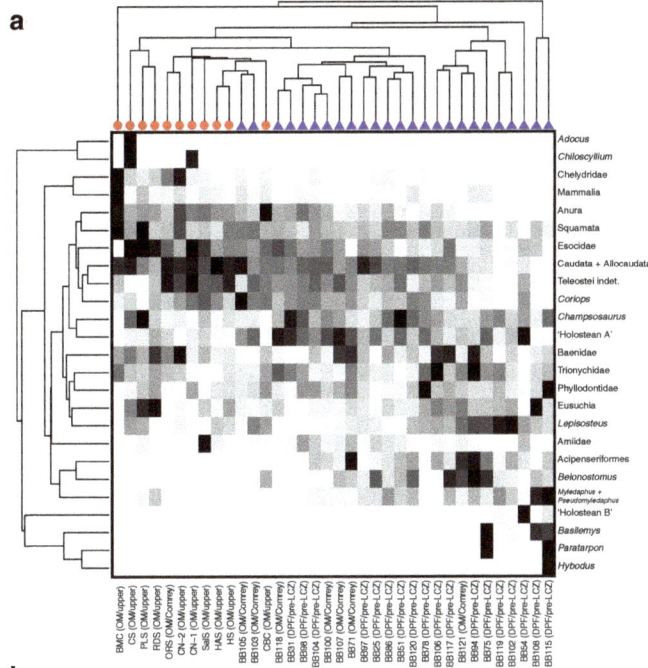

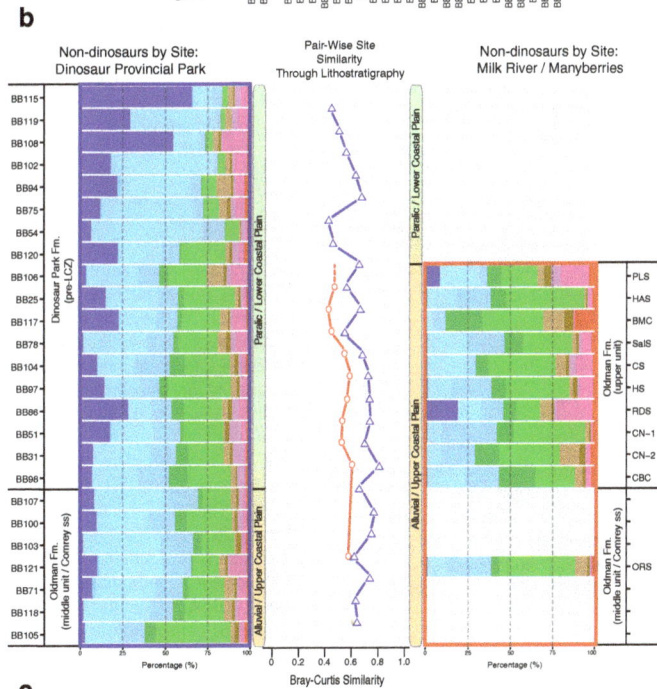

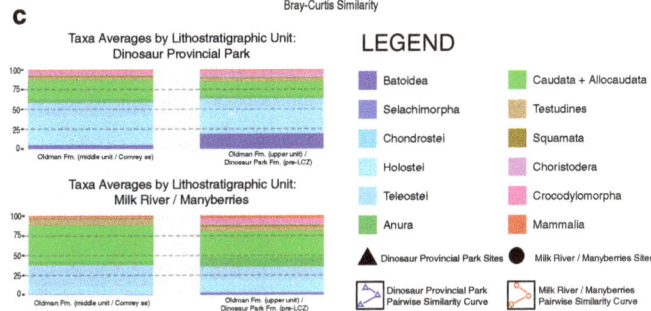

this question, but it is possible that consistently high rates of evolution and niche partitioning among species within each of the sampled dinosaur families played a greater role in the observed high diversity and frequent turnovers in dinosaur taxa throughout the Late Cretaceous of North America.

Additional files

Additional file 1. Taxon abundance table of dinosaurs for sites from time-equivalent Oldman and Dinosaur Park formations. Milk River/Manyberries source data consistent with Table 1, while Dinosaur Provincial Park data derived from targeted dinosaur re-sampling of Brinkman et al [8].

Additional file 2. Environmental factors associated with each Belly River Group microfossil site, for use in redundancy analysis.

Abbreviations
BRG: Belly River Group; DPF: Dinosaur Park Formation; DPP: Dinosaur Provincial Park; LCZ: Lethbridge Coal Zone; MRM: Milk River/Manyberries; OM: Oldman Formation; RDA: redundancy analysis; UPGMA: unweighted pair group method with arithmetic mean.

Authors' contributions
TMC and DCE conceived and designed the project; TMC assembled the data and performed the analyses; TMC made the figures and tables. TMC and DCE wrote the manuscript. All authors read and approved the final manuscript.

Author details
[1] Department of Ecology and Evolutionary Biology, University of Toronto, Toronto, ON, Canada. [2] Department of Natural History, Royal Ontario Museum, 100 Queen's Park, Toronto, ON M5S 2C6, Canada.

Acknowledgements
We would like to thank previous researchers on Belly River Group microsites whose data we utilized, and in particular Don Brinkman for his peerless contributions to vertebrate microfossil palaeoecology. David Eberth and Don Brinkman are thanked for their helpful comments on an earlier draft of this manuscript. Michael J. Ryan, Derek Larson, Karma Nanglu, Kentaro Chiba, Caleb Brown, Kirstin Brink, Jordan Mallon, and Nicolas Campione are thanked for additional discussions related to this research. We also thank A. Dunhill and T. Tortosa for their helpful reviewer comments.

Competing interests
The authors declare that they have no competing interests.

Funding
Funding provided to T.M.C through a Natural Sciences and Engineering Research Council Alexander Graham Bell Canada Graduate Scholarship (CGSD3-444747-2013), and to D.C.E through a Natural Sciences and Engineering Research Council Discovery Grant (RGPIN 355845).

References

1. Gates TA, Sampson SD, Zanno LE, Roberts EM, Eaton JG, Nydam RL, Hutchison JH, Smith JA, Loewen MA, Getty MA. Biogeography of terrestrial and freshwater vertebrates from the late Cretaceous (Campanian) Western Interior of North America. Palaeogeogr Palaeoclimatol Palaeoecol. 2010;291(3–4):371–87.
2. Lehman TM. Late Cretaceous dinosaur provinciality. In: Tanke D, Carpenter K, editors. Mesozoic vertebrate life. Ottawa: NRC Research Press; 2001. p. 310–28.
3. Sampson SD, Loewen MA, Farke AA, Roberts EM, Forster CA, Smith JA, Titus AL. New horned dinosaurs from Utah provide evidence for intracontinental dinosaur endemism. PLoS ONE. 2010;5(9):e12292.
4. Gates TA, Prieto-Márquez A, Zanno LE. Mountain building triggered late Cretaceous North American megaherbivore dinosaur radiation. PLoS ONE. 2012;7(8):e42135.
5. Horner J, Varricchio DJ, Goodwin MB. Marine transgressions and the evolution of Cretaceous dinosaurs. Nature. 1992;358:59–61.
6. Lehman TM. Late Maastrichtian paleoenvironments and dinosaur biogeography in the western interior of North America. Palaeogeogr Palaeoclimatol Palaeoecol. 1987;60:189–217.
7. Lehman TM. Late Campanian dinosaur biogeography in the western interior of North America. Dinofest Int. 1997;1997:223.
8. Brinkman DB, Ryan MJ, Eberth DA. The paleogeographic and stratigraphic distribution of ceratopsids (Ornithischia) in the Upper Judith River Group of Western Canada. Palaios. 1998;13(2):160–9.
9. Loewen MA, Irmis RB, Sertich JJ, Currie PJ, Sampson SD. Tyrant dinosaur evolution tracks the rise and fall of Cretaceous oceans. PLoS ONE. 2013;8(11):e79420.
10. Mallon JC, Evans DC, Ryan MJ, Anderson JS. Feeding height stratification among the herbivorous dinosaurs from the Dinosaur Park Formation (upper Campanian) of Alberta, Canada. BMC Ecology. 2013;13(1):1–15.
11. Carrano MT. Dinosaurian faunas of the later Mesozoic. In: Brett-Surman MK, Holtz Jr TR, Farlow JO, editors. The complete dinosaur. 2nd ed. Bloomington: Indiana University Press; 2012. p. 1003–26.
12. Brinkman DB. Paleoecology of the Judith River Formation (Campanian) of Dinosaur Provincial Park, Alberta, Canada: Evidence from vertebrate microfossil localities. Palaeogeogr Palaeoclimatol Palaeoecol. 1990;78:37–54.
13. Beavan NR, Russell AP. An elasmobranch assemblage from the terrestrial-marine transitional Lethbridge Coal Zone (Dinosaur Park Formation: Upper Campanian), Alberta. Canada. J Paleontol. 1999;73(3):494–503.
14. Brinkman DB, Russell AP, Eberth DA, Peng JH. Vertebrate palaeocommunities of the lower Judith River Group (Campanian) of southeastern Alberta, Canada, as interpreted from vertebrate microfossil assemblages. Palaeogeogr Palaeoclimatol Palaeoecol. 2004;213(3–4):295–313.
15. Fanti F, Miyashita T. A high latitude vertebrate fossil assemblage from the Late Cretaceous of west-central Alberta, Canada: evidence for dinosaur nesting and vertebrate latitudinal gradient. Palaeogeogr Palaeoclimatol Palaeoecol. 2009;275(1–4):37–53.
16. Cullen TM, Fanti F, Capobianco C, Ryan MJ, Evans DC. A vertebrate microsite from a marine-terrestrial transition in the Foremost Formation (Campanian) of Alberta, Canada, and the use of faunal assemblage data as a paleoenvironmental indicator. Palaeogeogr Palaeoclimatol Palaeoecol. 2016;444:101–14.
17. Butler RJ, Benson RB, Carrano MT, Mannion PD, Upchurch P. Sea level, dinosaur diversity and sampling biases: investigating the 'common cause' hypothesis in the terrestrial realm. Proc R Soc Lond B Biol Sci. 2011;278(1709):1165–70.
18. Dunhill AM, Bestwick J, Narey H, Sciberras J. Dinosaur biogeographical structure and Mesozoic continental fragmentation: a network-based approach. J Biogeogr. 2016;43:1691–704. doi:10.1111/jbi.12766.
19. Horner J. The Mesozoic terrestrial ecosystem of Montana. In: Montana Geological Society Field Conference; 1989.
20. Horner J. Three ecologically distinct vertebrate faunal communities from the Late Cretaceous two medicine formation of Montana, with discussion of evolutionary pressures induced by interior seaway fluctuations. In: Montana Geological Society Field Conference; 1984.
21. Bell PR, Currie PJ. A high-latitude dromaeosaurid, Boreonykus certekorum, gen. et sp. nov. (Theropoda), from the upper Campanian Wapiti Formation, west-central Alberta. J Vertebr Paleontol. 2016;36(1):e1034359.

22. Chiba K, Ryan MJ, Braman DR, Eberth DA, Scott EE, Brown CM, Kobayashi Y, Evans DC. Taphonomy of a monodominant Centrosaurus apertus (Dinosauria: Ceratopsia) bonebed from the upper Oldman Formation of southeastern Alberta. Palaios. 2015;30(9):655–67.

23. Evans D, Currie P, Eberth D, Ryan M. High-resolution lambeosaurine dinosaur biostratigirphy, Dinosaur Park Formation, Alberta: sexual dimorphism reconsidered. J Vertebr Paleontol. 2006;26(3):59A.

24. Evans DC, McGarrity CT, Ryan MJ. A skull of Prosaurolophus maximus from Southeastern Alberta and the spatiotemporal distribution of faunal zones in the Dinosaur Park Formation. In: Eberth D, Evans DC, editors. Hadrosaurs. Bloomington: Indiana University Press; 2014. p. 200–7.

25. Lucas SG. Dinosaur communities of the San Juan Basin: A case for lateral variations in the composition of Late Cretaceous dinosaur communities. In: Lucas SG, Rigby K, Kues BS, editors. Advances in San Juan Basin Paleontology. Albuquerque: University of New Mexico Press; 1981. p. 337–93.

26. Prieto-Marquez A, Wagner JR, Bell PR, Chiappe LM. The late-surviving 'duck-billed'dinosaur Augustynolophus from the upper Maastrichtian of western North America and crest evolution in Saurolophini. Geol Mag. 2015;152(02):225–41.

27. Rivera-Sylva HE, Hedrick BP, Dodson P. A Centrosaurine (Dinosauria: Ceratopsia) from the Aguja Formation (Late Campanian) of Northern Coahuila, Mexico. PLoS ONE. 2016;11(4):e0150529.

28. Sullivan RM, Lucas SG. The Kirtlandian land-vertebrate "age"-faunal composition, temporal position and biostratigraphic correlation in the nonmarine Upper Cretaceous of western North America. NM Mus Na Hist Sci Bull. 2006;35:7–30.

29. Béland P, Russell DA. Paleoecology of Dinosaur Provincial Park (Cretaceous), Alberta, interpreted from the distribution of articulated vertebrate remains. Can J Earth Sci. 1012;1978:15.

30. Coe MJ, Dilcher DL, Farlow JO, Jarzen DM, Russell DA. Dinosaurs and land plants. In: Friis EM, Chaloner WG, Crane PR, editors. The origins of Angiosperms and their biological consequences. Cambridge: Cambridge University Press; 1987.

31. Currie PJ, Dodson P. Mass death of a heard of ceratopsian dinosaurs. In: Reif WE, Westphal F, editors. Third symposium of Mesozoic terrestrial ecosystems. Tübingen: Attempto Verlag; 1984.

32. du Toit JT, Cumming DHM. Functional significance of ungulate diversity in African savannas and the ecological implications of the spread of pastoralism. Biodivers Conserv. 1999;8:1643.

33. Farlow JO. A consideration of the trophic dynamics of a Late Cretaceous large-dinosaur community (Oldman Formation). Ecology. 1976;57:841.

34. Farlow JO, Dodson P, Chinsamy A. Dinosaur biology. Annu Rev Ecol Syst. 1995;26:445.

35. Horner JR, Makela R. Nest of juveniles provides evidence of family structure among dinosaurs. Nature. 1979;282:296–8. doi:10.1038/282296a0.

36. Ryan MJ, Russell AP, Eberth DA, Currie PJ. The taphonomy of a Centrosaurus (Ornithischia: Ceratopsidae) bone bed from the Dinosaur Park Formation (Upper Campanian), Alberta, Canada, with comments on cranial ontogeny. PALAIOS. 2001;16:482.

37. Varricchio DJ, Horner JR. Hadrosaurid and lambeosaurid bone beds from the Upper Cretaceous Two Medicine Formation of Montana: taphonomic and biologic implications. Can J Earth Sci. 1993;30:997.

38. Bakker RT. Dinosaur feeding behaviour and the origin of flowering plants. Nature. 1978;274:661–3. doi:10.1038/274661a0.

39. Campione NE, Evans DC. Cranial growth and variation in Edmontosaurus (Dinosauria: Hadrosauridae): implications for latest Cretaceous megaherbivore diversity in North America. PLoS ONE. 2011;6:e25186.

40. Carrano MT, Janis CM, Sepkoski JJ. Hadrosaurs as ungulate parallels: lost lifestyles and deficient data. Acta Palaeontol Pol. 1999;44:237.

41. Dodson P. A faunal review of the Judith River (Oldman) Formation, Dinosaur Provincial Park, Alberta. Mosasaur. 1983;1:89.

42. Fricke HC, Pearson DA. Stable isotope evidence for changes in dietary niche partitioning among hadrosaurian and ceratopsian dinosaurs of the Hell Creek Formation, North Dakota. Paleobiology. 2008;34(4):534–52.

43. Mallon JC, Evans DC, Ryan MJ, Anderson JS. Megaherbivorous dinosaur turnover in the Dinosaur Park Formation (upper Campanian) of Alberta, Canada. Palaeogeogr Palaeoclimatol Palaeoecol. 2012;350–352:124–38.

44. Ryan MJ, Evans DC. Ornithischian Dinosaurs. In: Currie PJ, Koppelhus EB, editors. Dinosaur Provincial Park: a spectacular ancient ecosystem revealed. Bloomington: Indiana Univeristy Press; 2005. p. 312–48.

45. Sampson SD, Loewen MA. Unraveling a radiation: a review of the diversity, stratigraphic distribution, biogeography, and evolution of horned dinosaurs (Ornithischia: Ceratopsidae). In: Ryan MJ, Chinnery-Allgeier BJ, Eberth DA, editors. New perspectives on horned dinosaurs: the royal tyrrell museum Ceratopsian symposium. Bloomington: Indiana University Press; 2010.

46. Longrich NR. Judiceratops tigris, a New Horned Dinosaur from the Middle Campanian Judith River Formation of Montana. Bull Peabody Mus Nat Hist. 2013;54(1):51–65.

47. Evans DC, Reisz RR. Anatomy and relationships of Lambeosaurus magnicristatus, a crested hadrosaurid dinosaur (Ornithischia) from the Dinosaur Park Formation, Alberta. J Vertebr Paleontol. 2007;27(2):373–93.

48. Currie PJ, Koppelhus EB, editors. Dinosaur Provincial Park: a spectacular ancient ecosystem revealed. Bloomington: Indiana University Press; 2005.

49. Eberth DA. The Geology. In: Currie PJ, Koppelhus EB, editors. Dinosaur Provincial Park: a spectacular ancient ecosystem revealed. Bloomington: Indiana University Press; 2005. p. 54–82.

50. Eberth DA, Hamblin AP. Tectonic, stratigraphic, and sedimentologic significance of a regional discontinuity in the upper Judith River Group (Belly River wedge) of southern Alberta, Saskatchewan, and northern Montana. Can J Earth Sci. 1993;30:174–200.

51. Ogunyomi O, Hills LV. Depositional environments, Foremost Formation (Late Cretaceous), Milk River area, southern Alberta. Bull Can Pet Geol. 1977;25(5):929–68.

52. Eberth DA. Stratigraphy and sedimentology of vertebrate microfossil sites in the uppermost Judith River Formation (Campanian), Dinosaur Provincial Park, Alberta, Canada. Palaeogeogr Palaeocl. 1990;78:1–36.

53. Brinkman DB, Braman DR, Neuman AG, Ralrick PE, Sato T. A vertebrate assemblage from the marine shales of the Lethbridge Coal Zone. In: Currie PJ, Koppelhus EB, editors. Dinosaur Provincial Park: a spectacular ancient ecosystem revealed. Bloomington: Indiana University Press; 2005. p. 486–500.

54. Braman DR, Koppelhus EB. Campanian palynomorphs. In: Currie PJ, Koppelhus EB, editors. Dinosaur Provincial Park: a spectacular ancient ecosystem revealed. Bloomington: Indiana University Press; 2005. p. 101–30.

55. Larson DW, Brinkman DB, Bell PR. Faunal assemblages from the upper Horseshoe Canyon Formation, an early Maastrichtian cool-climate assemblage from Alberta, with special reference to the Albertosaurus sarcophagus bonebed This article is one of a series of papers published in this Special Issue on the theme Albertosaurus. Can J Earth Sci. 2010;47(9):1159–81.

56. Sankey JT, Brinkman D, Guenther M, Currie PJ. Small theropod and bird teeth from the late Cretaceous (late Campanian) Judith River Group, Alberta. J Paleontol. 2002;76(4):751–63.

57. Eberth D, Shannon M, Noland B. A bonebeds database: classification, biases, and patterns of occurrence. In: Rogers R, Eberth D, Fiorillo A, editors. Bonebeds: genesis, analysis, and paleobiological significance. Chicago: University of Chicago Press; 2007. p. 103–220.

58. Peng J, Russell AP, Brinkman DB. Vertebrate microsite assemblages (exclusive of mammals) from the Foremost and Oldman Formations of the Judith River Group (Campanian) of southeastern Alberta: an illustrated guide. Prov Mus Alberta Nat His Occ Papers. 2001;25:1–54.

59. Rogers RR, Brady ME. Origins of microfossil bonebeds: insights from the Upper Cretaceous Judith River Formation of north-central Montana. Paleobiology. 2010;36(1):80–112.

60. Rogers RR, Kidwell SM. A conceptual framework for the genesis and analysis of vertebrate skeletal concentrations. In: Rogers RR, Eberth DA, Fiorillo AR, editors. Bonebeds: genesis, analysis, and paleobiological significance. Chicago: University of Chicago Press; 2007. p. 1–63.

61. Frampton EK. Taphonomy and palaeoecology of mixed invertebrate-vertebrate fossil assemblage in the Foremost Formation (Cretaceous, Campanian), Milk River Valley, Alberta. Calgary: University of Calgary; 2005.

62. Bell PR, Snively E. Polar dinosaurs on parade: a review of dinosaur migration. Alcheringa. 2008;32(3):271–84.

63. Kirkland JJ, Eaton JG, Brinkman DB. Elasmobranchs from Upper Cretaceous freshwater facies in southern Utah. In: Loewen MA, Titus AL, editors. At the top of the Grand Staircase—the Late Cretaceous of southern Utah. Bloomington: Indiana University Press; 2013. p. 153–94.

64. Newbrey MG, Murray AM, Brinkman DB, Wilson MVH, Neuman AG. A new articulated freshwater fish (Clupeomorpha, Ellimmichthyiformes) from the Horseshoe Canyon Formation, Maastrichtian, of Alberta, Canada. Can J Earth Sci. 2010;47(9):1183–96.

65. Oksanen J, Blanchet FG, Kindt R, Legendre P, Minchin PR, O'Hara RB, Simpson GL, Solymos P, Stevens MH, Wagner H. Vegan: Community Ecology Package. In., 2.0-10 edn; 2013.

66. De Cáceres M, Legendre P, He F. Dissimilarity measurements and the size structure of ecological communities. Methods Ecol Evol. 2013;4:1167–77.

67. Wickham H. ggplot2: elegant graphics for data analysis. New York: Springer; 2009.

68. Wickham H. Package 'reshape2, flexibly reshape data: a reboot of the reshape package. In., 1.4.1 edn; 2014.

69. Vavrek MJ. Package 'fossil', palaeoecological and palaeogeographical analysis tools. In., 0.3.7 edn; 2012.

70. Ryan MJ, Evans DC, Shepherd KM. A new ceratopsid from the Foremost Formation (middle Campanian) of Alberta. Can J Earth Sci. 2012;49(11):1251–62.

71. Brown CM, Evans DC, Campione NE, O'Brien LJ, Eberth DA. Evidence for taphonomic size bias in the Dinosaur Park Formation (Campanian, Alberta), a model Mesozoic terrestrial alluvial-paralic system. Palaeogeogr Palaeoclimatol Palaeoecol. 2013;372:108–22.

72. Christman SP. Geographic variation for salt water tolerance in the frog Rana sphenocephalia. Copeia. 1974;3:773–8.

73. Smith HW. Water regulation and its evolution in the fishes. Q Rev Biol. 1932;7(1):1–26.

74. Vavrek MJ, Larsson HC. Low beta diversity of Maastrichtian dinosaurs of North America. Proc Natl Acad Sci USA. 2010;107(18):8265–8.

9

Diet segregation in American bison (*Bison bison*) of Yellowstone National Park (Wyoming, USA)

John L. Berini[1]* and Catherine Badgley[2]

Abstract

Background: Body size is a major factor in the nutritional ecology of ruminant mammals. Females, due to their smaller size and smaller rumen, have more rapid food-passage times than males and thereby require higher quality forage. Males are more efficient at converting high-fiber forage into usable energy and thus, are more concerned with quantity. American bison are sexually dimorphic and sexually segregate for the majority of their adult lives, and in Yellowstone National Park, they occur in two distinct subpopulations within the Northern and Central ranges. We used fecal nitrogen and stable isotopes of carbon and nitrogen from American bison to investigate sex-specific differences in diet composition, diet quality, and dietary breadth between the mating season and a time period spanning multiple years, and compared diet indicators for these different time periods between the Northern and Central ranges.

Results: During mating season, diet composition of male and female American bison differed significantly; females had higher quality diets, and males had greater dietary breadth. Over the multi-year period, females had higher quality diets and males, greater dietary breadth. Diet segregation for bison in the Central Range was more pronounced during the mating season than for the multi-year period and females had higher quality diets than males. Finally, diet segregation in the Northern Range was more pronounced during the multi-year period than during the mating season, and males had greater dietary breadth.

Conclusions: Female bison in Yellowstone National Park have higher quality diets than males, whereas males ingest a greater diversity of plants or plants parts, and bison from different ranges exhibited more pronounced diet segregation during different times. Collectively, our results suggest that diet segregation in bison of Yellowstone National Park is associated with sex-specific differences in nutritional demands. Altogether, our results highlight the importance of accounting for spatial and temporal heterogeneity when conducting dietary studies on wild ungulates.

Background

Body size is a major factor in the nutritional ecology of ruminant mammals, as mass-specific energy demands generally decrease with increasing body mass [1]. On average, the ability of ungulates to convert high-fiber, low-quality forage into usable energy increases with body size, a trend known as the Jarman–Bell principle [2, 3]. Although the Jarman–Bell principle was first used to explain dietary differences among African ruminants of varying sizes [2, 3], evidence suggests that the principles upon which the Jarman–Bell principle is founded hold true for dietary differences within species as well, and therefore, may help explain sexual segregation in sexually dimorphic ruminants [4]. Applying the Jarman–Bell principle to size-dimorphic ruminants suggests that smaller individuals should be more selective feeders and have higher quality diets than larger individuals [2–4]. In sexual segregation theory, the forage-selection hypothesis suggests that females, due to their smaller size and smaller rumen, have more rapid food-passage times than males and thereby require higher quality forage in order to maintain body weight. Males,

*Correspondence: beri0015@umn.edu
[1] Department of Fisheries, Wildlife, and Conservation Biology, University of Minnesota, 135 B Skok Hall, 2003 Upper Buford Circle, St. Paul, MN 55108-1052, USA
Full list of author information is available at the end of the article

with larger size and longer food-retention times, are more efficient at converting high-fiber forage into usable energy and thus, are more concerned with quantity [5–7]. Although numerous hypotheses have been invoked to explain sexual segregation in ruminants (for a review see [8]), this phenomenon is best characterized as the differential use of space, habitat, or food by males and females.

Sexual dimorphism and segregation in large mammals are typically associated with sex-specific differences in energetic requirements and digestion [9]. For example, the activity budget hypothesis assumes that females are less efficient at digesting forage than males, and therefore spend more time foraging, whereas males spend more time ruminating [10]. While the forage-selection hypothesis also assumes that females are less efficient at digesting forage, this hypothesis predicts group formation based on dietary preferences as opposed to activity budget. American bison (*Bison bison*) are highly size-dimorphic, with the average male (800 kg) weighing roughly 350 kg more than the average female (450 kg; [11]). Reproductively active females have higher energy demands than males due to gestation and lactation, with energetic requirements peaking during early to mid-summer [12, 13]. The energetic demands of males also peak during early to mid-summer, when mature individuals are replenishing energy stores in preparation for the rut, which occurs during late summer [12, 14]. During the rut, male and female bison spend approximately 1 month in large mating groups, with males tending to potential mates in an effort to maximize breeding potential [15]. Throughout the rest of the year, males and females remain segregated. Collectively, the anatomical, behavioral, and physiological differences between male and female American bison make them an ideal species for studies of diet segregation. Previous studies of sexual segregation in American bison provide evidence for both the forage-selection hypothesis and the predation-risk hypothesis. The latter assumes that females and their calves are more vulnerable to predation than males and therefore select habitats that provide increased safety over increased forage quality [5, 16].

In Yellowstone National Park (YNP), the bison population consists of two geographically and ecologically distinct subpopulations. In the Northern and Central ranges of YNP (Fig. 1), bison populations differ in fetal growth rates [17], median calving dates [17], and tooth-wear patterns [18]. Differences in mtDNA haplotype frequencies between ranges provide evidence for assortative breeding [19], and since their reintroduction in 1995, wolf predation rates on bison have been higher in the Central Range than in the Northern Range [20, 21]. Additionally, higher

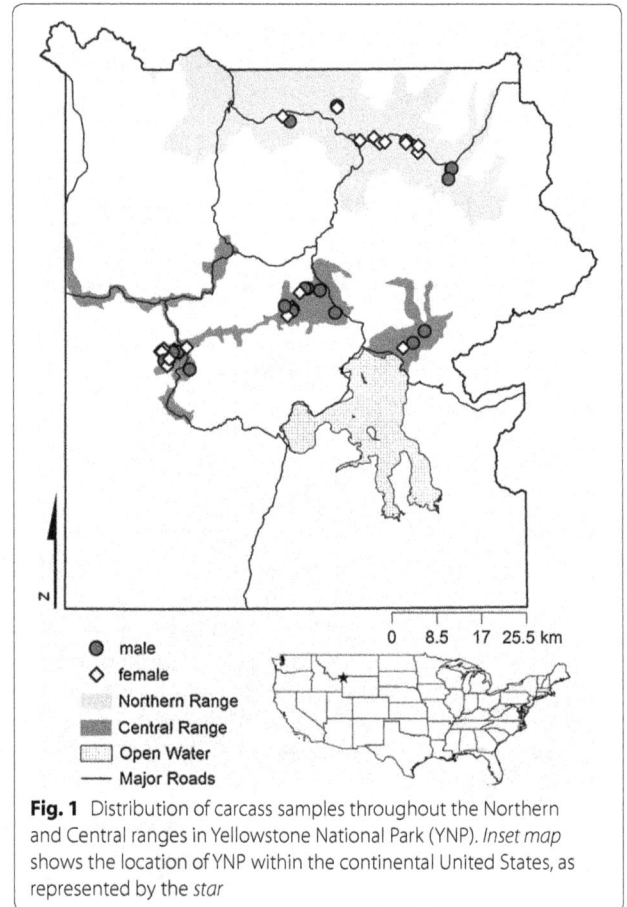

Fig. 1 Distribution of carcass samples throughout the Northern and Central ranges in Yellowstone National Park (YNP). *Inset map* shows the location of YNP within the continental United States, as represented by the *star*

altitude and colder temperatures in the Central Range result in deeper snow cover, which remains significantly longer than in the Northern Range [22, 23]. The Central Range also has higher levels of geothermal activity than the Northern Range, resulting in greater abundance of winter forage [24]. However, the quality of forage located on geothermal patches tends to be of relatively low quality, with elevated levels of fluoride and silica [25]. The two ranges also differ in bedrock geology and soil type [23]. The Northern Range is dominated by andesitic bedrock, which generates soils that are richer in plant-available nutrients and organic carbon compared to soils generated by rhyolitic bedrock, which is prevalent across the Central Range [23]. Population growth of bison in YNP is thought to be contingent on herbivore density and winter-forage availability [26]. In general, Yellowstone National Park is considered to be forage-limited [26], and population densities of both elk and bison are greater on the Northern Range compared to those of the Central Range [27, 28]. Spatial and temporal variations in the biotic and abiotic environments should be reflected in the quality, abundance, and distribution of forage available to bison throughout YNP and therefore, also reflected in the composition and quality of their diets.

If different sexes face different nutritional demands, then differences in forage quality and density-dependent influences on forage availability and selection should be reflected in how similar (or different) the diets of males and females are within each range. Moreover, the degree of diet segregation between males and females within each range should vary between the mating and non-mating seasons. In the Central Range, bison and elk densities are relatively low throughout the year [27, 28]. As a result, animals in this region of the park likely experience strong density-dependent effects during late summer, when animals form large, mixed-sex mating groups. Thus, diet segregation in the Central Range should be most pronounced during this time, and should be in sharp contrast to the degree of diet segregation we might observe throughout the rest of the year. In the Northern Range, bison and elk densities are considerably higher, and as a result, animals in this region of the park likely experience strong density-dependent effects throughout the entire year. While ungulate densities in the Northern Range may also peak during the mating season, differences in the degree of diet segregation between the mating and non-mating seasons are likely less pronounced than that which we might observe on the Central Range.

Using fecal nitrogen and stable isotopes of carbon ($\delta^{13}C$) and nitrogen ($\delta^{15}N$), we evaluated whether male and female bison in YNP exhibit diet segregation and whether the degree of diet segregation varies within each range during mating season and across a time period spanning multiple years. To determine whether diet segregation of bison in YNP is associated with sex-specific nutritional demands and whether these demands are met differently in each range, we tested four hypotheses. (1) Male and female bison have distinct diet compositions during the mating season and (2) across multiple years. (3) Male and female bison from the Central Range differ more in diet composition (i.e., show more pronounced diet segregation) during the mating season than across multiple years. (4) Male and female bison from the Northern Range have different diet compositions, regardless of season. Regardless of time period or range, we expect females to have higher quality diets than males. Finally, during the mating season, males are most concerned with tending potential mates in an effort to maximize breeding potential, and thus, the variety of forages available to both males and females should be similar during this time. As a result, we expect males and females to have similar dietary breadths during the mating season, but statistically distinct dietary breadths throughout the multi-year period, regardless of range.

Methods

Study area

Yellowstone National Park occupies 891,000 ha of primarily forested habitat in northwestern Wyoming, USA, and ranges from 1500 to 3300 m in elevation [27]. During the summer months, the park is home to an estimated 3700 bison [29] that reside principally in two areas, the Central Range and the Northern Range (Fig. 1). As winter approaches, bison move to lower elevations, including the lower Yellowstone River drainage and Blacktail Deer Plateau in the Northern Range, where snow pack is shallower and the growth of spring vegetation begins earlier [30]. As noted above, distinct biotic and abiotic differences between the two ranges influence the abundance and distribution of high-quality forage for bison in YNP [23].

Analytical methods

Fecal nitrogen is an established measure of dietary quality [31], and stable isotopes, especially of carbon and nitrogen, are commonly used to investigate dietary differences both within and among mammal populations [32]. The primary source of nitrogen in feces of mammalian herbivores is dietary protein, although plant secondary metabolites that bind to proteins as well as microbial sources of nitrogen may influence estimates of protein in feces [31]. Regardless, absolute forage intake [33], forage digestibility, and dietary protein are positively correlated with fecal N in grazing ungulates [34]. Stable isotopes of carbon found in the biogenic materials of herbivores (e.g., feces and collagen) reflect the average $\delta^{13}C$ of plants ingested and assimilated during the formation of these materials [35, 36]. Thus, $\delta^{13}C$ is commonly used to distinguish diet compositions of large herbivores using mathematical mixing models [37]. Even without the use of mixing models, however, both the mean and variance of stable isotopes provide information about dietary habits. Differences in mean $\delta^{13}C$ have been used to analyze population differences in diet [38], and individual variation in $\delta^{13}C$ from biogenic materials (e.g., feces and bone) provides a measure of dietary breadth within a population. As animals ingest a greater range of plant species and plant parts, the variance of $\delta^{13}C$ increases [39, 40]. In general, the mean $\delta^{15}N$ of animal tissues reflects the protein content of the animal's diet [41, 42], with a negative correlation between the nitrogen content of ingested plants and $\delta^{15}N$ values of herbivore bone collagen ($\delta^{15}N_{collagen}$; [43]).

Stable isotopes of carbon and nitrogen from bone collagen and feces, along with fecal N content, provide information about diet over different temporal windows. Bone formation is a relatively continuous process [44], and bone collagen has a turnover rate on the order of years

[45, 46]. Thus, the isotopic composition of bone collagen reflects the average foraging trends over much of the lifetime of sampled individuals (i.e., multi-year). Stable isotope values from bison fecal samples, in contrast, reflect the isotopic composition of forage ingested during the previous 24–48 h [47].

Sample collection

We collected 60 fecal samples from bison in the Northern and Central ranges (15 males and 15 females from each range) over a period of 10 days during the mating season of early August 2009. We monitored animals along the periphery of the herd and when defecation occurred, we recorded a compass bearing in the direction of that potential sample, as well as any landmarks useful in determining the precise location of that sample (e.g., tree, bushes, flowers, mounds, etc.). No two samples were collected within close spatial proximity of one another (i.e., approximately 20 m) within the same herd, and samples were collected within 2 h of elimination. All fecal samples came from animals of reproductive age (≥6 years for males and ≥2 years for females); one fecal sample from the Central Range had to be discarded due to molding. We determined whether animals were of reproductive age based on size, head morphology, and pelage (mature males have well-defined manes that are absent on females and immature males; [48]).

The YNP scientific staff provided the locations of bison carcasses from adult animals that had died over the previous 36 months. Sample collection took place between 15 July and 7 August 2009. We collected 22 hemi-mandibles and 18 vertebrae from 40 carcasses. Twenty-five of the sampled carcasses were from the Central Range (13 females, 12 males), and 15 (11 females, 4 males) were from the Northern Range (Fig. 1). We estimated age at time of death using patterns of cheek–tooth eruption and wear [49] and the degree of fusion of suture lines on the frontal bones of the skull [50]. We determined the sex of carcasses based on sex-specific morphological differences in the skull [51] and pelvic girdle [52, 53]. For stable isotope analysis of bone collagen, we collected hemi-mandibles; when a hemi-mandible could not be located, we collected a single vertebra.

Sample preparation and analysis

We dried fecal samples at 60 °C until stable weights were achieved, then homogenized them in a SPEX Certiprep 8000D ball mill. We weighed 0.3 ± 0.05 mg of each sample and wrapped it in a Costech 5×9 mm tin capsule for analysis of %N and $\delta^{13}C$. For stable isotope analysis of bone collagen, we demineralized approximately 50 mg of bone in 0.5 N HCl for 72 h at 4 °C. We then rinsed samples with de-ionized water and placed them in 6 ml

of chloroform/methanol lipid-extraction solution [54]. Finally, we lyophilized the samples for 24 h and then wrapped 0.3 ± 0.05 mg of collagen sample in a Costech 5×9 mm tin capsule for analysis of $\delta^{13}C$ and $\delta^{15}N$ [55].

Samples were analyzed in the Nadelhoffer Stable Isotope Laboratory at the University of Michigan Biological Station (Pellston, Michigan). Fecal and collagen samples were analyzed on a Thermo Delta Plus XL isotope ratio mass spectrometer via combustion in a Costech CHN analyzer, Model 4010. Analytical precision was estimated via multiple analyses of a caffeine standard (mean $\delta^{13}C = -49.35$, SD = 0.16; mean $\delta^{15}N = -1.7‰$, SD = 0.12). Stable isotope values of C and N were reported to the nearest 0.01 and 0.1‰, respectively, and fecal-N content ($\%N_{feces}$) was reported to the nearest 0.01%. Data acquisition and instrument control were accomplished using ISODAT 2.0. Isotopic values of feces and bone collagen were reported relative to international standards, Vienna Pee Dee Belemnite (VPDB) for carbon and atmospheric N for nitrogen.

Statistical analysis

We used multivariate analysis of variance (MANOVA) to determine whether diets differ throughout YNP as a function of sex and range. Specifically, we evaluated six different models (Table 1) to determine whether diets vary as a function of sex and range during the mating season (hypothesis 1, model 1) and over a multi-year period (hypothesis 2, model 2) and whether diets of males and females differ during the mating season and over a multi-year period in both the Central (hypothesis 3, models 3.1 and 3.2, respectively) and Northern ranges (hypotheses 4, models 4.1 and 4.2, respectively). For any tests resulting in $P < 0.10$, we applied a univariate ANOVA to determine whether any of the dependent variables were significantly influenced by our independent variables (i.e., sex and range).

We also investigated differences in dietary breadth using Bartlett's test for homogeneity of variance [56]. We compared variance of $\delta^{13}C_{feces}$ to determine if dietary breadth varied as a function of sex and range during the mating season and compared variance of $\delta^{13}C_{collagen}$ to determine whether dietary breadth varied as a function of sex across a multi-year time period. While our sample sizes may seem small when subdivided by sex and range, Clementz and Koch [57] reported that a sample size of $n = 5$ resulted in a standard error of 0.01‰ for the mean $\delta^{13}C$ value of the study population and recommended ≥5 as suitable sample size for stable isotope studies of populations. However, one of our collagen test groups, Northern Range males, had a sample size just below this threshold. To determine the influence of this small sample size, we calculated 95% confidence intervals for each sex-range

Table 1 Structure and sample size for MANOVA models

H_a	x	y	Time period	Sample size		Model ID
				Female	Male	
1	Sex	$\delta^{13}C_{feces}$, $\%N_{feces}$	Mating season	30	29	1
2	Sex	$\delta^{13}C_{collagen}$, $\delta^{15}N_{collagen}$	Multi-year	24	16	2
3	Sex, CR	$\delta^{13}C_{feces}$, $\%N_{feces}$	Mating season	15	14	3.1
		$\delta^{13}C_{collagen}$, $\delta^{15}N_{collagen}$	Multi-year	13	12	3.2
4	Sex, NR	$\delta^{13}C_{feces}$, $\%N_{feces}$	Mating season	15	15	4.1
		$\delta^{13}C_{collagen}$, $\delta^{15}N_{collagen}$	Multi-year	11	4	4.2

Models tested whether diet varies as a function of sex and range between mating season and a multi-year time period. To test hypotheses 3 and 4, we subset the data for each range (CR, Central Range; NR, Northern Range), and analyzed the effect of sex on proxy measures for diet within each range. Columns labeled "x" and "y" denote the independent and dependent variables, respectively, for each model

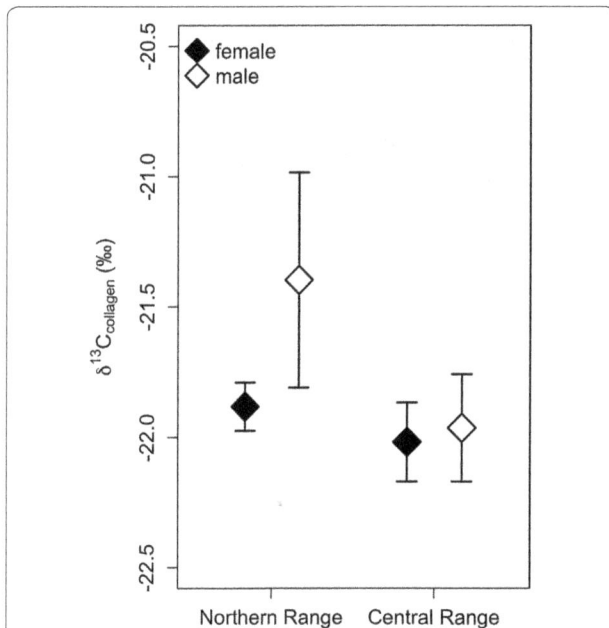

Fig. 2 Mean and 95% confidence intervals for each combination of sex and range for $\delta^{13}C$ of collagen ($\delta^{13}C_{collagen}$). *Points* represent mean values for $\delta^{13}C_{collagen}$ and *bars* represent 95% confidence intervals. Sample sizes for Northern Range bison are n = 11 for females and n = 4 for males, and sample sizes for Central Range bison are n = 13 for females and n = 12 for males

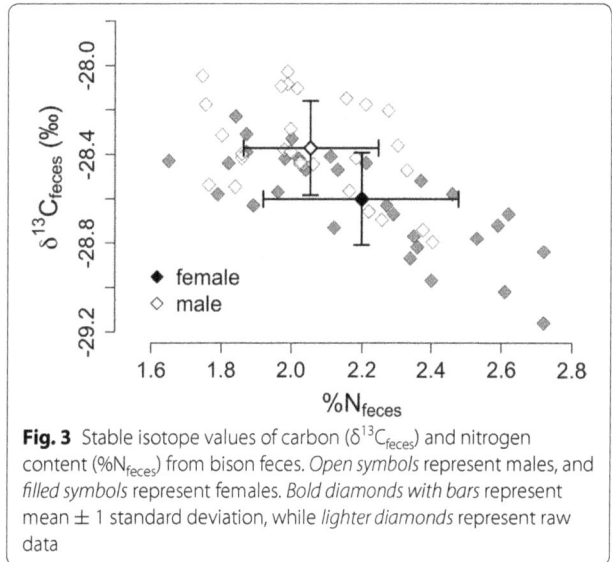

Fig. 3 Stable isotope values of carbon ($\delta^{13}C_{feces}$) and nitrogen content ($\%N_{feces}$) from bison feces. *Open symbols* represent males, and *filled symbols* represent females. *Bold diamonds with bars* represent mean ± 1 standard deviation, while *lighter diamonds* represent raw data

combination (Fig. 2). Additionally, we performed an F test to compare the variance of Northern Range males (n = 4) to Central Range males (n = 12) and found the variance between the two groups to be statistically indistinguishable (F = 1.3463, P = 0.6193). All statistical analyses were performed using base packages in R 3.1.2 [58].

Results

Male and female bison throughout YNP had distinctly different diets during the mating season (Fig. 3; Table 2). While there was no difference in dietary breadth during this time (variance of $\delta^{13}C_{feces}$, Table 2), mean values of

Table 2 Results for single-factor MANOVA tests and Bartlett's tests for homogeneity of variance

Model ID	MANOVA		P	K^2	Bartlett's test P
	df	F			
1	1,57	8.698	<0.001*	0.008	0.931
2	1,38	2.991	0.063	7.175	0.007*
3.1	1,27	11.915	<0.001*	1.718	0.190
3.2	1,23	1.434	0.260	0.781	0.377
4.1	1,28	2.025	0.153	1.430	0.232
4.2	1,13	5.403	0.021*	5.076	0.024*

Statistically significant results (P < 0.05) for both tests are identified with an asterisk. Bartlett's test for homogeneity of variance was used to analyze differences in variance of $\delta^{13}C_{feces}$ (mating season) and $\delta^{13}C_{collagen}$ (multi-year)

$\delta^{13}C_{feces}$ were significantly different between sexes indicating distinct diet compositions. Females had greater fecal nitrogen ($\%N_{feces}$) than males (Fig. 3; Table 3),

Table 3 Univariate ANOVA results for MANOVA tests with P < 0.10

Model ID	x	y	Mean values		df	SS	F	P
			Female	Male				
1	Sex	$\delta^{13}C_{feces}$	−28.61	−28.36	1,57	0.904	17.570	<0.001*
		$\%N_{feces}$	2.20	2.04	1,57	0.357	5.292	0.025*
2	Sex	$\delta^{13}C_{collagen}$	−21.96	−21.82	1,38	0.171	1.536	0.223
		$\delta^{15}N_{collagen}$	3.94	4.46	1,38	2.583	4.905	0.033*
3.1	Sex, CR	$\delta^{13}C_{feces}$	−28.76	−28.38	1,27	1.056	18.670	<0.001*
		$\%N_{feces}$	2.46	2.14	1,27	0.713	20.410	<0.001*
4.2	Sex, NR	$\delta^{13}C_{collagen}$	−21.88	−21.40	1,13	0.691	11.550	0.005*
		$\delta^{15}N_{collagen}$	3.66	3.82	1,13	0.076	0.228	0.641

Significant difference in $\delta^{13}C$ values between males and females indicates different dietary sources during each respective time period. Higher $\%N_{feces}$ indicates higher quality diets during the mating season, whereas lower $\delta^{15}N_{collagen}$ indicates higher quality diet during the multi-year time period. Stable isotope values are reported to the nearest 0.01‰, and $\%N_{feces}$ is reported to the nearest 0.01%. Columns labeled "x" and "y" denote the independent and dependent variables, respectively, for each model

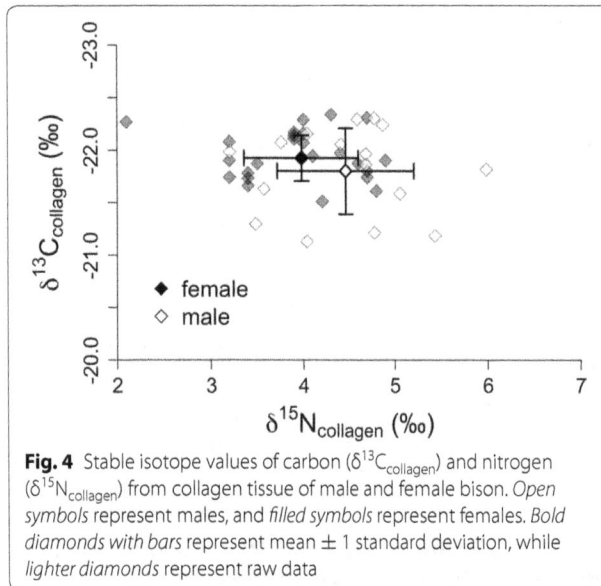

Fig. 4 Stable isotope values of carbon ($\delta^{13}C_{collagen}$) and nitrogen ($\delta^{15}N_{collagen}$) from collagen tissue of male and female bison. *Open symbols* represent males, and *filled symbols* represent females. *Bold diamonds with bars* represent mean ± 1 standard deviation, while *lighter diamonds* represent raw data

indicating higher quality diets during the mating season. Throughout the multi-year period, there was no difference in mean $\delta^{13}C_{collagen}$ (Fig. 4; Table 3) between sexes, indicating that males and females ingest similar plants or plant parts, on average. Despite having similar mean diet compositions, as a group, males had greater dietary breadth (variance of $\delta^{13}C_{collagen}$, Table 2) during the multi-year period, whereas females had higher quality diets, as indicated by lower $\delta^{15}N_{collagen}$ (Fig. 4; Table 3).

In the Central Range, male and female bison had significantly different diets during the mating season but not across the multi-year period (Table 2). Mean values of $\delta^{13}C_{feces}$ differed significantly between sexes in the Central Range, indicating distinct diet compositions during the mating season. Females had greater $\%N_{feces}$ compared

to males (Table 3), indicating higher quality diets; however, there was no difference in dietary breadth between sexes during this time (variance $\delta^{13}C_{feces}$, Table 2). During the multi-year period, male and female diets in the Central Range were statistically indistinguishable in all regards, indicating diets of similar composition ($\delta^{13}C_{collagen}$), quality ($\delta^{15}N_{collagen}$), and breadth (variance $\delta^{13}C_{collagen}$, Table 2).

In contrast to what we observed in the Central Range, male and female bison in the Northern Range had significantly different diets during the multi-year period but not during the mating season (Table 2). During mating season, male and female diets in the Northern Range were statistically indistinguishable in all regards, indicating diets of similar composition, quality, and breadth (Table 2). However, during the multi-year period, $\delta^{13}C_{collagen}$ differed significantly between sexes, indicating sex-specific differences in mean diet composition (Table 3). Finally, although males had greater dietary breadth (variance $\delta^{13}C_{collagen}$, Table 2) during the multi-year period in the Northern Range, there was no difference in diet quality ($\delta^{15}N_{collagen}$, Table 3).

Discussion

We studied whether diet segregation of bison in YNP is associated with sex-specific nutritional demands and whether these demands are met differently in each range. We tested four hypotheses and found complete or partial support for each. Specifically, we found that mean diet composition of male and female bison during the mating season differs significantly, with females having higher quality diets and males having greater dietary breadth (hypothesis 1). Further, while mean diet composition for male and female bison throughout multiple years is statistically indistinguishable, females have higher quality diets and males have greater dietary breadth (hypothesis 2). Additionally, diet segregation for bison in the Central Range was more pronounced during the mating season than across the multi-year period; while females had higher quality diets than males during this time, there was no difference in dietary breadth (hypothesis 3). Finally, diet segregation in the Northern Range was more pronounced across the multi-year period than during the mating season; while males had greater dietary breadth during this time, there was no difference in diet quality (hypothesis 4).

Our results suggest that diet segregation of bison in YNP is associated with sex-specific nutritional demands and density-dependent influences associated with meeting these demands. During mating season, the diets of male and female bison are composed of different plants or plant parts (based on differences in $\delta^{13}C_{feces}$), with females consuming higher quality diets on average (higher %N_{feces}). Although the difference in $\delta^{13}C_{feces}$ between sexes is small (0.25‰), it represents more than 20% of the total range of $\delta^{13}C_{feces}$ values. Over the multi-year time period, male and female bison, on average, consume a similar array of plants or plant parts throughout the park (based on similarity of $\delta^{13}C_{collagen}$). Despite the similarity in average diet composition, however, females have higher quality diets (lower $\delta^{15}N_{collagen}$) than males and appear to be more consistent in their use of available forages (lower variance in $\delta^{13}C_{collagen}$). Collectively, these results suggest that females exhibit more selective feeding behavior throughout the majority of the year compared to males. Numerous field studies of ungulates have reported that, while different sexes may forage in the same habitats, their diets can differ significantly, with females usually selecting forage of higher quality [7, 59, 60]. For example, dietary overlap of male and female white-tailed deer decreases during periods of aggregation (e.g., mating season) compared to periods of segregation [61], and fecal nitrogen of females is greater than that of males, regardless of season [62].

Despite the fact that the two bison populations studied are separated by only tens of kilometers, we found evidence of opposing responses of sex-specific diet segregation in the two ranges. In the Central Range, diet segregation in bison was apparent during the mating season but not during the multi-year period, whereas in the Northern Range, diet segregation was apparent during the multi-year period but not during the mating season. In the Central Range, although males and females obtained a majority of their forage from different plants or plant parts during the mating season, with females ingesting higher quality forage, there was no difference in dietary breadth. Over the multi-year period, males and females from the Northern Range obtained a majority of their forage from different plants or different plant parts. While there was no sex-specific difference in dietary

quality for Northern Range bison during this time, males consumed a greater diversity of dietary items compared to females. Opposing responses of diet segregation in each range may result from differences in the abundance and distribution of high-quality forage and the varying degree of competition for this forage across different ranges and different time periods.

While the Yellowstone ecosystem is considered to be forage-limited for its ungulate herbivores, competition for forage is likely greater on the Northern Range than on the Central Range [26]. The Northern Range offers high-quality forage throughout much of the year, and population densities of elk and bison are greater on the Northern Range, especially during the mating season [27, 28]. Although competition for forage peaks during the mating season for bison in both the Northern and Central ranges, range-specific differences in the availability of high-quality forage may help explain the differences in diet segregation.

Another important consideration is the abundance and distribution of potential predators. While predation rates on bison vary throughout the park, elk are the primary prey of wolves in Yellowstone. In the Northern Range, where elk are most abundant, bison make up only about 4% of wolf kills, with elk comprising approximately 80% of kills [21]. Elk are much less abundant in the Central Range, and as a result, wolf predation rates on bison are higher there than anywhere else in the park, with approximately 10% of wolf kills in this region consisting of bison and around 40% of kills consisting of elk [20]. Moreover, while bison populations without predators have been characterized by the forage selection hypothesis [5, 16], populations with predators have been characterized by the predation risk hypothesis [60], in which females select habitats that offer increased safety at the cost of decreased food quality [63, 64]. Evidence suggests that the presence of potential predators can result in decreased foraging activity [65, 66], thereby directly influencing diet quality of prey [67]. Altogether, our results suggest that bison in Yellowstone National Park are best characterized by the forage selection hypothesis.

Although we interpret our results in the context of the forage selection hypothesis, our results are also consistent with the niche trade-off framework described by Bowyer [8], in which overlap on one niche axis (e.g., space) is accompanied by avoidance on another (e.g., diet). During mating season in YNP, females outnumber males by about 2–1 [68, 69]. Thus, males may be at a competitive disadvantage for those plants that females prefer but that make up at least a portion of male diets during the rest of the year. However, intraspecific competition as a driver of diet segregation in ruminants has been rejected several times in the literature [70, 71], and testing this hypothesis is beyond the scope of this study. Although higher densities of females during the mating season may exclude males from consuming certain forages, these differences are best explained by morphological differences between sexes resulting in different physiologies and therefore different nutritional demands. Specifically, it has been hypothesized that a significant increase in the ingestion of high-quality forage by males could lead to mal-absorption, bloat, and ultimately, decreased digestive efficiency [72]. For comparison, in southern Texas, adult white-tailed deer shifted to lower quality diets as levels of intraspecific competition increased, with dietary shifts of males being more pronounced than those of females [61]. However, the authors rejected competition as a cause for sexual segregation and suggested that predator avoidance by females with young best explained the observed pattern [61]. Similarly, bighorn sheep in Sierra Nevada, California exhibited stronger niche differentiation during periods of spatial overlap than during periods of spatial separation [73]; however, the authors noted that these differences are most likely explained either by morphological differences or by how different sexes manage risk of predation [73]. It is also important to note that male ruminants tend to restrict feeding behavior during mating season [74], potentially resulting in decreased dietary breadth and lower variance of $\delta^{13}C$ during periods of aggregation.

Finally, many studies investigating isotopic variation among conspecifics on different dietary regimes or in different regions have found only minor differences in isotope values between test groups. For example, several populations of mule deer with significant differences in fecal N content exhibited only small differences in mean $\delta^{15}N_{feces}$, ranging from −0.54 to 1.10‰ [75]. Similarly, a study investigating the isotopic composition of hair in cattle (*Bos taurus*) from different confinement and dietary practices found that six of 13 test groups had mean $\delta^{13}C$ values within a range of less than 1‰ [76]. Cattle, elk, and mule deer in Starkey Experimental Forest, Oregon, had a collective $\delta^{13}C$ range of less than 2‰ and a collective $\delta^{15}N$ range of less than 4‰ [39]. In YNP, Feranec reported that mean $\delta^{13}C$ values of feces collected over a 1-month period from bison, elk, mule deer, and white-tailed deer (*O. virginianus*) fell within a range of less than 1‰, and mean $\delta^{13}C$ values of collagen for bison, elk, and mule deer lay within a range of about 2‰ [77]. Moreover, of the large herbivores resident to YNP, bison had the smallest $\delta^{13}C$ range for feces and collagen, both of which were 0.9‰. Thus, while differences in forage type, feeding strategy (e.g., browser vs. grazer), and ecological setting may result in only small differences in isotopic values, these differences may correspond to biologically significant variations in diet [78]. Finally, while some of

our sample sizes may seem small for groups divided by sex and range, all but one sample size was well over the recommended threshold for stable-isotopic characterization of populations [53]. For the sample that lay below this threshold, Northern Range males, the variance is statistically indistinguishable from that of Central Range males, which has three times the number of samples. Regardless, this particular result should be considered preliminary. Also, the threshold value for sample size was determined for enamel carbonate samples, not for collagen and fecal isotopes; however, more recent studies corroborate these findings using collagen isotopes [79].

Conclusions

Ecological theory suggests that smaller-bodied, female bison should displace males from high-quality foraging habitats, since females forage in large groups and deplete resources more rapidly than do males [2, 3, 7]. Based on our results, female bison in YNP have higher quality diets than males do, suggesting that diet segregation is associated with sex-specific nutritional demands. Additionally, range-specific differences in the abundance and distribution of high-quality forage, in conjunction with seasonal variation in population density of bison and elk, may influence spatial and temporal differences in diet segregation. Altogether, our results highlight the importance of accounting for spatiotemporal heterogeneity when conducting dietary studies on wild ungulates.

Authors' contributions
JLB and CB formulated the study idea and developed the study methods. JLB conducted fieldwork, laboratory work, and analyzed the data. JLB and CB wrote the manuscript. Both authors read and approved the final manuscript.

Author details
[1] Department of Fisheries, Wildlife, and Conservation Biology, University of Minnesota, 135 B Skok Hall, 2003 Upper Buford Circle, St. Paul, MN 55108-1052, USA. [2] Department of Ecology and Evolutionary Biology, University of Michigan, 1109 Geddes Avenue, Ann Arbor, MI 48109-1079, USA.

Acknowledgements
We thank S. Gunther, C. Hendrix, C. Curry, and B. Guild for providing logistical support and R. Wallen and K. Gunther for familiarizing us with Yellowstone National Park and its wildlife populations. We thank A. Pfrimmer for help in collecting carcass and fecal samples and J. LeMoine and M. Grant for guidance during stable isotope analysis. The suggestions of P. Myers and K. Nadelhoffer significantly improved this study. R. Feranec, D. L. Fox, and anonymous reviewers provided constructive comments on the manuscript.

Competing interests
The authors declare that they have no competing interests.

Funding
The University of Michigan provided financial support via the Rackham Graduate Student Research Grant, the Department of Ecology and Evolutionary Biology Block Grants, and the Frontiers Master's Fellowship Program in the Department of Ecology and Evolutionary Biology.

References

1. Mysterud A. The relationship between ecological segregation and sexual body size dimorphism in large herbivores. Oecologia. 2000;124:40–54.
2. Bell RH. A grazing ecosystem in the Serengeti. Sci Am. 1971;225:86–93.
3. Jarman P. The social organisation of antelope in relation to their ecology. Behaviour. 1974;48:215–67.
4. Perez-Barberia FJ, Pérez-Fernández E, Robertson E, Alvarez-Enriquez B. Does the Jarman–Bell principle at intra-specific level explain sexual segregation in polygynous ungulates? Sex differences in forage digestibility in Soay sheep. Oecologia. 2008;157:21–30.
5. Post DM, Armbrust TS, Horne EA, Goheen JR. Sexual segregation results in differences in content and quality of bison (Bos bison) diets. J Mammal. 2001;82:407–13.
6. Ruckstuhl KE. Foraging behaviour and sexual segregation in bighorn sheep. Anim Behav. 1998;56:99–106.
7. Du Toit JT. Sex differences in the foraging ecology of large mammalian herbivores. In: Ruckstuhl K, Neuhaus P, editors. Sexual segregation in vertebrates: ecology of the two sexes. Cambridge: Cambridge University Press; 2006. p. 35–52.
8. Bowyer RT. Sexual segregation in ruminants: definitions, hypotheses, and implications for conservation and management. J Mammal. 2004;85:1039–52.
9. Main MB, Du Toit JT. Sex differences in reproductive strategies affect habitat choice in ungulates. In: Ruckstuhl K, Neuhaus P, editors. Sexual segregation in vertebrates: ecology of the two sexes. Cambridge: Cambridge University Press; 2006. p. 148–62.
10. Ruckstuhl KE, Neuhaus P. Sexual segregation in ungulates: a comparative test of three hypotheses. Biol Rev Camb Philos Soc. 2002;77:77–96.
11. Berger J, Peacock M. Variability in size–weight relationships of Bison bison. J Mammal. 1988;69:618–24.
12. Main MB, Weckerly FW, Bleich VC. Sexual segregation in ungulates: new directions for research. J Mammal. 1996;77:449–61.
13. Robbins C. Wildlife feeding and nutrition. Atlanta: Elsevier; 2012.
14. Forsyth DM, Duncan RP, Tustin KG, Gaillard J-M. A substantial energetic cost to male reproduction in a sexually dimorphic ungulate. Ecology. 2005;86:2154–63.
15. Larter NC. Diet and habitat selection of an erupting wood bison population. Master's thesis. University of British Columbia, Vancouver; 1988.
16. Mooring MS, Reisig DD, Osborne ER, Kanallakan AL, Hall BM, Schaad EW, et al. Sexual segregation in bison: a test of multiple hypotheses. Behaviour. 2005;142:897–927.
17. Gogan PJP, Podruzny KM, Olexa EM, Pac HI, Frey KL. Yellowstone bison fetal development and phenology of parturition. J Wildl Manag. 2005;69:1716–30.
18. Christianson DA, Gogan PJP, Podruzny KM, Olexa EM. Incisor wear and age in Yellowstone bison. Wildl Soc Bull. 2005;33:669–76.
19. Gardipee FM. Development of fecal DNA sampling methods to assess genetic population structure of Greater Yellowstone bison. Master's thesis. The University of Montana, Missoula; 2007.
20. Becker MS, Garrott RA, White PJ, Gower CN, Bergman EJ, Jaffe R. Wolf prey selection in an elk–bison system: choice or circumstance? Terr Ecol. 2008;3:305–37.
21. Smith D, Stahler D, Albers E, Metz M, Williamson L, Ehlers N, et al. Yellowstone wolf project: annual report 2008 (YCR-2009-03). National Park Service, Yellowstone center for resources; 2009.
22. Despain DG. Two climates of Yellowstone National Park. Proc Mont Acad Sci. 1987;47:11–9.
23. Despain DG. Yellowstone vegetation: consequences of environment and history in a natural setting. Boulder: Roberts Rinehart Publishers; 1990.
24. Fournier RO. Geochemistry and dynamics of the Yellowstone National Park hydrothermal system. Annu Rev Earth Planet Sci. 1989;17:13–53.
25. Garrott RA, Eberhardt LL, Otton JK, White PJ, Chaffee MA. A geochemical trophic cascade in Yellowstone's geothermal environments. Ecosystems. 2002;5:659–66.
26. Gates CC, Stelfox B, Muhly T, Chowns T, Hudson RJ. The ecology of bison movements and distribution in and beyond Yellowstone National Park: a critical review with implications for winter use and transboundary population management. The United States National Park Service; 2005.
27. Smith DW, Mech LD, Meagher M, Clark WE, Jaffe R, Phillips MK, et al. Wolf–bison interactions in Yellowstone National Park. J Mammal. 2000;81:1128–35.
28. Smith DW, Drummer TD, Murphy KM, Guernsey DS, Evans SB. Winter prey selection and estimation of wolf kill rates in Yellowstone National Park, 1995–2000. J Wildl Manag. 2004;68:153–66.
29. McDonald T, Carlson E, Patterson A, Zaluski M, Jones A, McDonald K, et al. 2012 Annual report of the interagency bison management plant. 2012.
30. White PJ, Wallen RL, Hallac DE, Jerrett JA, editors. Yellowstone bison—conserving an american icon in modern society. Yellowstone National Park: Yellowstone Association; 2015.
31. Leslie DM, Bowyer RT, Jenks JA. Facts from feces: nitrogen still measures up as a nutritional index for mammalian herbivores. J Wildl Manag. 2008;72:1420–33.
32. Cerling T, Harris J, Leakey M. Browsing and grazing in elephants: the isotope record of modern and fossil proboscideans. Oecologia. 1999;120:364–74.
33. Stallcup OT, Davis GV, Shields L. Influence of dry matter and nitrogen intakes on fecal nitrogen losses in cattle. J Dairy Sci. 1975;58:1301–7.
34. Lancaster RJ. Estimation of digestibility of grazed pasture from faeces nitrogen. Nature. 1949;163:330–1.
35. Cerling T, Harris J. Carbon isotope fractionation between diet and bioapatite in ungulate mammals and implications for ecological and paleoecological studies. Oecologia. 1999;120:347–63.
36. Deniro M, Epstein S. Influence of diet on the distribution of carbon isotopes in animals. Geochim Cosmochim Acta. 1978;42:495–506.
37. Phillips DL. Converting isotope values to diet composition: the use of mixing models. J Mammal. 2012;93:342–52.
38. Angerbjörn A, Hersteinsson P, Lidén K, Nelson E. Dietary variation in arctic foxes (Alopex lagopus)—an analysis of stable carbon isotopes. Oecologia. 1994;99:226–32.
39. Stewart KM, Bowyer RT, Kie JG, Dick BL, Ben-David M. Niche partitioning among mule deer, elk, and cattle: do stable isotopes reflect dietary niche? Ecoscience. 2003;10:297–302.
40. Newsome SD, Tinker MT, Monson DH, Oftedal OT, Ralls K, Staedler MM, et al. Using stable isotopes to investigate individual diet specialization in California sea otters (Enhydra lutris nereis). Ecology. 2009;90:961–74.
41. Ambrose SH. Effects of diet, climate and physiology on nitrogen isotope abundances in terrestrial foodwebs. J Archaeol Sci. 1991;18:293–317.
42. Schoeninger MJ, DeNiro MJ. Nitrogen and carbon isotopic composition of bone collagen from marine and terrestrial animals. Geochim Cosmochim Acta. 1984;48:625–39.
43. Sealy JC, van der Merwe NJ, Thorp JAL, Lanham JL. Nitrogen isotopic ecology in southern Africa: implications for environmental and dietary tracing. Geochim Cosmochim Acta. 1987;51:2707–17.
44. Glimcher MJ. Bone: nature of the calcium phosphate crystals and cellular, structural, and physical chemical mechanisms in their formation. Rev Mineral Geochem. 2006;64:223–82.
45. Koch PL. Isotopic study of the biology of modern and fossil vertebrates. In: Michener R, Lajtha K, editors. Stable isotopes in ecology and environmental science. 2nd ed. Malden: Blackwell Publishing Ltd; 2008. p. 99–154.
46. Rubenstein D, Hobson K. From birds to butterflies: animal movement patterns and stable isotopes. Trends Ecol Evol. 2004;19:256–63.
47. Rutley BD, Hudson RJ. Seasonal energetic parameters of free-grazing bison (Bison bison). Can J Anim Sci. 2000;80:663–71.
48. Reynolds HW, Gates CC, Glaholt RD. Bison. Wild mammals of North America: biology, management, and conservation. 2nd ed. Baltimore: The Johns Hopkins University Press; 2003. p. 1009–60.
49. Frison GC, Reher CA. Age determination of buffalo by teeth eruption and wear. Plain Anthropol. 1970;15:46–50.
50. Skinner MF, Kaisen OC. The fossil bison of Alaska and preliminary revision of the genus. Bull Am Mus Nat Hist. 1947;89:1–154.
51. Shackleton DM, Hills LV, Hutton DA. Aspects of variation in cranial characters of plains bison (Bison bison bison Linnaeus) from Elk Island National Park, Alberta. J Mammal. 1975;56:871–87.
52. Edwards JK, Marchinton RL, Smith GF. Pelvic girdle criteria for sex determination of white-tailed deer. J Wildl Manag. 1982;46:544–7.

53. Tyler NJC. Sexual dimorphism in the pelvic bones of Svalbard reindeer, *Rangifer tarandus platyrhynchus*. J Zool. 1987;213:147–52.

54. Bligh EG, Dyer WJ. A rapid method of total lipid extraction and purification. Can J Biochem Physiol. 1959;37:911–7.

55. Burton RK, Koch PL. Isotopic tracking of foraging and long-distance migration in northeastern Pacific pinnipeds. Oecologia. 1999;119:578–85.

56. Sokal RR, Rohlf FJ. Bartlett's test of homogeneity of variances. Biometry. San Francisco: WH Freeman and Co.; 1969. p. 370–89.

57. Clementz MT, Koch PL. Differentiating aquatic mammal habitat and foraging ecology with stable isotopes in tooth enamel. Oecologia. 2001;129:461–72.

58. R Core Team. R: A language and environment for statistical computing. R Foundation for Statistical Computing; 2013. http://www.R-project.org. Accessed 7 Mar 2014.

59. Clutton-Brock TH, Guinness FE, Albon SD. Red deer: behavior and ecology of two sexes. Chicago: University of Chicago Press; 1982.

60. Komers PE, Messier F, Gates CC. Group structure in wood bison: nutritional and reproductive determinants. Can J Zool. 1993;71:1367–71.

61. Kie JG, Bowyer RT. Sexual segregation in white-tailed deer: density-dependent changes in use of space, habitat selection, and dietary niche. J Mammal. 1999;80:1004–20.

62. Gallina S, Sánchez-Rojas G, Buenrostro-Silva A, López-González CA. Comparison of faecal nitrogen concentration between sexes of white-tailed deer in a tropical dry forest in southern Mexico. Ethol Ecol Evol. 2015;27:103–15.

63. Festa-Bianchet M. Seasonal range selection in bighorn sheep: conflicts between forage quality, forage quantity, and predator avoidance. Oecologia. 1988;75:580–6.

64. Conradt L. Definitions, hypotheses, models and measures in the study of animal segregation. In: Ruckstuhl K, Neuhaus PJP, editors. Sexual segregation in vertebrates: ecology of the two sexes. Cambridge: Cambridge University Press; 2006. p. 11–32.

65. Lima SL, Dill LM. Behavioral decisions made under the risk of predation: a review and prospectus. Can J Zool. 1990;68:619–40.

66. Altendorf KB, Laundré JW, López González CA, Brown JS. Assessing effects of predation risk on foraging behavior of mule deer. J Mammal. 2001;82:430–9.

67. Hernández L, Laundré JW. Foraging in the "landscape of fear" and its implications for habitat use and diet quality of elk *Cervus elaphus* and bison *Bison bison*. Wildl Biol. 2005;11:215–20.

68. Wolff JO. Breeding strategies, mate choice, and reproductive success in American bison. Oikos. 1998;83:529–44.

69. Bowyer RT, Bleich VC, Manteca X, Whiting JC, Stewart KM. Sociality, mate choice, and timing of mating in American bison (*Bison bison*): effects of large males. Ethology. 2007;113:1048–60.

70. Conradt L, Gordon IJ, Clutton-Brock TH, Thomson D, Guinness FE. Could the indirect competition hypothesis explain inter-sexual site segregation in red deer (*Cervus elaphus* L.)? J Zool. 2001;254:185–93.

71. Conradt L, Clutton-Brock TH, Thomson D. Habitat segregation in ungulates: are males forced into suboptimal foraging habitats through indirect competition by females? Oecologia. 1999;119:367–77.

72. Barboza PS, Bowyer RT. Sexual segregation in dimorphic deer: a new gastrocentric hypothesis. J Mammal. 2000;81:473–89.

73. Schroeder CA, Bowyer RT, Bleich VC, Stephenson TR. Sexual segregation in Sierra Nevada bighorn sheep, *Ovis canadensis sierrae*: ramifications for conservation. Arct Antarct Alp Res. 2010;42:476–89.

74. Miquelle DG. Why don't bull moose eat during the rut? Behav Ecol Sociobiol. 1990;27:145–51.

75. Darimont CT, Paquet PC, Reimchen TE. Stable isotopic niche predicts fitness of prey in a wolf–deer system. Biol J Linn Soc. 2007;90:125–37.

76. Schwertl M, Auerswald K, Schäufele R, Schnyder H. Carbon and nitrogen stable isotope composition of cattle hair: ecological fingerprints of production systems? Agric Eco Environ. 2005;109:153–65.

77. Feranec RS. Stable carbon isotope values reveal evidence of resource partitioning among ungulates from modern C3-dominated ecosystems in North America. Palaeogeogr Palaeoclimatol Palaeoecol. 2007;252:575–85.

78. Singer FJ, Norland JE. Niche relationships within a guild of ungulate species in Yellowstone National Park, Wyoming, following release from artificial controls. Can J Zool. 1994;72:1383–94.

79. Fox-Dobbs K, Bump JK, Peterson RO, Fox DL, Koch PL. Carnivore-specific stable isotope variables and variation in the foraging ecology of modern and ancient wolf populations: case studies from Isle Royale, Minnesota, and La Brea. Can J Zool. 2007;85:458–71.

10

Loss and conservation of evolutionary history in the Mediterranean Basin

S. Veron*⑩, P. Clergeau and S. Pavoine

Abstract

Background: Phylogenetic diversity and evolutionary distinctiveness are highly valuable components of biodiversity, but they are rarely considered in conservation practices. Focusing on a biodiversity hotspot, the Mediterranean Basin, we aimed to identify those areas where evolutionary history is highly threatened and range-restricted in the region. Using null models, we first compared the spatial distributions of three indices: two measured threatened evolutionary history—Expected PD*loss* and Heightened Evolutionary distinctiveness and Global Endangerment—and one measured endemic evolutionary history—Biogeographically Evolutionary Distinctiveness. We focused on three vertebrate groups with high proportions of endemic, threatened species: amphibians, squamates and terrestrial mammals. Second, we estimated the spatial overlap of hotspots of threatened and endemic evolutionary history within the network of protected areas under several conservation scenarios.

Results: Areas that concentrate evolutionary history of conservation interest greatly differed among taxa and indices, although a large proportion of hotspots were identified in the Maghreb, in the East of the Mediterranean Basin as well as in islands. We found that, in a minimum conservation scenario, there was a significant proportion of hotspots for amphibians and squamates that were protected but not for terrestrial mammals. However, in a strong conservation scenario, only few hotspots overlapped with protected areas and they were significantly less protected than in a model where hotspots were chosen randomly.

Conclusions: Some sites concentrate highly threatened and range-restricted evolutionary history of the Mediterranean basin and their conservation could be much improved. These sites are relevant for conservation studies aimed at designing new conservation actions to preserve evolutionary history and the option values it represents.

Keywords: Amphibians, Evolutionary distinctiveness, Endemism, Mammals, Mediterranean basin, Phylogenetic diversity, Protected areas, Squamates

Background

Due to human activities, species are going extinct at such high rates that a sixth mass extinction crisis has probably begun [1]. In the future extinction risks are expected to intensify. However, not all species can be saved and conservationists have to make a choice about how to best protect biodiversity [2]. Basing conservation on species richness or threatened species, as is usually the case, may be an inadequate strategy to conserve the diversity of life because it considers all species as equal

[3]. A more valuable strategy, in which there is increasing interest, is to protect species evolutionary history. One main benefit of evolutionary history over species richness is that it may capture future diversity and provide future unexpected services for humans and ecosystems [3]. Other benefits are ethical: helping to protect Earth's evolutionary heritage [4]; aesthetical: humans may appreciate a variety of living forms [5]; and evolutionary: providing possibilities for future evolution [6] (but see [7]). Loss of evolutionary history could be much higher than species richness loss when extinctions are phylogenetically clustered (thus threatening not only terminal branches but also deep branches shared by the species at risk), when evolutionarily distinct species go extinct

*Correspondence: sveron@edu.mnhn.fr
Centre d'Ecologie et des Sciences de la Conservation (CESCO UMR7204), Sorbonne Universités, MNHN, CNRS, UPMC, CP51, 43-61 rue Buffon, 75005 Paris, France

and when the phylogenetic tree is unbalanced, i.e. the extent to which some branches lead to many species (or higher taxa) while their sister branches lead only to a few [8–10]. However, evolutionary history is rarely included in conservation programs and is poorly represented in protected areas [10]. Depending on those factors, evolutionary history is more threatened in some regions of the world than in others [10]. In this study, we are interested in the risks of losing evolutionary history in the Mediterranean Basin, one of the richest regions of biodiversity on Earth and where many endemic and threatened species live.

The Mediterranean Basin is situated at the junction of Europe, Africa and Asia. It extends eastward from Morocco to Turkey and southward from northern Italy to the Canary Islands. Countries of the Mediterranean Basin share a common climate [11], which is characterized by hot, dry summers and cool, humid winters. This climate strongly influences the wildlife of the Mediterranean Basin such that many species are found nowhere else on Earth [12]. Moreover, intense human activity has resulted in landscapes characterized by complex patchworks of habitats, generating a high diversity of species [11]. The Mediterranean Basin has been identified as a hotspot in terms of its diversity and its high ratio of endemism in plants [13, 14]. It also shelters a rich but threatened diversity of marine and terrestrial animals that includes many endemic species [15]. Among Mediterranean species, 26 % of mammals, 48 % of squamates and 64 % of amphibians are endemic to the region [11].

The degradation of habitats, climate change, invasive species, overexploitation of natural resources and pollution are the most significant threats to biodiversity in the region, causing extensive damage to ecosystems, fauna and flora [11]. These threats, particularly habitat degradation and climate change, are expected to intensify in the future [15]. Moreover the recent growth of tourism activities increases the risks of losing biodiversity [16]. Of the 1912 species evaluated by the International Union for the Conservation of Nature (IUCN) in the Mediterranean region (including amphibians, squamates, birds, mammals, crayfish and crabs, cartilaginous fishes, endemic fresh water fishes and dragonflies), 19 % are threatened with extinction [11].

The IUCN has analysed the threats to biodiversity in the Mediterranean Basin and advocated the protection of those endemic species that capture unique phylogenetic information [11]. To date, few studies have considered evolutionary history or its conservation in the region, although several studies have explored fish evolutionary history [17, 18]. A valuable strategy to measure the evolutionary history of conservation interest is to use phylogenetic diversity (PD) and evolutionary distinctiveness

(ED) metrics. The PD of a subset of taxa is measured as the sum of the branch lengths of the minimum path that joins those taxa on a phylogenetic tree [19]. ED quantifies the number of relatives a species has and how phylogenetically distant they are [19, 20]. PD and ED are complementary measures for conservation. PD identifies the amount of shared evolutionary history of the species present in an area and may capture functional diversity and future benefits [3, 19, 21]. ED enables us to prioritize species according to their phylogenetic isolation (which decreases with the number of relatives and increases with the phylogenetic distance to those relatives) and may capture rare features important for ecosystem services [22, 23]. Preserving ED species may also help to conserve PD when all species which maximize PD cannot be protected [24]. Due to the high rates of endemism and the threats faced by species in the Mediterranean basin, our objectives were to identify those terrestrial areas where PD and ED are threatened or range-restricted and to analyse the effectiveness of protected areas in conserving those hotspots. To match with conservation policies we identified hotspots according to the Aïchi target defined by the Convention on Biological Diversity in 2011 [25] which aims to protect 17 % of land areas. We considered protected areas in categories I, II, and IV as their main management objective is to directly protect species [26]. We focused on the vertebrate groups that are known to have high proportions of endemic and threatened species in the region: amphibians, squamates and mammals.

Results

We searched for hotspots that concentrate threatened evolutionary history according to three indices:

1. Expected PD*loss* [27], which calculates at-risk PD: branch lengths of the phylogeny are weighted by the extinction probabilities of the species they support and Expected PD*loss* is the sum of those weighted branches. The use of this metric was highly recommended to measure the total branch length at risk because it accounts for the phylogenetic complementarities of extinction risks, i.e. the extinction probability of a deep branch depends on the probability that all its descendant species go extinct. Expected PD*loss* identifies the amount of threatened evolutionary history of the species present in an area.

2. Heightened Evolutionary Distinctiveness and Global Endangerment (HEDGE) [28], which calculates at-risk ED in a probabilistic framework where the branch lengths of the phylogeny are also weighted by the extinction probabilities of the species they support. Similarly to Expected PD*loss*, its use was recommended because it is based on the phylogenetic complementarities of extinction risks. However, contrary to Expected PD*loss*, it gives a score to each species.

3. Biogeographically weighted Evolutionary Distinctiveness (BED) [29], which identifies species that encompass high amounts of ED and are also the most spatially restricted. Like HEDGE, it considers phylogenetic complementarity (but with range sizes instead of extinction probabilities) and it gives a score to each species.

Hotspots depend on indices and taxa

For mammals, 24 % of the hotspots were identified by all three indices, whereas 22 and 15 % of hotpots were shared by the three indices in squamates and amphibians, respectively.

For mammals, considering the Aïchi target of 17 % of protected territory, the areas that were expected to lose disproportionate amounts of PD (according to expected PD) and which harbour top HEDGE species were situated in the Maghreb, Turkey, the Balkans, Israel and the Canary islands (Fig. 1a, b). Some key BED areas were found in southern Morocco, Israel, Lebanon, Jordan, northern Egypt, Turkey and southern France (Fig. 1c). For example, *Macaca sylvanus*, the endangered and only primate of the Mediterranean region, is found only in northern Morocco and Algeria. The Equidae species *Equus hemonius* is also highly threatened, with only one small, reintroduced population in Israel (Fig. 4a).

For squamates, the hotspots based on Expected PD*loss* and HEDGE were identified in Israel, Lebanon, central Spain, northern Maghreb and islands: the Baleares, Canary Islands, Crete, Cyprus, and north of Sicily (Fig. 2a, b). Key BED sites were mainly located in the Canary Islands, Israel, Lebanon, Jordan, and Greece (Fig. 2c). Several top HEDGE species were highly restricted and had high BED scores (Fig. 4b). In particular, the lizard genus *Ibolacerta* represents an important evolutionary radiation in the Mediterranean Basin, with a high proportion of endemic and threatened species.

For amphibians, HEDGE, Expected PD*loss* and BED hotspots were identified in Sardinia, the Spanish Pyrenees, Morocco, northern Algeria, the Balkans, Crete and southern Turkey (Fig. 3a–c). Some species, in particular those from the genus *Lyciasalamandra* and the critically endangered *Hyla heinzsteinitzi*, ranked high in both BED and HEDGE scores (Fig. 4c).

For all groups, the identified hotspots captured high proportions of the regional Expected PD*loss* as well as high proportions of accumulated regional HEDGE and BED values (sum of species HEDGE and BED values). These proportions were significantly higher than the proportions obtained when hotspots were chosen randomly: all p values < 0.01 except HEDGE and Expected PD*loss* for mammals (p < 0.1) (Table 1).

When choosing hotspots independently of the 17 % Aïchi threshold, we identified fewer hotspots than 17 % of total cells, except for mammal HEDGE hotspots which, in that case, covered more than 17 % of the land (Additional files 1, 2, 3). In mammals, many additional hotspots with high HEDGE values were found in the North of Maghreb and in Turkey (Additional file 1).

We found several moderate correlations between the distribution of species richness and PD, Expected PD*loss*, HEDGE and BED values (Table 2; see also maps of species richness in Additional files 4, 5, 6). Places where species are highly threatened are likely to be hotspots of evolutionary history at risk. Especially in amphibians, sites in southern Turkey, in Israel and in Sardinia have many threatened species and are hotspots for Expected PD*loss*, HEDGE and BED. Yet a species richness approach also misses some sites where threatened evolutionary history concentrate. For example, the Canary Islands are hotspots of terrestrial mammal Expected PD*loss* although relatively few threatened species are found there (Fig. 1; Additional file 4).

Poorly protected hotspots

The number of hotspots that overlapped with protected areas varied among groups. Nonetheless, the degree of protection was low for all groups; i.e. only a few hotspots were protected on more than half of their surface, particularly in the categories I, II and IV of protected areas (Fig. 5).

Coverage by all Mediterranean protected areas

In the minimum protection scenario for mammals, approximately 60 % of BED hotspots intersected with at least one protected area; however, these sites were significantly under-protected (significance refers to the frequency to which the proportion of hotspots protected was higher than if priority grid cells were distributed randomly (F_{PA}): significantly under-protected means $F_{PA} \leq 0.25$, significantly over-protected means $0.75 < F_{PA} \leq 1$; Fig. 5a). Both Expected PD*loss* and HEDGE mammal hotspots were significantly under-protected, and 45 and 55 % of these sites, respectively, contained at least one protected area (Fig. 5a). Around 50 % of squamate hotspots intersected with at least one protected area for all indices (Fig. 5a). These proportions were similar to those expected if hotspots were randomly distributed for BED and Expected PD*loss* indices and greater than expected for the HEDGE index (Fig. 5a). For amphibians, 65 % of HEDGE and Expected PD*loss* hotspots intersected with protected areas and they were as protected as expected from a random distribution of hotspots. BED hotspots were significantly over-protected and 80 % of them intersected with at least one protected area (Fig. 5a).

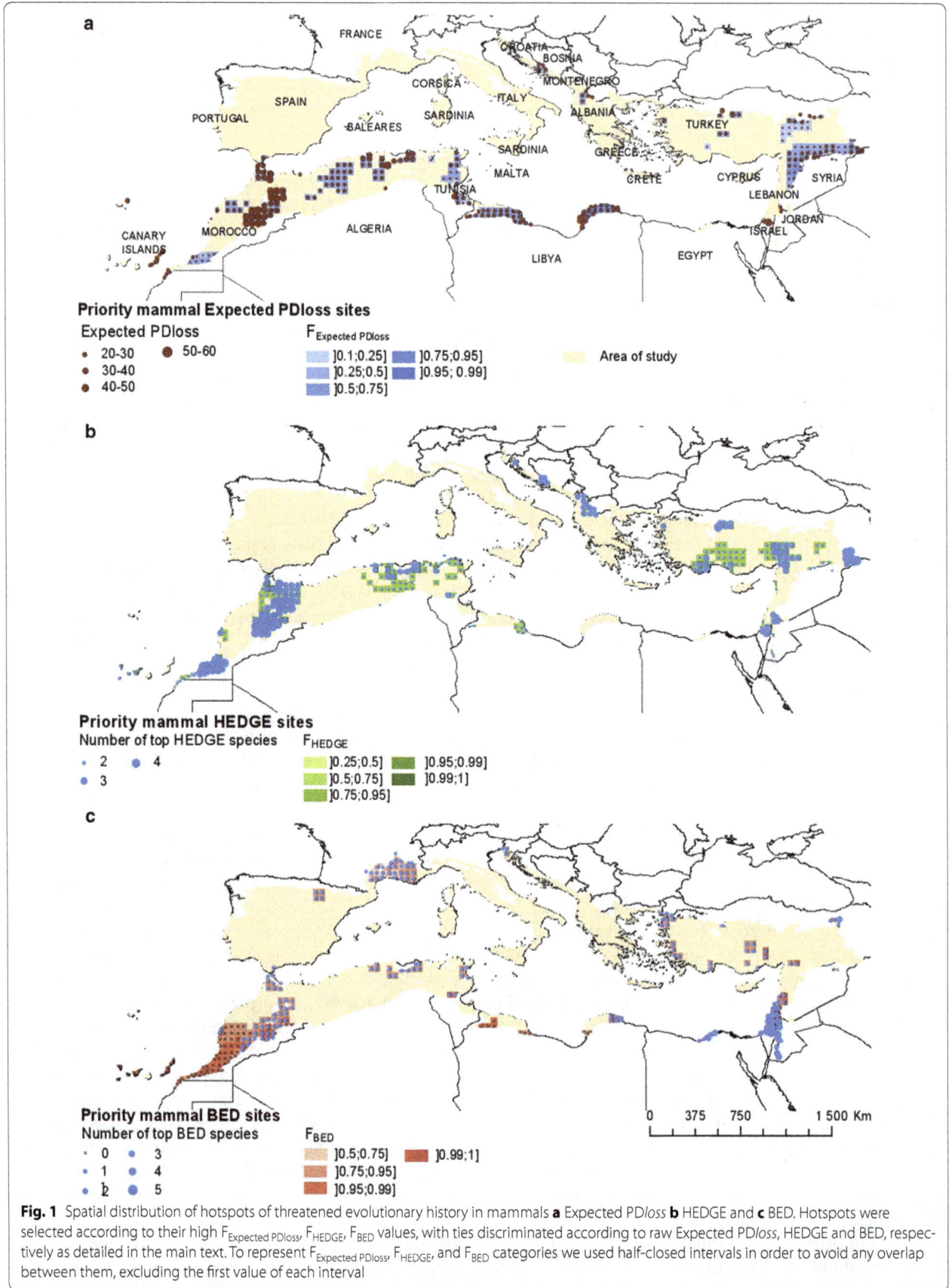

Fig. 1 Spatial distribution of hotspots of threatened evolutionary history in mammals **a** Expected PD*loss* **b** HEDGE and **c** BED. Hotspots were selected according to their high $F_{Expected PDloss}$, F_{HEDGE}, F_{BED} values, with ties discriminated according to raw Expected PD*loss*, HEDGE and BED, respectively as detailed in the main text. To represent $F_{Expected PDloss}$, F_{HEDGE}, and F_{BED} categories we used half-closed intervals in order to avoid any overlap between them, excluding the first value of each interval

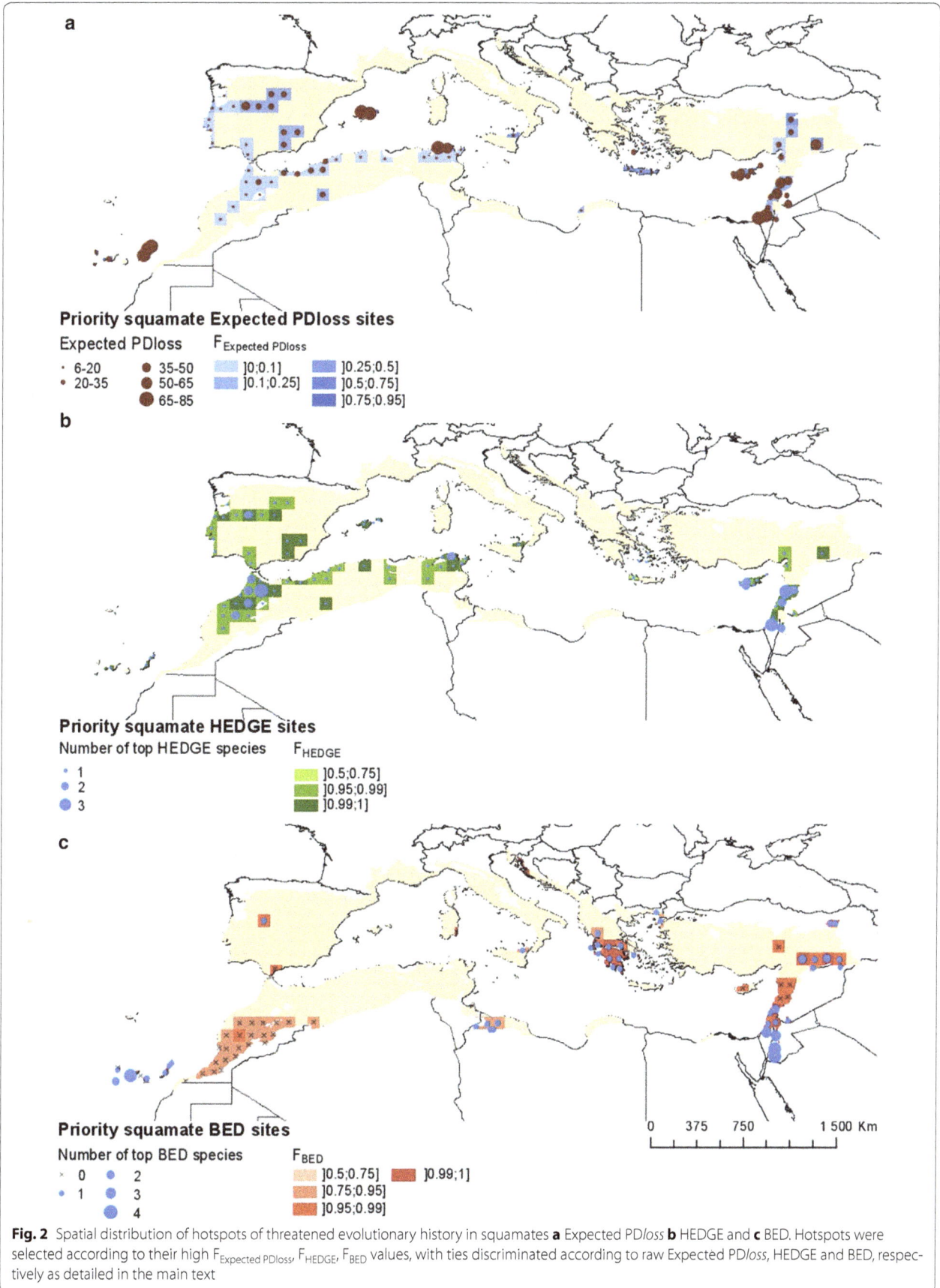

Fig. 2 Spatial distribution of hotspots of threatened evolutionary history in squamates **a** Expected PD*loss* **b** HEDGE and **c** BED. Hotspots were selected according to their high $F_{Expected\ PDloss}$, F_{HEDGE}, F_{BED} values, with ties discriminated according to raw Expected PD*loss*, HEDGE and BED, respectively as detailed in the main text

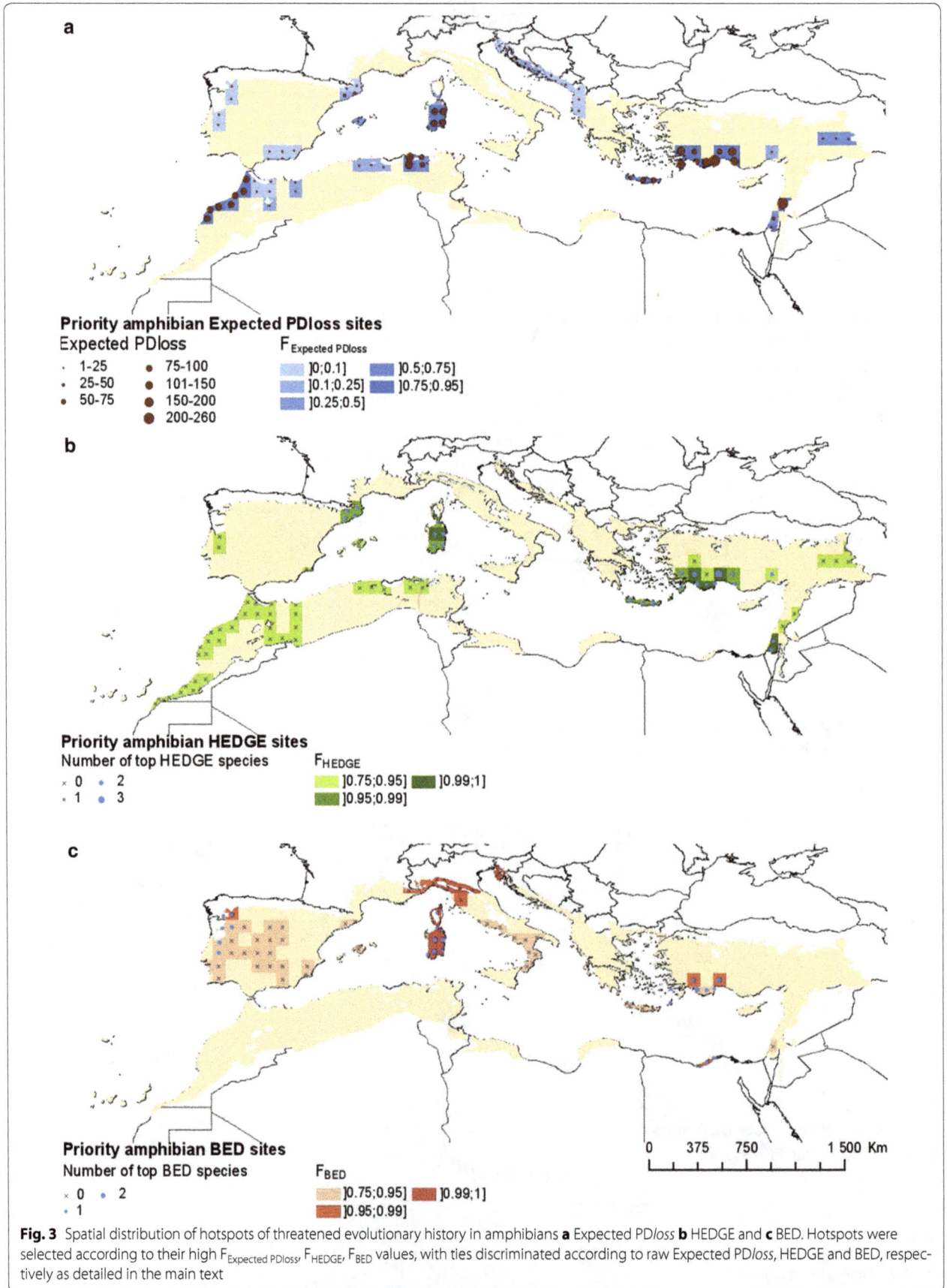

Fig. 3 Spatial distribution of hotspots of threatened evolutionary history in amphibians **a** Expected PD*loss* **b** HEDGE and **c** BED. Hotspots were selected according to their high $F_{Expected PDloss}$, F_{HEDGE}, F_{BED} values, with ties discriminated according to raw Expected PD*loss*, HEDGE and BED, respectively as detailed in the main text

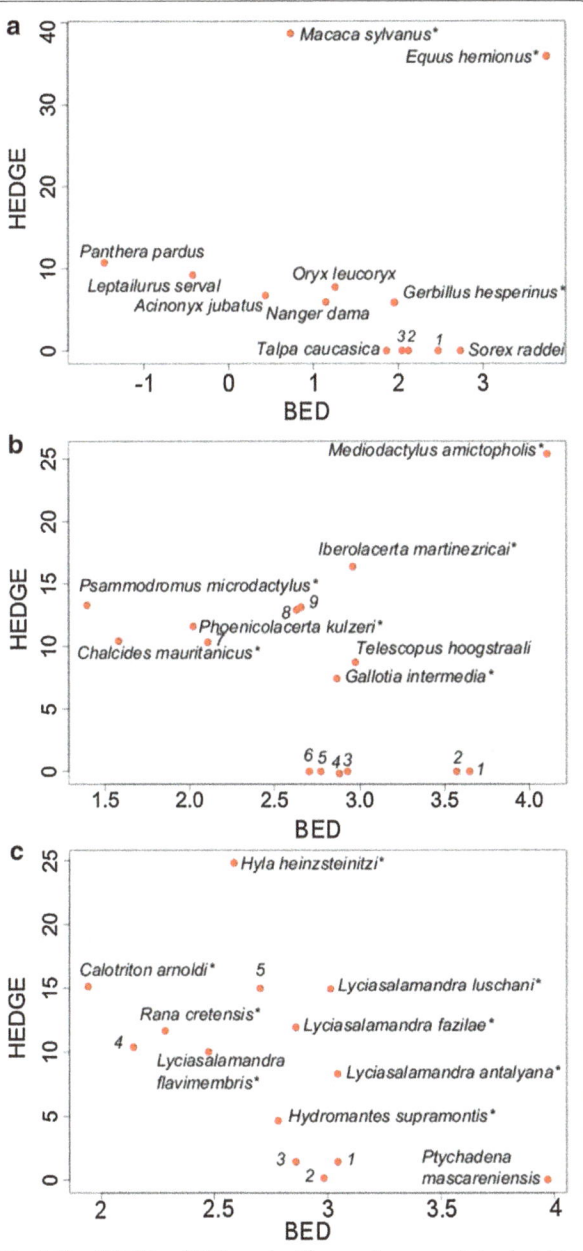

Fig. 4 Top HEDGE and BED species. The *graph* represents top heightened evolutionary distinctiveness globally endangered (HEDGE) scores against top biogeographically evolutionary distinctiveness (BED) scores for squamate, amphibian and terrestrial mammal species. BED scores are represented with a logarithm scale. A *means that the species is endemic from the Mediterranean Basin. **a** Top ten HEDGE and top ten BED mammal species *1 Gerbillus floweri, 2 Gerbillus cheesmani, 3 Arvicanthis niloticus.* **b** Top 10 % HEDGE and top 10 % BED squamate species. *1 Pristurus rupestris 2 Platyceps sinai 3 Asaccus elisae 4 Tarentola gomerensis* 5 Bunopus tuberculatus 6 Acanthodactylus tilburyi 7 Algyroides marchi* 8 Podarcis raffonei* 9 Acanthodactylus beershebensis** **c** Top 10 % HEDGE and top 10 % BED amphibian species. *1 Hydromantes genei* 2 Discoglossus montalentii* 3 Lyciasalamandra helverseni* 4 Euproctus platycephalus* 5 Lyciasalamandra billae**

Table 1 Unique threatened and endemic evolutionary history represented in priority grid cells

	Terrestrial mammals	Squamates	Amphibians
Expected PD*loss*	88.59*	97.18**	81.7**
HEDGE	90.96*	94.87**	87.16**
BED	95.71**	86.92**	73.15**

The table represents the proportion of unique Expected PD*loss*, HEDGE and BED captured by the corresponding hotspots. To calculate p values we determined how often the proportion of unique Expected PD*loss*, HEDGE and BED, captured by a random selection of hotspots, was higher than or equal to the observed values

Significance represented with the symbol * corresponds to marginal significance ($p \leq 0.1$) and ** to significance ($p \leq 0.01$)

Table 2 Correlations between a phylogenetic and a species richness approach

	Terrestrial mammals	Squamates	Amphibians
Correlations with species richness			
PD	0.88**	0.94**	0.95**
Correlations with richness in threatened species			
Expected PD*loss*	0.65**	0.47**	0.68**
HEDGE	0.70**	0.38**	0.71**
Correlations with richness in range-restricted species			
BED	0.37**	0.46**	0.27**

Spearman correlation between evolutionary history and species richness for PD, richness in threatened species (species classified as critically endangered, endangered, or vulnerable) for Expected PD*loss* and HEDGE and richness in range-restricted species (number of species in the top 10 % species with the smallest range size) for BED

Significance represented with the symbol * corresponds to marginal significance (p value ≤ 0.1) and ** to significance (p value ≤ 0.01)

However, in the strong protection scenario, where sites were considered protected when more than half of their area was covered by protected areas, the proportion of protected hotspots decreased (Fig. 5b). For terrestrial mammals, hotspots were significantly under-protected for Expected PD*loss* and HEDGE indices but as protected as expected from random for BED hotspots (Fig. 5b). In amphibians, HEDGE hotspots were significantly under-protected whereas BED and Expected PD*loss* hotspots were as protected as expected from random (Fig. 5b). In squamates hotspots were significantly under-protected for all indices (Fig. 5b).

Coverage by Mediterranean protected areas of category I, II and IV

We repeated the same analyses with only the categories I, II and IV of protected areas, i.e., protected areas specifically dedicated to species conservation or with stringent

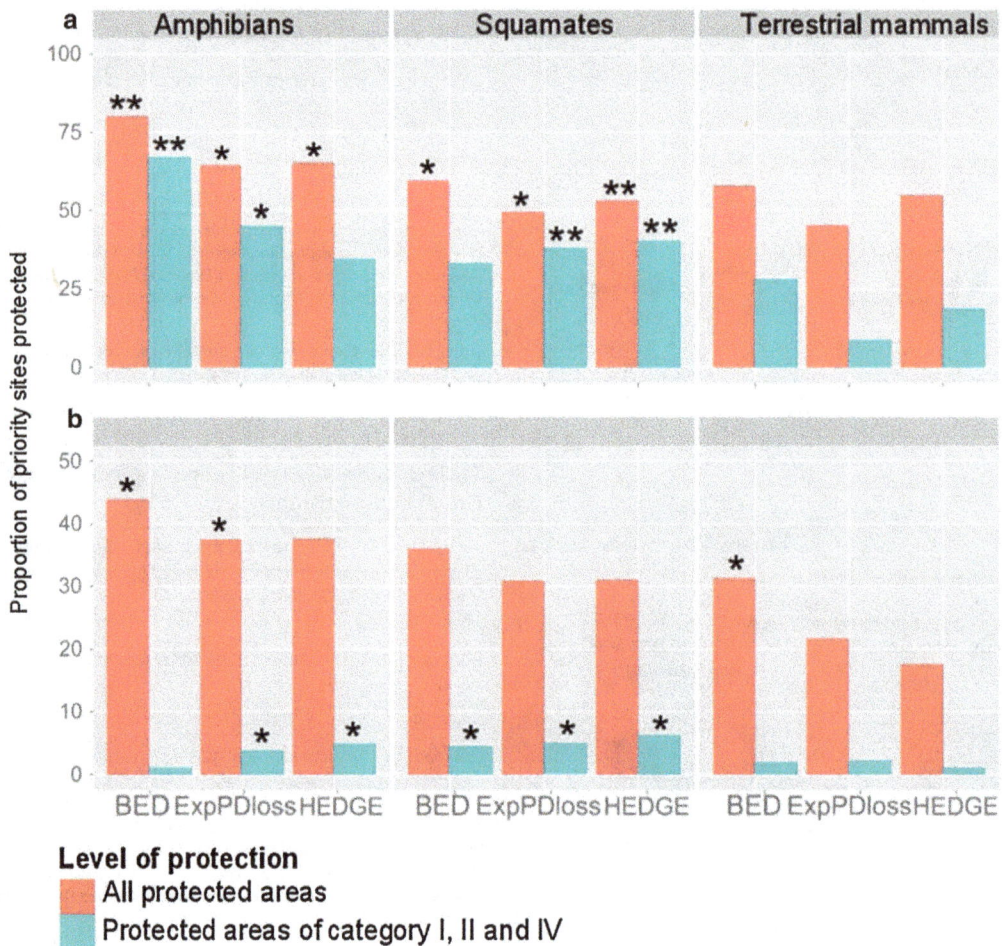

Fig. 5 Proportion of hotspots of threatened and endemic evolutionary history protected in two conservation scenarios. Each graph represents the proportion of HEDGE, BED and Expected PD*loss* hotspots protected. Red bars correspond to the degree of protection if all protected areas are included and blue bars the degree of protection if only protected areas of categories I, II and IV are included. Star symbols correspond to the frequency to which the proportion of hotspots protected was higher than if priority grid cells were distributed randomly (F_{PA}): no star means $F_{PA} \leq 0.25$; * means $0.25 < F_{PA} \leq 0.75$; **$0.75 < F_{PA} \leq 1$. We ran analysis for **a** a scenario of minimum protection in which hotspots were safe if they intersected at least one protected area; **b** a scenario of strong protection where a site was considered safe if it was protected on more than half of its area

regulations. As expected, for all groups, the proportion of protected hotspots decreased compared with that observed where all protected areas were considered (Fig. 5). In amphibians, protected areas of category I, II and IV overlapped with as many Expected PD*loss* hotspots as expected from random. Amphibian BED hotspots were significantly over-protected but HEDGE hotspots were significantly under-protected (Fig. 5a). In squamates only BED hotspots were significantly under-protected whereas, in terrestrial mammals, hotspots were significantly under-protected according to all indices. Very few sites had more than half of their area covered by protected areas of categories I, II and IV; nonetheless, there were as many such sites as under a random distribution of hotspots in squamates and amphibians, except

for amphibian BED hotspots, whereas in mammals BED, HEDGE and Expected PD*loss* hotspots were significantly under-protected (Fig. 5b). Some species with high BED and HEDGE scores were indeed not found in any protected area (Additional file 9). In mammals 13 and 2 species from the top 10 % BED (28 species) and HEDGE species (23 species), respectively, were not found in any protected area. In squamates, 8 and 5 species from the top 10 % BED (24 species) and HEDGE species (23 species), respectively were not found in any protected area. As for amphibians, 3 of the top 10 % BED species (11 species) and 4 of the top 10 % HEDGE species (11 species) were not found in protected areas (Additional file 10).

When hotspots were not defined according to Aïchi targets, hotspots were globally over protected compared to a random arrangement of protected areas for squamates, whereas hotspots for amphibians were as protected as expected from random and hotspots for mammals were significantly under-protected. In a scenario where a higher degree of protection was required, only few hotspots were covered on more than half of their surface by protected areas of category I, II or IV and they were as protected or less protected than random (Additional file 7).

Discussion

Localization of hotspots differed between taxa but many were identified in the Maghreb, in eastern countries (Israel, Lebanon, Jordan and Turkey) and in islands. For mammals, 24 % of hotspots were common to the three indices, whereas 22 and 15 % of hotpots were shared by the three indices in squamates and amphibians, respectively. Differences between groups may be due to a higher number of top HEDGE species which are also in the top BED species in mammals and to the narrower distribution of threatened and/or endemic amphibians with high evolutionary distinctiveness. These top hotspots supported by all indices revealed areas that capture both high amounts of threatened and range-restricted evolutionary history. For example, in amphibians, some sites in southern Turkey and in Israel were identified as hotspots for all indices and harbour top HEDGE species which also had high BED scores, e.g. *Lyciasalamandra billae*, *Hyla heinzsteinitzi* (Fig. 4; but see the discussion about the taxonomic uncertainty of *Hyla heinzsteinitzi* [30]). Recently, *L. billae* was also identified as one of the top 15 vertebrate species with the highest probability of extinction [31], indicating the importance of conservation efforts in the Mediterranean Basin. However, this overlapping of priority zones did not always occur, and each index provided unique information emphasizing particular conservation requirements. We also found several moderate correlations between the distribution of species richness and Expected PD*loss*, HEDGE and BED values, challenging the use of surrogates among indices [32]. PD and species richness distribution are expected to be, at least partially, correlated (Table 2). However, they are also expected to differ under some conditions, including when phylogenies are unbalanced, closely related species tend to be found near to each other, old species have smaller geographical distributions on average than young species, and old species are found in species-poor areas [33]. The correlation with a species richness approach was lower for the indices we used (Expected PD*loss*, HEDGE and BED) than for PD probably because these indices include information about threat status or range size of deep branches based on phylogenetic complementarity, i.e. the fact that the risk to lose a deep branch

depends on the probabilities of extinctions of all the species it supports. For example, even if a site is occupied by a threatened species, deep phylogenetic branches can be secured if non threatened descendants of these branches also live there. As for BED the range size of a deep branch depends on the size of the union of the range of the species it supports.

Our method was based on a traditional, widely used hotspot approach [14, 34–36]. Our aim was to identify areas that contain disproportionate amounts of threatened and endemic evolutionary history, even if some threatened and endemic branches may be present in several hotspots [37, 38]. Similarly, several priority hotspots for the conservation of the evolutionary history of marine mammals at risk were identified in the Mediterranean sea [38]. An alternative to the hotspot approach is the network approach that specifically analyses how many species or how much evolutionary history is shared by sites. The network approach searches for a minimum set of areas that capture as many species or as much evolutionary history in a region as possible [39, 40]. The network approach is of particular interest at a regional scale but we believe the hotspot approach enables us to identify sites whose ecosystem resilience may be threatened and where "option values", captured by evolutionary history, are at risk [41, 42]. Option values are the values of preserving the option to use services in the future [25]. They are wholly unanticipated and because evolutionary history is expected to capture genotypic, phenotypic and functional diversity it may be the best measure to preserve those as-yet-unexpected services and to promote system resilience in a changing world [27, 43]. An advantage of the hotspot approach is thus to identify sites where option values are at risk at a local scale whereas a network approach would have prioritized sites relevant for conservation at the regional scale. Regional and local conservation needs thus differ, yet local hotspots also contain a very high proportion of Mediterranean threatened and endemic evolutionary history especially within the squamates and amphibians (as shown in Table 1). At the species level, evolutionary distinctiveness may also capture unique features and current and future benefits [44, 45]. Species with high HEDGE and BED values are thus of conservation interest because they may represent at risk option values (Additional file 11). Mouillot et al. [46] showed that rare species represent a large proportion of unique feature diversity which will potentially help to maintain ecosystems that are resilient to threats such as climate change. By capturing rarity and evolutionary history, BED species may thus be key species for the preservation of option values. In this study, we used regional and global data on extinction risks. When available for future studies, adding information about local threat would enable us to refine the definition of Mediterranean hotspots. Moreover an increasing knowledge on the biodiversity of the region and the range distribution of species will improve our comprehension of the risks to lose phylogenetic diversity. Especially, in the future, initiatives such as the global assessment of reptile distribution [47] will help to have more data on the distribution of squamates which were missing in our study. Yet because of the threats, endemism and evolutionary distinctiveness of the species already present in our data we believe the hotspots we identified are important areas for conservation and that more data will contribute to the identification of new hotspots.

Our initial results suggested that hotspots were well covered by protected areas, especially for squamates and amphibians. However, these results were based on the inclusion of all categories of protected areas and the consideration of a cell as protected if it intersected at least one protected area. A more stringent conservation strategy would require greater coverage of sites by protected areas and include management objectives explicitly directed toward species conservation. When such criteria were accounted for, the number of hotspots protected was low. In addition, some species ranking among the species with the highest BED and HEDGE scores were not found in any protected area (e.g. *Gerbillus hesperinus* in mammals, *Acanthodactylus harranensis* in squamates or *Lyciasalamandra antalyana* in amphibians). This gap can potentially be explained by the low coverage of the land by protected areas (4.3 %) [48], more common networks of protected areas in the North such as Natura 2000, lack of protected areas in arid zones (whereas nearly 400 sites have been designed as RAMSAR sites for the protection of wetlands [49]) and more numerous protected areas dedicated to bird conservation (e.g. Bird Directive 2009/147/EC).

Previous studies have highlighted the poor conservation of evolutionary history. At a global scale, Safi et al. [34] found that only 15.6 and 4.7 % of evolutionary distinct and globally endangered amphibian and mammal priority zones, respectively, intersected with protected areas. In Europe, terrestrial mammal, squamate and bird PD and ED are less protected than expected if protected areas were randomly distributed [50, 51]. At the Mediterranean scale, Guilhaumon et al. [18] showed that protected areas in the Mediterranean Sea did not reach conservation targets for fish PD (see also [17]). An approach to conserving Mediterranean PD and ED would be to consider evolutionary history in the definition of Key Biodiversity Areas [52]. Key Biodiversity Areas are sites of global significance for biodiversity conservation and are identified using standard criteria such as vulnerability and irreplaceability. They are a basis for

conservation planning and are important to the maintenance of viable species populations [53]. Evolutionary history criteria are not yet included in the standards that define Key Biodiversity Areas, but Brooks et al. [54] proposed the inclusion of at-risk phylogenetic endemism and evolutionary distinctiveness in the standards. By measuring the spatial distribution of BED, Expected PD*loss* and HEDGE, we identified hotspots that may inform the establishment of new Key Biodiversity Areas. However, further research is needed to develop an approach that combines the advantages of the hotspot and of the network approaches by considering local and regional conservation needs while also considering other essential principles for reserve design, such as costs, flexibility and irreplaceability. It would also be meaningful to estimate the coverage of hotspots with already defined Key Biodiversity Areas. Another possible measure to make the conservation of PD more stringent would be to dedicate some already protected areas specifically to the conservation of species and of their evolutionary history. Indeed, some hotspots exclusively contain protected areas of category III, V and VI, which do not directly aim to conserve species. For example the management objective of protected areas classified under the category V is to protect landscapes. Conservation areas are not the only way to conserve biodiversity; many species exist beyond protected areas, and a key goal should be to preserve the quality of their habitats [31]. Evolutionary history criteria should also be included in prioritization lists of species [55]. This may be an important complementary approach to species conservation, as conservation objectives within protected areas are not always met [56].

Conclusion
We conducted the first study that identifies those areas where at-risk and range-restricted evolutionary history is concentrated in the Mediterranean Basin. Hotspots were mainly found south and east of the Mediterranean Sea and in islands but were poorly covered by protected areas. We also showed that some species representing the threatened endemic evolutionary history of the region were not found in any protected area. Thus, not only local sites but the region itself are at risk of losing large amounts of phylogenetic diversity. Underwood et al. [57] stressed the importance of expanding the network of Mediterranean protected areas; we showed that new protected areas should be delineated to avoid that some sites lose too much of their phylogenetic diversity. We thus encourage practitioners to consider evolutionary history criteria in their efforts to protect habitats, ecosystems, species and their related benefits to societies. We recommend to use both a site and a species approach and not only to consider species threat status but also endemism,

while accounting for the different categories of protected areas. We advise the use of indices which consider the phylogenetic complementarities of extinction risks or range restrictions, meaning that a branch in the phylogeny may be lost only if all its descending species are lost and that it is range-restricted only if its descending species are all endemic to the same restricted area. We also encourage the use of null models, as they enable us to identify sites at risk independently of species richness. As the resolution and completeness of phylogenies improve, the use of phylogenetic diversity in conservation is becoming increasingly reliable and meaningful. Future assessments of conservation needs at different scales as well as assessments of data-deficient species status could enhance our knowledge of the risks of losing evolutionary history.

Methods
Data
For the mammal phylogeny, we used a maximum clade credibility tree [58]. We used recently established phylogenies for squamates [59] and amphibians [60], both phylogenies being fully resolved and dated. Yet, some Mediterranean amphibian and squamate species were not present in those phylogenies. Some of them were highly threatened and/or endemic of the basin and could represent evolutionary history of conservation interest (3 mammals, 1 squamate and 4 amphibians among missing species were threatened and 8 mammals, 4 squamates, 6 amphibians missing in the phylogenies were endemic to the region) we thus included them by creating polytomies with species belonging to the same genus. Fifteen mammals, 15 squamates and 5 amphibians were added as polytomies. It was shown that the effect of polytomies on PD and ED may be very low [10]. To test whether the placement of missing species had little impact on the identification of hotspots we randomized all species for each genus in the phylogenies and found that observed HEDGE, Expected PD*loss* and BED rankings of grid cells were highly correlated with rankings obtained under the randomization procedure (Spearman correlation test; rho > 0.95 and p ≪ 0.001 for all indices)

Spatial data were mapped using ArcGIS 9.3 software. We used the ecoregion limits of the Mediterranean Basin [61]. We have delimited the Mediterranean Basin as the biome "Mediterranean forests, woodlands, and scrub" of the Palearctic Realm. Distribution ranges of species were downloaded from the IUCN extent of occurrence maps [62]. All maps were projected in a World Behrman projection.

The conservation status of each species was downloaded from the IUCN Red List assessments [62]. We used the regional assessments for terrestrial mammals

[63] but used the global assessments for amphibians and squamates because no Mediterranean assessments exist for these two groups [62]. We removed data-deficient (DD) species to calculate the HEDGE and Expected PD*loss* indices but not the BED index, as this latter index does not rely on threat categories (see the next section for a definition of these indices). For all indices, we removed extinct and regionally extinct species, and in mammals, we removed those species for which a regional assessment was not applicable. Our final data set comprised 229 terrestrial mammals, 107 amphibians and 230 squamates for HEDGE and Expected PD*loss* and 258 terrestrial mammals, 107 amphibians, and 238 squamates for BED; higher numbers for BED were due to the integration of species classified data-deficient in the IUCN Red List (Additional file 10). All known Mediterranean terrestrial mammals and amphibians were included whereas some information about the distribution of squamates was missing (342 squamates were assessed in the region; [62]). More justification on the data used can be found in the Additional file 8.

Protected areas were downloaded from the most recent updates of the world database on protected areas [64]. Some shape information was missing for some protected areas; these areas were thus represented as points. We added buffer areas around these points that corresponded to the respective sizes of the areas. Protected areas for which both shape and size data were missing (24 protected areas) were excluded from analysis. The final data set included 9093 protected areas.

Metrics
We first assessed Expected PD*loss* [27] and HEDGE metrics [28], using probabilities of extinction within 50 years based on the transformation of the IUCN categories into extinction probabilities as described by [65]. We then calculated BED from the range size of species [29]. We used R [66] and the most recent versions of the picante [67] and phylobase [68] packages for analysis.

Identifying hotspots where evolutionary history is at risk
We first defined and mapped hotspots of Expected PD*loss*, HEDGE and BED values according to the following procedure. We applied a 1° × 1° resolution grid to the Mediterranean Basin map for squamate and amphibian analyses (corresponding to approximately 100 × 100 km and 477 grid cells) and 0.5° × 0.5° for mammal analyses (50 km × 50 km and 1489 grid cells). We made this choice as a compromise between having a sufficient number of species per grid cell, having an accurate resolution and decreasing omission errors. Moreover it was shown that, for mammals, at a 0.5° × 0.5° resolution distribution data from the IUCN and atlas data were similar [69]

showing the reliability of data at this scale. We made a complementary analysis at a $1° \times 1°$ resolution for mammals and found that hotspots were identified in the same region and the proportion protected was similar to the results found at a $0.5° \times 0.5°$ resolution (unpublished result). Similarly Safi et al. [34] found that the resolution of grid cells did not affect the identification of sites where threatened evolutionary history concentrate. We then calculated the Expected PD*loss* value, the number of top-priority HEDGE species, and the number of top-priority BED species for each cell and species group.

We defined "top" species as the 10 % of species with the highest scores for a given index (either HEDGE or BED). This approach depends on the number of species in each cell (Table 2; see also [34]). We thus defined a second criterion to identify hotspots by using null models as follows.

For Expected PD*loss*, we identified those areas with greater losses than expected if extinction risks were randomly distributed. We randomized extinction risks among species in the phylogeny one thousand times, calculated the new value of Expected PD*loss* in each grid cell for each randomization, and then determined the frequency (termed $F_{\text{Expected PDloss}}$) with which the observed Expected PD*loss* value in a given grid cell was higher than the simulated values. For the HEDGE and BED values, we randomized species identities within each grid cell one thousand times while maintaining the species richness of each grid cell constant. We then calculated the sum of species HEDGE and BED values per grid cell for each randomization and calculated the F_{HEDGE} and F_{BED} values as the frequencies with which the observed HEDGE and BED values in a given grid cell were higher than the simulated values. We thus used two classes of null models. Because Expected PD*loss* is measured as a characteristic of a given site, we fixed the composition of each cell and randomized species extinction risks. In contrast, because HEDGE and BED values are measured as characteristics of each species independently of the cell in which they occur, we shuffled species across cells. The two approaches thus complement each other to reveal hotspots for conservation.

Hotspots were identified based on Aïchi targets defined in the Strategic Plan for Biodiversity 2011–2020 and aiming to protect 17 % of the total land area [25]. We ranked grid cells in increasing order of $F_{\text{Expected PDloss}}$, F_{HEDGE} and F_{BED}. Ranks for ties were determined using the raw Expected PD*loss*, HEDGE or BED values. For example, grid cells with equal $F_{\text{Expected PDloss}}$ values were ranked in increasing order of Expected PD*loss*. Then, we selected the 17 % of grid cells with the highest ranks. We also proposed an alternative selection of hotspots independent

of Aïchi targets in Additional files 1, 2, 3 and 7. This alternative strategy enables to identify either the most threatened and range-restricted hotspots which do not necessarily cover 17 % of the territory or, on the contrary, additional sites which were threatened but not identified due to the 17 % threshold. Expected PD*loss* alternative hotspots were defined as areas where Expected PD*loss* was higher than the mean value of all sites and areas where Expected PD*loss* was higher than under a random distribution of threats ($F_{\text{Expected PDloss}} \geq 0.5$). HEDGE and BED alternative hotspots were defined as areas where HEDGE and BED, respectively, contained at least one species from the 10 % of species with the highest HEDGE and BED scores and areas where HEDGE and BED, respectively, was higher than under a random distribution of threats ($F_{\text{HEDGE/BED}} \geq 0.5$).

To assess the extent to which the selected hotspots complemented one another, we examined the unique Expected PD*loss* values, and the sum of species HEDGE and BED values of combined hotspots (considering the pool of species occurring in at least one of the hotspots) and compared these values to those expected if the hotspots were randomly distributed. A p value was defined as the frequency with which random values were higher or equal than the observed value. Finally, to test whether a species richness approach would have identified similar hotspots [33], we used Spearman correlations to assess the correlations between species richness and PD, the richness in threatened species (i.e., species classified as critically endangered, endangered or vulnerable in the IUCN Red List) and either Expected PD*loss* or HEDGE, and the richness in range-restricted species (number of species in the top 10 % species with the smallest range size) and BED.

Are hotspots for the conservation of Mediterranean evolutionary history well protected?

For each group and metric, we calculated the observed proportion of hotspots of threatened and endemic evolutionary history that were protected. We then randomly designated 17 % of the grid cells as hotspots and calculated the number of those simulated priority grid cells that were protected. Finally, we calculated the frequency (termed F_{PA}) with which the observed proportion of priority grid cells that were protected was greater than that in simulations. We considered the conservation of hotspots to be more efficient than if randomly distributed when the F_{PA} value was greater than 0.75, as efficient when the F_{PA} value ranged between 0.25 and 0.75, and less efficient when the F_{PA} value was less than 0.25. Due to the low coverage of the Mediterranean basin by protected areas we considered that, if the proportion of protected hotspots was higher than in our null models in more than 75 % of the simulations, hotspots were well protected.

We performed the calculations for all of the protected areas first and then for only the categories I, II and IV of protected areas. Categories I, II and IV require a higher level of protection, and their management objectives are specifically dedicated to species protection, whereas other categories may focus on other aspects of conservation, such as the sustainable use of resources [26]. Other authors have not included category IV protected areas as their regulation may not be as stringent as that of categories I and II [50, 51]. However, we included them in the present study because category IV encompasses different designations in the Mediterranean Basin, some of which are highly regulated (e.g., natural reserves, national parks) [26]. Out of the 9093 protected areas, 1837 were of categories I, II or IV.

We repeated this method for two protection scenarios: a minimum protection scenario, in which a grid cell was considered protected if it intersected at least one protected area; and a strong protection scenario, in which a grid cell had to have more than 50 % of its area covered by protected areas to be considered protected [70]. The latter scenario did not include protected areas for which range size information was missing (see Data section).

Additional files

Additional file 1. Identification of priority sites independently of Aïchi targets in terrestrial mammals. Data. A. Identification of Expected PD*loss* priority sites independently of Aïchi targets. Expected PD*loss* hotspots were defined as areas where Expected PD*loss* was higher than the mean value of all sites and areas where Expected PD*loss* was higher than under a random distribution of threats ($F_{Expected\ PDloss} \geq 0.5$). B and C. Identification of HEDGE and BED priority sites independently of Aïchi targets: HEDGE and BED hotspots were defined as areas where HEDGE and BED, respectively, contained at least one species from the 10 % of species with the highest HEDGE and BED scores and areas where HEDGE and BED, respectively, was higher than under a random distribution of threats ($F_{HEDGE/BED} \geq 0.5$).

Additional file 2. Identification of priority sites independently of Aïchi targets in squamates. Data. A. Identification of Expected PDloss priority sites independently of Aïchi targets. Expected PDloss hotspots were defined as areas where Expected PDloss was higher than the mean value of all sites and areas where Expected PDloss was higher than under a random distribution of threats ($F_{Expected\ PDloss} \geq 0.5$). B and C. Identification of HEDGE and BED priority sites independently of Aïchi targets: HEDGE and BED hotspots were defined as areas where HEDGE and BED, respectively, contained at least one species from the 10 % of species with the highest HEDGE and BED scores and areas where HEDGE and BED, respectively, was higher than under a random distribution of threats ($F_{HEDGE/BED} \geq 0.5$).

Additional file 3. Identification of priority sites independently of Aïchi targets in amphibians. Data: A. Identification of Expected PD*loss* priority sites independently of Aïchi targets. Expected PD*loss* hotspots were defined as areas where Expected PD*loss* was higher than the mean value

of all sites and areas where Expected PD*loss* was higher than under a random distribution of threats ($F_{Expected\ PDloss} \geq 0.5$). B and C. Identification of HEDGE and BED priority sites independently of Aïchi targets: HEDGE and BED hotspots were defined as areas where HEDGE and BED, respectively, contained at least one species from the 10 % of species with the highest HEDGE and BED scores and areas where HEDGE and BED, respectively, was higher than under a random distribution of threats ($F_{HEDGE/BED} \geq 0.5$).

Additional file 4. Spatial distribution of species richness in mammals. A. Species richness. B. Threatened species richness (number of species whose threat status is CR, EN or VU). C. Endemic species richness (number of species among the 10 % of species with the smallest range).

Additional file 5. Spatial distribution of species richness in squamates. A. Species richness. B. Threatened species richness (number of species whose threat status is CR, EN or VU). C. Endemic species richness (number of species among the 10 % of species with the smallest range).

Additional file 6. Spatial distribution of species richness in amphibians. A. Species richness. B. Threatened species richness (number of species whose threat status is CR, EN or VU). C. Endemic species richness (number of species among the 10 % of species with the smallest range).

Additional file 7. conservation scenarios. Each graph represents the proportion of HEDGE, BED and Expected PD*loss* priority sites protected. Red bars correspond to the degree of protection if all protected areas are included and blue bars the degree of protection if only protected areas of categories I, II and IV are included. Star symbols correspond to the frequency to which the proportion of hotspots protected was higher than if priority grid cells were distributed randomly (F_{PA}): no star means $F_{PA} \leq 0.25$; * means $0.25 < F_{PA} \leq 0.75$; ** means $0.75 < F_{PA} \leq 1$. We ran analysis for **A.** a scenario of minimum protection in which hotspots were safe if they intersected at least one protected area; **B.** a scenario of strong protection where a site was considered safe if it was protected on more than half of its area.

Additional file 8. Justification of the data used.

Additional file 9. Species not found in any protected areas and their respective HEDGE and BED scores. Data: Species not found in any protected areas of category I, II and IV, a * means that the species is not present in any protected area.

Additional file 10. Data-deficient species and their BED scores. Data-deficient mammal and squamate species and their BED scores (there are no amphibian DD species in the Mediterranean basin).

Additional file 11. HEDGE and BED scores of all Mediterranean species. HEDGE and BED scores of all Mediterranean species.

Abbreviations
BED: biogeographic weighted evolutionary distinctiveness; ED: evolutionary distinctiveness; DD: data-deficient; Expected PDloss: expected loss of phylogenetic diversity; HEDGE: heightened evolutionary distinctiveness and global endangerment; IUCN: international union for conservation of nature; PD: phylogenetic diversity.

Authors' contributions
SV and SP conceived analyses, SV performed analyses, SV, PC and SP wrote the manuscript. All authors read and approved the final manuscript.

Acknowledgements
We thank Daniel Faith and Ana Rodrigues for their useful comments and advice on our study. We are grateful to Florian Kirchner from the IUCN France for providing information about threat status and range distribution of species. We gratefully acknowledge support from the CNRS/IN2P3 Computing Center (Lyon/Villeurbanne-France), for providing a significant amount of the computing resources needed for this work.

Competing interests
The authors declare that they have no competing interests.

Fundings
This study has been supported by the French State through the Research National Agency under the LabEx ANR-10-LABX-0003-BCDiv, within the framework of the program 'Investing for the future' (ANR-11-IDEX-0004-02).

References
1. Ceballos G, Ehrlich PR, Barnosky AD, García A, Pringle RM, Palmer TM. Accelerated modern human–induced species losses: entering the sixth mass extinction. Sci Adv. 2015;1:e1400253.
2. Vane-Wright RI, Humphries CJ, Williams PH. What to protect? Systematics and the agony of choice. Biol Cons. 1991;55:235–54.
3. Faith DP, Magallón S, Hendry AP, Conti E, Yahara T, Donoghue MJ. Evosystem services: an evolutionary perspective on the links between biodiversity and human well-being. Curr Opin Env Sust. 2010;2:66–74.
4. Davies TJ, Buckley LB. Phylogenetic diversity as a window into the evolutionary and biogeographic histories of present-day richness gradients for mammals. Phil Trans Roy Soc B Biol Sci. 2011;366:2414–25.
5. Mooers AO, Heard SB, Chrostowski E. Evolutionary heritage as a metric for conservation. Phyl Cons. 2005;1:120–38.
6. Morlon H, Potts MD, Plotkin JB. Inferring the dynamics of diversification: a coalescent approach. PLoS Biol. 2010;8:e1000493.
7. Rolland J, Cadotte MW, Davies J, Devictor V, Lavergne S, Mouquet N, Pavoine S, Rodrigues A, Thuiller W, Turcati L, et al. Using phylogenies in conservation: new perspectives. Biol Lett. 2012;8:692–4.
8. Purvis A, Agapow PM, Gittleman JL, Mace GM. Nonrandom extinction and the loss of evolutionary history. Science. 2000;288:328–30.
9. von Euler F. Selective extinction and rapid loss of evolutionary history in the bird fauna. P Roy Soc B Biol Sci. 2001;268:127–30.
10. Veron S, Davies TJ, Cadotte MW, Clergeau P, Pavoine S. Predicting loss of evolutionary history: where are we? Biol Rev Camb Philos Soc. 2015. doi:10.1111/brv.12228 .
11. Cuttelod A, García N, Abdul Malak D, Temple H, Katariya V. The Mediterranean: a biodiversity hotspot under threat. In: Vié JC, Hilton-Taylor C, Stuart SN, editors. The 2008 Review of the IUCN red list of threatened species. Gland: IUCN; 2008.
12. Blondel J, Aronson J. Biology and wildlife of the Mediterranean region. Oxford: University Press; 1999.
13. Myers N, Mittermeier RA, Mittermeier CG, Da Fonseca GAB, Kent J. Biodiversity hotspots for conservation priorities. Nature. 2000;403:853–8.
14. Mittermeier RA, Turner WR, Larsen FW, Brooks TM, Gascon C. Global biodiversity conservation: the critical role of hotspots. In: Zachos FE, Habel JC, editors. Biodiversity hotspots. Berlin: Springer; 2011. p. 3–22.
15. Coll M, Piroddi C, Steenbeek J, Kaschner K, Lasram FBR, Aguzzi J, Ballesteros E, Bianchi NC, Corbera J, Dailanis T, Danovaro R, Voultsiadou E. The biodiversity of the Mediterranean Sea: estimates, patterns, and threats. PLoS ONE. 2010;5:e11842.
16. Apostolopoulos Y, Loukissas P, Leontidou L. Mediterranean tourism: facets of socioeconomic development and cultural change. London: Routledge; 2014.
17. Mouillot D, Albouy C, Guilhaumon F, Lasram FBR, Coll M, Devictor V, Meynard CN, Pauly D, Tomasini JA, Troussellier M, et al. Protected and threatened components of fish biodiversity in the Mediterranean Sea. Curr Biol. 2011;21:1044–50.
18. Guilhaumon F, Albouy C, Claudet J, Velez L, Ben Rais Lasram F, Tomasini JA, Douzery EJP, Meynard CN, Mouquet N, Troussellier M, Mouillot D. Representing taxonomic, phylogenetic and functional diversity: new challenges for Mediterranean marine-protected areas. Div Dist. 2015;21:175–87.
19. Faith DP. Conservation evaluation and phylogenetic diversity. Biol Cons. 1992;61:1–10.
20. Forest F, Grenyer R, Rouget M, Davies TJ, Cowling RM, Faith DP, Savolainen V. Preserving the evolutionary potential of floras in biodiversity hotspots. Nature. 2007;445:757–60.
21. Redding DW, Mooers AØ. Incorporating evolutionary measures into conservation prioritization. Cons Biol. 2006;20:1670–8.
22. Redding DW, DeWolff CURT, Mooers AØ. Evolutionary distinctiveness, threat status, and ecological oddity in primates. Cons Biol. 2010;24:1052–8.
23. Gascon C, Brooks TM, Contreras-MacBeath T, Heard N, Konstant W, Lamoreux J, Launay F, Maunder M, Russel A, Mittermeier SM. The importance and benefits of species. Curr Biol. 2015;25:R431–8.
24. Redding DW, Hartmann K, Mimoto A, Bokal D, DeVos M, Mooers AØ. Evolutionarily distinctive species often capture more phylogenetic diversity than expected. J Theor Biol. 2008;251:606–15.
25. Convention on Biological Diversity. https://www.cbd.int/sp/targets/. Accessed 4th Dec 2015.
26. Ornat A, Reynés AP. Use of the protected areas management categories in the Mediterranean region. Gland: IUCN; 2004.
27. Faith DP. Threatened species and the potential loss of phylogenetic diversity: conservation scenarios based on estimated extinction probabilities and phylogenetic risk analysis. Cons Biol. 2008;22:1461–70.
28. Steel M, Mimoto A, Mooers AØ. Hedging our bets: the expected contribution of species to future phylogenetic diversity. Evol Bioinform Online. 2007;3:237–44.
29. Cadotte MW, Davies JT. Rarest of the rare: advances in combining evolutionary distinctiveness and scarcity to inform conservation at biogeographical scales. Div Dist. 2010;16:376–85.
30. Stöck M, Dubey S, Klütsch C, Litvinchuk SN, Scheidt U, Perrin N. On tree frog cryptozoology and systematics–response to Y Werner. Mol Phyl Evol. 2010;57:957–8.
31. Conde DA, Colchero F, Güneralp B, Gusset M, Skolnik B, Parr M, Byers O, Johnson K, Young G, Flesness N, et al. Opportunities and cost for preventing vertebrate extinctions. Curr Biol. 2015;25:219–21.
32. Davies TJ, Fritz SA, Grenyer R, Orme CDL, Bielby J, Bininda-Emonds OR, Cardillo M, Jones KE, Gittleman JL, Mace GM, Purvis A. Phylogenetic trees and the future of mammalian biodiversity. Proc Natl Acta Sci USA. 2008;105:11556–63.
33. Rodrigues ASL, Brooks TM, Gaston KJ. Integrating phylogenetic diversity in the selection of priority areas for conservation: does it make a difference. In: Purvis A, Gittleman JL, Brooks T, editors. Biodiversity hotspots. Cambridge: Cambridge University Press; 2005. p. 101–19.
34. Safi K, Armour-Marshall K, Baillie JE, Isaac NJ. Global patterns of evolutionary distinct and globally endangered amphibians and mammals. PLoS ONE. 2013;8:e63582.
35. Kati V, Devillers P, Dufrêne M, Legakis A, Vokou D, Lebrun P. Hotspots, complementarity or representativeness? Designing optimal small-scale reserves for biodiversity conservation. Biol Cons. 2004;120:471–80.
36. Shriner SA, Wilson KR, Flather CH. Reserve networks based on richness hotspots and representation vary with scale. Eco App. 2006;16:16601673.
37. Faith DP, Reid CAM, Hunter J. Integrating phylogenetic diversity, complementarity, and endemism for conservation assessment. Cons Biol. 2004;18:255–61.
38. May-Collado RJ, Zambrana-Torrelio C, Agnarsson I. Global spatial analysis of phylogenetic conservation priorities for aquatic mammals. In: Pellens R, Grandcolas P, editors. Biodiversity conservation and phylogenetic systematics: preserving our evolutionary history in an extinction crisis. Berlin: Springer; 2016. p. 305–318.
39. Pressey RL, Humphries CJ, Margules CR, Vane-Wright RI, Williams PH. Beyond opportunism: key principles for systematic reserve selection. Trends Ecol Evol. 1993;8:124–8.
40. Margules CR, Pressey RL. Systematic conservation planning. Nature. 2000;405:243–53.
41. Millennium Ecosystem Assessment. Ecosystems and human well-being: synthesis. Washington: Island Press; 2005.
42. Faith DP, Magallón S, Hendry AP, Conti E, Yahara T, Donoghue MJ. Evosystem services: an evolutionary perspective on the links between biodiversity and human well-being. Curr Opin Env Sust. 2010;2:66–74.
43. Lean C, MacLaurin J. The value of phylogenetic diversity. In: Pellens R, Grandcolas P, editors. Biodiversity conservation and phylogenetic systematics: preserving our evolutionary history in an extinction crisis. Berlin: Springer; 2016. p. 19–38.
44. Pavoine S, Ollier S, Dufour AB. Is the originality of a species measurable? Eco Lett. 2005;8:579–86.

45. Collen B, Turvey ST, Waterman C, Meredith HM, Kuhn TS, Baillie JE, Isaac NJ. Investing in evolutionary history: implementing a phylogenetic approach for mammal conservation. Phil Trans Roy Soc B Biol Sci. 2011;366:2611–22.

46. Mouillot D, Bellwood DR, Baraloto C, Chave J, Galzin R, Harmelin-Vivien M, Kulbicki M, Lavergne S, Lavorel S, Mouquet N, et al. Rare species support vulnerable functions in high-diversity ecosystems. PLoS Biol. 2013;11:e1001569.

47. Global Assessment of Reptile Distribution. http://www.gardinitiative.org/. Accessed 24th June 2016.

48. Cox RL, Underwood EC. The importance of conserving biodiversity outside of protected areas in Mediterranean ecosystems. PLoS ONE. 2011;6:e14508.

49. Mediterranean Wetland Initiatives. http://medwet.org/aboutwetlands/ramsarmedsites/. Accessed 24th June 2016.

50. Thuiller W, Maiorano L, Mazel F, Guilhaumon F, Ficetola GF, Lavergne S, Renaud L, Roquet C, Mouillot D. Conserving the functional and phylogenetic trees of life of European tetrapods. Phil Trans Roy Soc B Biol Sci. 2015;370:20140005.

51. Zupan L, Cabeza M, Maiorano L, Roquet C, Devictor V, Lavergne S, Mouillot D, Mouquet N, Renaud J, Thuiller W. Spatial mismatch of phylogenetic diversity across three vertebrate groups and protected areas in Europe. Div Dist. 2014;20:674–85.

52. Eken G, Bennun L, Brooks TM, Darwall W, Fishpool LD, Foster M, Knox D, Langhammer D, Matiku P, Radford D, et al. Key biodiversity areas as site conservation targets. Bioscience. 2004;51:1110–8.

53. Langhammer PF. Identification and gap analysis of key biodiversity areas: targets for comprehensive protected area systems. Gland: IUCN; 2007.

54. Brooks TM, Cuttelod A, Faith DP, Garcia-Moreno J, Langhammer P, Pérez-Espona S. Why and how might genetic and phylogenetic diversity be reflected in the identification of key biodiversity areas? Phil Trans Roy Soc B Biol Sci. 2015;370:20140019.

55. Hidasi-Neto J, Loyola RD, Cianciaruso MV. Conservation actions based on Red Lists do not capture the functional and phylogenetic diversity of birds in Brazil. PLoS ONE. 2013;8:e73431.

56. Hockings M, Phillips A. How well are we doing? Some thoughts on the effectiveness of protected areas. Parks. 1999;9:5–14.

57. Underwood EC, Klausmeyer KR, Cox RL, Busby SM, Morrison SA, Shaw MR. Expanding the global network of protected areas to save the imperiled Mediterranean biome. Cons Biol. 2009;23:43–52.

58. Rolland J, Condamine FL, Jiguet F, Morlon H. Faster speciation and reduced extinction in the tropics contribute to the mammalian latitudinal diversity gradient. PLoS Biol. 2014;12:e1001775.

59. Pyron RA, Burbrink FT. Early origin of viviparity and multiple reversions to oviparity in squamate reptiles. Ecol Lett. 2014;17:13–21.

60. Pyron RA, Wiens JJ. Large-scale phylogenetic analyses reveal the causes of high tropical amphibian diversity. P Roy Soc B Biol Sci. 2013;2013:16–22.

61. Olson DM, Dinerstein E, Wikramanayake ED, Burgess ND, Powell GV, Underwood EC, D'amico JA, Itoua I, Strand HE, Morrison JC, et al. Terrestrial ecoregions of the world: a new map of life on earth. Bioscience. 2001;51:933–8.

62. IUCN. The IUCN red list of threatened species. Version 2015.2. http://www.iucnredlist.org. Accessed 19th Mar 2015.

63. Temple HJ, Cuttelod A. The status and distribution of mediterranean mammals. Gland: IUCN; 2009.

64. World Data Base on Protected Areas. www.protectedplanet.net/. Accessed 4th Dec 2015.

65. Mooers AØ, Faith DP, Maddison WP. Converting endangered species categories to probabilities of extinction for phylogenetic conservation prioritization. PLoS ONE. 2008;3:e3700.

66. R Development Core Team. R: a language and environment for statistical computing. Vienna: R Foundation for Statistical Computing; 2001. SBN 3-900051-07-0, Available at: http://www.R-project.org/.

67. Kembel SW, Cowan PD, Helmus MR, Cornwell WK, Morlon H, Ackerly DD, Blomberg SP, Webb CO. Picante: R tools for integrating phylogenies and ecology. Bioinformatics. 2015;26:1463–4.

68. Bolker B, Butler M, Cowan P, de Vienne D, Eddelbuettel D, Holder M, Jombart T, Kembel S, Michonneau F, Orme B, et al. Phylobase: base package for phylogenetic structures and comparative data, R package version 0.8.0. 2015.

69. Márcia Barbosa A, Estrada A, Márquez AL, Purvis A, Orme CDL. Atlas versus range maps: robustness of chorological relationships to distribution data types in European mammals. J Biol. 2012;39:1391–400.

70. Araújo MB. Matching species with reserves–uncertainties from using data at different resolutions. Biol Cons. 2004;118:533–8.

11

Importance of latrine communication in European rabbits shifts along a rural–to–urban gradient

Madlen Ziege[1*], David Bierbach[2], Svenja Bischoff[1], Anna-Lena Brandt[1], Mareike Brix[1], Bastian Greshake[3], Stefan Merker[4], Sandra Wenninger[1], Torsten Wronski[5] and Martin Plath[6]

Abstract

Background: Information transfer in mammalian communication networks is often based on the deposition of excreta in latrines. Depending on the intended receiver(s), latrines are either formed at territorial boundaries (*between-group communication*) or in core areas of home ranges (*within-group communication*). The relative importance of both types of marking behavior should depend, amongst other factors, on population densities and social group sizes, which tend to differ between urban and rural wildlife populations. Our study is the first to assess (direct and indirect) anthropogenic influences on mammalian latrine-based communication networks along a rural-to-urban gradient in European rabbits (*Oryctolagus cuniculus*) living in urban, suburban and rural areas in and around Frankfurt am Main (Germany).

Results: The proportion of latrines located in close proximity to the burrow was higher at rural study sites compared to urban and suburban ones. At rural sites, we found the largest latrines and highest latrine densities close to the burrow, suggesting that core marking prevailed. By contrast, latrine dimensions and densities increased with increasing distance from the burrow in urban and suburban populations, suggesting a higher importance of peripheral marking.

Conclusions: Increased population densities, but smaller social group sizes in urban rabbit populations may lead to an increased importance of *between-group communication* and thus, favor peripheral over core marking. Our study provides novel insights into the manifold ways by which man-made habitat alterations along a rural-to-urban gradient directly and indirectly affect wildlife populations, including latrine-based communication networks.

Keywords: Chemical communication, Communication center, Core marking, Localized defecation, Urban ecology

Background

Mammalian communication through localized defecation sites

The transmission of information in localized defecation sites (latrines) plays a central role in mammalian communication ([1–3], reviewed in [4]). Latrines deposited along territory boundaries are known to serve as a visual and olfactory fence, not only to indicate territorial occupancy, but also to signal the competitive ability of the territory owner(s), e.g., towards neighboring territory holders (*between-group communication*; seen in European badgers, *Meles meles* [5, 6]; lemurs [7]; meerkats, *Suricata suricatta* [8], and bushbuck, *Tragelaphus scriptus* [9]). Besides this peripheral marking behavior, several species also establish latrines in central parts of their home ranges—termed core marking—in order to support the monopolization of key resources, such as food, shelter, burrows, or nest sites (seen in European badgers [6, 10], lemurs [4, 7], and Arabian gazelles, *Gazella arabica* [11, 12]). Furthermore, latrines that are located in core areas of home ranges facilitate information exchange between the members of the same social group and thus, can enhance and maintain social bonds or dominance hierarchies (*within-group communication* [6, 13, 14]).

*Correspondence: madlen.ziege@mailbox.org
[1] Department of Ecology and Evolution, Goethe University Frankfurt, Max-von-Laue-Str. 13, 60439 Frankfurt am Main, Germany
Full list of author information is available at the end of the article

Relative importance of core vs. peripheral marking behavior

Dröscher and Kappeler [4] recently highlighted that we still have a limited understanding about how different ecological factors influence the structure and complexity of mammalian latrine-based communication networks. The relative importance of core vs. peripheral marking behavior seems to depend on population ecological variables; e.g., higher population densities increase competition for territorial space and thus, the necessity to indicate territorial occupancy. This, in turn, favors peripheral over core marking, as suggested for high density rural European badger populations [15, 16] (for European rabbits, *Oryctolagus cuniculus*, see also [17]).

Furthermore, economic considerations predict that the establishment, use, and maintenance of latrines depends on the time and energy animals can effectively invest in their marking behavior [3, 18]. If territory dimensions exceed a certain size, peripheral marking is likely to be replaced by the less time-consuming core marking behavior [3, 4, 18]. Likewise, if the number of individuals that contribute to peripheral marking is low and/or animals need to allocate a considerable proportion of their time to other behaviors—e.g., because they spend more time avoiding predators or human disturbance—latrine distribution patterns should become less complex, and a shift towards core marking would be predicted.

Effects of urbanization on latrine-based communication networks

Population densities of some mammalian species are higher in urban habitats compared to rural areas ([19–21], reviewed in [22]). Moreover, changes in population densities can be accompanied by differences in social organization, such as smaller social group sizes (European rabbits: [23]) or a less coherent social organization in urban and suburban populations (European badgers: [24–26]). Typical behavioral changes in some urban populations include a reduction in time spent foraging [27] and reduced territorial behavior [24–26], along with smaller territory dimensions (e.g., in raccoons, *Procyon lotor* [27]; European badgers [26]; or red foxes, *Vulpes vulpes* [28]; reviewed in [29]). While the aforementioned species are crepuscular and avoid human disturbance [5, 30], other species, like European rabbits, show extended activity rhythms and reduced anti-predator behavior in urban regions [31, 32], and so they are also unlikely to reduce territorial behavior.

Empirical studies considering the question of how urbanization affects latrine-based communication networks are largely restricted to European badgers [25, 26]. In rural areas, where badgers reached high population densities, both core—("hinterland marking" [5, 6,

10]) and peripheral marking behaviors were reported, but peripheral marking prevailed [15, 16]. Specifically, peripheral latrines were larger, more densely packed, and showed higher utilization frequencies [16]. By contrast, no peripheral latrines were found in a low-density suburban badger population in Bristol [25] and a high-density urban population in Brighton [26]. In case of the Bristol population, latrines accumulated close to the burrow, suggesting a role of latrines for communication within groups. A recent study by Domínguez-Cebrían and de Miguel [33] investigated the latrine-based communication network of a European rabbit population in a suburban forest of Madrid. Latrines deposited at the territorial periphery were previously hypothesized to signal territory occupancy in rabbits, whereas latrines situated in proximity to the burrow likely facilitate information exchange among group members [13, 14, 34–38]. Domínguez-Cebrían and de Miguel [33] found numbers of latrines to decrease with increasing distance from the burrow system and discuss that rabbits could face a higher predation risk when using peripheral latrines. However, no information was provided by the authors on population densities or social group sizes that would have allowed conclusions regarding the question of how (direct and indirect) effects of urbanization influence latrine-based communication networks in their study population.

Objectives of this study

European rabbits exchange information about individuals' age, sex, reproductive condition, and social status via secretions emanating from the anal and submandibular glands [14, 38, 39]. Rabbits deposit hard fecal pellets at latrines that are covered with anal gland secretions [36, 40] and smear secretions from the submandibular gland onto fecal pellets during so-called "chinning" behavior [14, 37, 39, 40]. It is thus well conceivable that latrines at territorial boundaries provide information about territorial occupancy to potential territory intruders (*between-group communication*) (e.g., [13, 14, 34–38]). In contrast, the common use of latrines located at core areas by different members of the same social rabbit group is probably mainly related to the establishment and maintenance of social group structures (*within-group communication*) [13, 14]. Previous studies were suggestive of a pattern in which peripheral marking is pronounced when population densities are high and distinct social groups are competing ([17], see also [15, 16] for European badger populations).

Population densities of European rabbits in rural areas of Europe are currently on decline [31, 41–44], while at the same time rabbits can reach high densities in urban and suburban areas (for Germany see [31, 43]) but tend to form much smaller social groups [23]. This trend is probably largely caused by intensified agricultural practices in rural areas, where the availability, e.g., of thickets for burrow construction is decreasing [23, 41–44]. Hence, European rabbits are an interesting species to compare population differences in latrine-based communication networks along a rural-to-urban gradient. The paucity of studies investigating the relative importance of core marking (*within-group communication*) vs. peripheral marking (*between-group communication*) in mammalian latrine-based communication networks further motivated our present study. We investigated rabbit populations along a rural-to-urban gradient. We located latrines at each site and established the distance of each latrine to the nearest burrow. We also assessed latrine dimensions and densities as indicators for long-term use, and numbers of fresh fecal pellets as an indicator for recent use. We further quantified direct and indirect anthropogenic impact at our study sites, including several (interrelated) variables describing human nuisance and anthropogenic landscape alterations (see 'degree of urbanity' [23, 31]). This allowed us to establish distribution patterns of latrines relative to the burrow, whereby a prevalence of *core marking* should be reflected by highest latrine densities, larger latrine dimensions, and more fecal pellets per latrine, close to the burrow compared to latrines afar from it. If *peripheral marking* prevails, this should lead to the opposite pattern.

Our predictions were derived from the observation that population densities of rabbits increase, while at the same time social group sizes decrease, along the rural-to-urban gradient considered here [23, 31]. We predicted that *peripheral marking* for territorial defense becomes more important in urbanized regions, as increasing population densities increase competition for space and other resources. Moreover, small group sizes at urban study sites should also favor peripheral over core marking behavior as the necessity to communicate within groups decreases. This should lead to a pattern where latrine densities, sizes, and utilization frequencies increase with increasing distance from the burrow towards the inner parts of the city, while the opposite pattern can be predicted for rural sites.

Methods
Selection of study sites
We studied rabbit populations in nine green spaces (measuring between 1 and 4.9 ha in size) in the city center of Frankfurt a.M. (Germany) that are highly fragmented and separated from each other by heavily used roads, in four parks at the periphery of the city (between 5.5 and 30.2 ha) and at two nearby rural study sites (both 36 ha; Table 1; Figs. 1, 2). Unfortunately, we were not able to include more study sites within the rural surrounding of Frankfurt a.M. due to difficulties in finding areas where a representative population density is still existent.

Table 1 Study sites

Study sites	Coordinates		Size [ha]	Degree of urbanity	Population density (rabbits/ha)	Mean social group size
Rural						
Bad Vilbel	N 50°9.418	E 8°41.820	36.00	−2.55	0.88	8.80
Maintal	N 50°8.653	E 8°49.094	36.00	−1.80	3.38	10.00
Suburban						
Ostpark	N 50°7.251	E 8°43.364	30.20	−0.45	19.14	9.50
Grüneburgpark	N 50°7.647	E 8°39.608	27.00	−0.43	0.26	3.50
Rebstockpark	N 50°6.674	E 8°36.773	21.10	−0.36	15.02	4.00
Miquelanlage	N 50°7.970	E 8°39.524	5.50	−0.04	2.27	2.83
Urban						
Site 1	N 50°6.999	E 8°41.503	4.90	0.47	8.16	2.90
Site 2	N 50°6.673	E 8°41.608	3.53	0.47	4.53	4.00
Site 3	N 50°6.723	E 8°40.220	3.64	0.50	9.07	4.00
Site 4	N 50°7.098	E 8°40.946	3.37	0.57	13.95	2.00
Site 5	N 50°7.160	E 8°41.198	2.18	0.59	15.60	3.40
Site 6	N 50°7.001	E 8°40.529	3.66	0.59	3.55	2.00
Site 7	N 50°6.865	E 8°40.263	1.33	0.76	9.02	1.50
Site 8	N 50°6.870	E 8°41.650	1.50	0.84	24.67	1.67
Site 9	N 50°6.606	E 8°40.323	1.00	0.85	5.00	2.00

Detail information for the 15 study sites situated along the rural-to-urban gradient in and around Frankfurt a.M., Germany

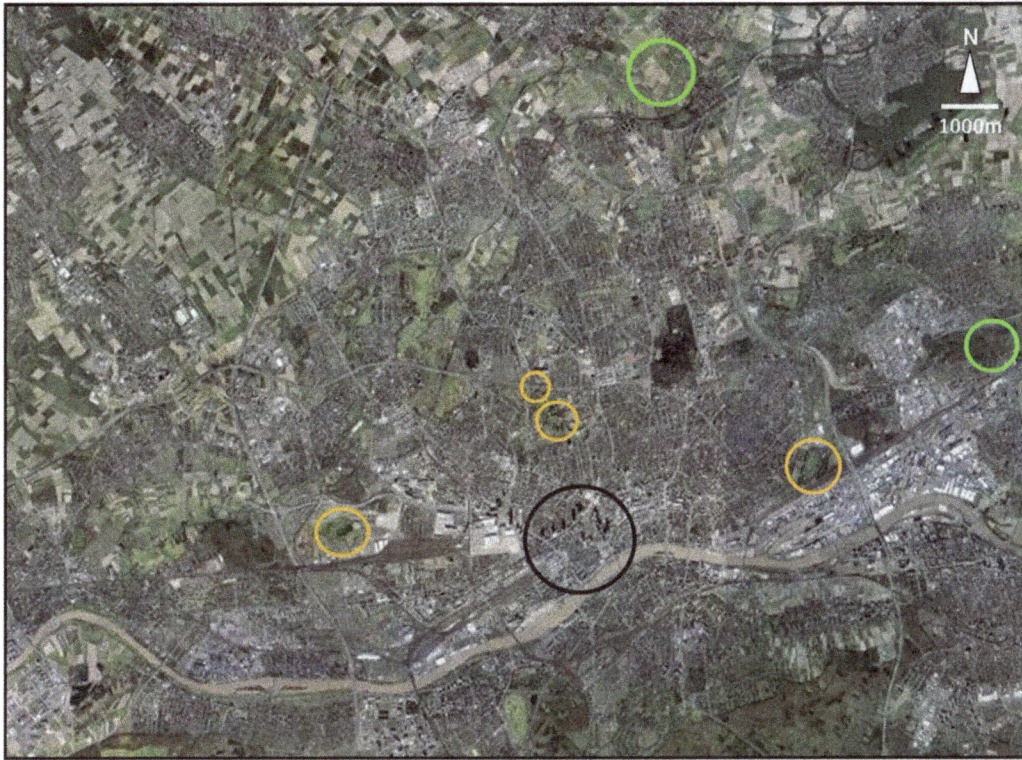

Fig. 1 Overview and location of study sites. Locations of all 15 study sites along the rural-to-urban gradient in and around Frankfurt a.M. *Black circles n = 9 urban study sites, orange circles n = 4 suburban study sites, green circles n = 2 rural study sites Source* Google Earth

In case of the suburban and urban study sites, short-cut meadows were the dominant landscape element (with a grass cutting regime of up to once a week during summer), and the dimensions of our study sites were clearly defined by park borders like streets or pathways. As comparable structures were lacking at both rural sites, we decided to selected quadrants of 600 × 600 m as our study sites, which were sufficiently large to include the outermost latrines afar from the burrow systems (Fig. 2). Here, open landscapes were dominated by agriculturally used areas where meadows (with a sheep grazing regime of two times per year), rape and wheat fields alternated. Between the meadows and fields, only few patches of thickets were present, mainly comprising blackberry bushes (Fig. 2).

Survey of latrine-based communication networks

We systematically mapped latrines and burrows by two persons walking line transects (app. 5 m apart) across the entire study area within all of our 15 study sites, starting in the early morning. We took GPS coordinates from the center of 3253 latrines and the center of 182 burrow systems using a Garmin 12 GPS [separate burrow systems were identified with the help of local hunters that use domesticated ferrets (*Mustelo putorius furo*) to chase rabbits out of the burrow within the framework of a regular hunting scheme, organized by the city of Frankfurt, hunting licence ID 1000250221]. We collected data during the reproductive season of rabbits, which in our latitude lasts from March to September, when territorial defense is strongest [36, 38]. Urban and suburban study sites as well as the rural study site Bad Vilbel were simultaneously sampled between May and September 2011, while the second rural study site (Maintal) was sampled between June and July 2012. Latrines were defined as an accumulation of at least 20 single fecal pellets within an area of 20 × 30 cm [44]. Based on the GPS coordinates we calculated distances of latrines to the nearest burrow system (see also [33, 35, 45]). We measured several variables for each latrine that are—according to previous studies on mammals, including European rabbits—suitable to characterize latrine-based communication networks [4, 6, 9, 11, 12, 33, 36]. Later we evaluated how those variables change with increasing distance of latrines from the respective burrow system (core vs. peripheral marking, see Statistical analyses). For example, if core marking prevails, latrines close to the burrow should be used more often by the members of the social group than peripheral ones, and this should be reflected by higher numbers of (fresh) fecal pellets compared to latrines that are less often used.

Fig. 2 Example of latrine distribution patterns. Detailed aerial photograph of the study site Bad Vilbel. *White triangles* indicate rabbit burrows, *white dots* indicate rabbit latrines *Source* Google Earth

We excluded $n = 10$ burrow systems with less than three latrines from our statistical analyses as those burrows did not show signs of regular use. Moreover, by doing so, we followed the methodological approach of another recent study on latrine distribution patterns of European rabbits in a suburban area [33] so that we were able to discuss our results in comparison to that study.

(a) Indicators of long-term latrine use

As one indicator of long-term latrine use, we established latrine sizes by measuring the maximum width and length of the area that fell into our definition of a latrine (see above). We approximated latrine dimensions [m^2] using a rectangular formula. We also determined numbers of fecal pellets per latrine as another estimate of latrine size. Accurately counting fecal pellets in all latrines through total clearing would have caused an enormous work load, and so we decided to estimate numbers of fecal pellets by eye (see [36]). This estimation method had been practiced before data collection at sites outside of our study area and was confirmed through total clearing after the test trials. As latrine sizes and numbers of fecal pellets both describe latrine dimensions, we log-transformed

and subjected both to a factor reduction (principal component analysis, PCA). We retrieved a single PC with an Eigenvalue >1 (1.50) that explained 75.3 % of the total variance, henceforth referred to as 'latrine dimension'.

Another variable that was used in previous studies to describe the relative importance of core vs. peripheral marking was the latrine density (e.g., latrines were more densely packed at the territorial periphery in a high-density urban badger population [16]). We expressed latrine densities by calculating the mean distance of each latrine to the nearest two neighboring latrines [11, 12].

(b) Indicator of recent latrine use

As an indicator of recent latrine use, we noted whether fresh fecal pellets were present ('0' no fresh fecal pellets present, '1' fresh fecal pellets present) and if present, we accurately counted them once during the process of latrine mapping in the early morning (see [36]).

(c) Indicator of territorial behavior at latrines

We noted whether rabbit paw-scrapings—signs of male territorial behavior [46, 47]—were present at latrines ('0' no paw-scrapings present, '1' paw-scrapings present).

However, we were unable to accurately quantify actual numbers of paw-scrapings.

(d) Effect of woody vegetation on latrine distribution

Finally, we also determined the distance of each latrine to the next woody vegetation (either shrubs or a tree), as this ecological variable is known to affect the placement and utilization frequency of latrines in European rabbits [33, 35].

Estimating the impact of urbanization

In order to relate (direct and indirect) anthropogenic influences to potential differences in latrine-based communication networks we calculated the 'degree of urbanity' for each of our 15 study sites following previous studies [23, 31]. In brief, we assessed the proportion of artificial ground cover (e.g., streets, play grounds) and numbers of anthropogenic objects per ha (e.g., benches, street lamps) at each study site, reflecting the availability of continuous living space. Information on the direct intensity of disturbance by humans (pedestrians and bikers) and leashed or unleashed dogs (per min and per ha) that rabbits were exposed to during their main activity periods at dusk and dawn was obtained through transect counts (for more details see [23, 31]). Additionally, we obtained data on numbers of human residents located within a radius of 500 m from the borders of the study sites from the registration office of Frankfurt a.M. (*Einwohnermeldeamt*, updated: 31.10.2010). These data provide an estimation of overall/peak numbers of visitors in the park areas, as residents tend to walk in nearby city parks.

We subjected the four (log-transformed) variables to PCA. A single principal component was retrieved (henceforth referred to as the PC 'degree of urbanity', Table 1) with an Eigenvalue >1 (3.44) that explained 85.9 % of the total variance (Table 2a). For display purpose only, study sites were categorized as rural ('degree of urbanity' values ≤ −0.5), suburban (> −0.5 and ≤0.5) and urban (>0.5), while the main statistical analyses were performed using continuous data (see below).

To establish a variable characterizing rabbit population dynamics, we relied on previously published data on rabbit densities (numbers of individuals per ha, assessed by direct census counts along pre-defined transects during dusk and dawn in September/October 2011; Table 1; [31]) and burrow densities [23, 31]. Moreover, we included data on social group sizes, obtained through behavioral observations and augmented by the use of ferrets to drive all members of a social group out of their burrow (Table 1; [23]). Again, we log-transformed the three variables and subjected them to PCA. A single principal component was retrieved with an Eigenvalue >1

Table 2 Degree of urbanity and rabbit population dynamics

	Axis loading
(a) Urbanization-related variables	
Proportion of artificial ground cover at each study site	0.84
Numbers of anthropogenic objects per ha at each study site	0.93
Intensity of disturbance by humans and leashed/unleashed dogs min^{-1} ha^{-1}	0.97
Numbers of human residents located within a radius of 500 m	0.96
(b) Variables related to population dynamics	
Population density	0.89
Burrow density	0.94
Social group size	−0.58

Axis loadings of two separated principal component analyses on variables related to (a) urbanization effects (explaining 85.9 % of the total variance) and (b) rabbit population dynamics, respectively (explaining 66.7 % of the total variance)

(2.00) that explained 66.7 % of the total variance (PC 'population dynamics'; Table 2b). As both principal components, the 'degree of urbanity' and 'population dynamics', were highly correlated (Spearman rank correlation: $r = 0.74, p = 0.002, n = 15$; see also [23, 31]), we decided to include only the 'degree of urbanity' in our statistical analyses. Running independent analytical models (see below) with different combinations of both covariates (e.g., 'population dynamics' and 'degree of urbanity'), however, yielded qualitatively very similar results (results not shown).

Statistical analyses

(a) Relative distance of latrines to the nearest burrow (d_{rel})

To compare the spatial distribution of latrines between sites, we first corrected for variation in the sizes of areas marked by latrines around burrow systems, e.g., different home range sizes. Unfortunately, radio-tracking and capture-mark-recapture approaches to establish exact home range dimensions were not feasible for all rabbit groups at our 15 study sites. By using the following approach we were still able to account for variation in home range sizes:

First, based on a distance matrix for all latrines and all burrows at a given study site, each latrine was assigned to the closest burrow (see also [33, 35]). Second, for each burrow we defined the perimeter in which 95 % of all latrines that had been assigned to this burrow were located. Third, we determined the mean distance of the two outermost latrines to the rabbit burrow within this 95 % perimeter (d_{max}) and used this value to calculate the dimensions of the latrine-marked area (A [ha]) around each rabbit burrow, assuming the burrow to be the center ($A = \pi \times d_{max}^2$). For every latrine belonging to this burrow system we corrected its absolute distance to the center of the burrow (d_{abs}) by d_{max} and thus obtained the relative distance of a latrine as $d_{rel} = d_{abs}/d_{max}$. Our approach was justified by the observation that we found latrines that were located close to the respective burrow system and afar from it in all cases, representing cases of core- and peripheral marking (see also [33]). Where we provide descriptive statistics, we categorized latrines depending on d_{rel}-values as ≤0.25 (e.g., around the burrow), 0.25–0.50, 0.50–0.75, or ≥0.75 (periphery), while all statistical tests were conducted using continuous data.

In our first approach, we used arcsine (square root)-transformed d_{rel}-values as the dependent variable in a linear mixed model (LMM, 'mixed' procedure in SPSS 13). We used 'burrow ID' as subject-grouping factor with random intercepts specified for each burrow and the 'degree of urbanity' as the explaining variable (covariate). A similar approach was used to investigate a potential effect of increasing urbanity on latrine-marked areas around rabbit burrows.

(b) Latrine characteristics in relation to the distance to the nearest burrow

In our second approach, we tested whether latrine dimensions and densities, numbers of fresh fecal pellets and distances to the next woody vegetation differed from the core to the periphery of the latrine-marked area, and if this pattern changes along the rural-to-urban gradient. We ran four LMMs using the respective variables (all log-transformed) and again included random intercepts for every burrow system ('burrow ID'), while 'd_{rel}'-values and the 'degree of urbanity' were used as explaining variables (covariates).

We included the interaction term 'd_{rel} × degree of urbanization' in the initial model and step-wise removed all non-significant explaining variables from the reduced model starting with the interaction effect. In case of significant interaction terms, we refrained from interpreting main effects and concentrated on the interaction effects. To analyze the binary variables 'presence of fresh fecal pellets' and 'presence of paw-scrapings' we ran logistic regressions each including 'd_{rel}', the 'degree of urbanity', and their interaction as the explaining variables. Non-significant effects were excluded in a step-wise backwards elimination procedure.

Results

Relative distance of latrines to the nearest burrow (d_{rel})

The 'degree of urbanity' had a significant effect on mean distances of latrines to the next burrow system (d_{rel}; Table 3a), reflecting that distribution patterns of latrines shifted from core- to more periphery-biased along the rural-to-urban gradient. At rural sites, 13.5 ± 0.6 % of all

latrines (mean proportion ± SE) were located in the core section close to the burrow ($d_{rel} \leq 0.25$) and 25.3 ± 1.6 % at the relative periphery ($d_{rel} \geq 0.75$). By contrast, only 3.4 ± 1.1 % of latrines were established within the core section at urban study sites, while 34.6 ± 7.0 % of latrines was found at the periphery of the latrine-marked area. At suburban study sites, 11.7 ± 2.1 % of latrines were located in the core section and 33.2 ± 4.7 % at the periphery.

We also detected a significant effect of the 'degree of urbanity' on the dimensions of the latrine-marked area around rabbit burrows ('Latrine-marked area'; Table 3b), which decreased from 2.73 ± 0.48 ha at rural sites, over 2.11 ± 0.27 ha at suburban sites, to 0.87 ± 0.25 ha at urban study sites.

Latrine characteristics in relation to their distance to the nearest burrow

(a) Indicators of long-term latrine use

Latrine dimensions were affected by the 'degree of urbanity' and the interaction term 'd_{rel} × degree of urbanity' ('Latrine dimension'; Table 3c). While latrine dimensions at rural study sites became smaller with increasing distance from the next burrow (Fig. 3a), the opposite pattern was observed at urban study sites: latrines that were located at the relative periphery of the latrine-marked area were larger than those located close to the burrow (Fig. 3c). Regarding suburban sites, latrine sizes showed

Table 3 Univariate linear mixed models

Fixed effects	F	df_1, df_2	P
(a) d_{rel}			
'Degree of urbanity'	11.13	1, 93	0.001
(b) Latrine-marked area (A)			
'Degree of urbanity'	25.49	1, 126	<0.001
(c) Latrine dimension (PC on latrine size and numbers of fecal pellets)			
'Degree of urbanity'	3.04	1, 531	<0.001
'd_{rel}'	0.29	1, 2960	0.589
'd_{rel} × degree of urbanity'	5.33	1, 2870	<0.001
(d) Latrine density			
'Degree of urbanity'	10.67	1, 190	0.001
'd_{rel}'	34.74	1, 2953	<0.001
'd_{rel} × degree of urbanity'	5.26	1, 2900	0.022
(e) Numbers of fresh fecal pellets			
'Degree of urbanity'	0.77	1, 269	0.38
'd_{rel}'	0.91	1, 295	0.34
'd_{rel} × degree of urbanity'	0.98	1, 521	0.32
(f) Distance to next woody vegetation			
'Degree of urbanity'	11.31	1, 2973	0.001
'd_{rel}'	354.29	1, 2853	<0.001

Results of univariate LMMs using (a) 'd_{rel}', (b) 'latrine-marked area (A)', (c) 'latrine dimension', (d) 'latrine density', (e) 'numbers of fresh fecal pellets' and (f) 'distance to next woody vegetation' as dependent variables

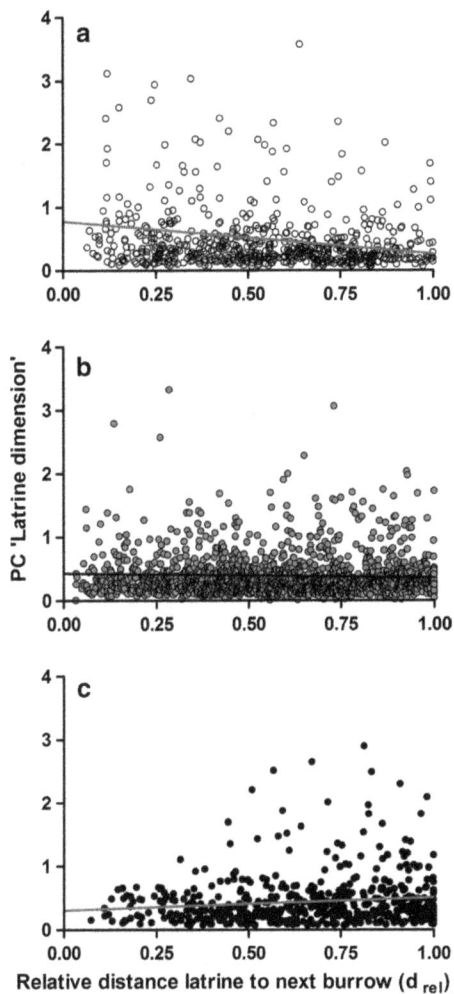

Fig. 3 Latrine dimension. Correlation between the PC 'latrine dimension' (incorporating the size of latrines [m²] and numbers of fecal pellets) and the relative distance to the next burrow (d_{rel}) at (**a**) rural sites with 'degree of urbanity' values ≤ -0.5 ($n = 547$ latrines), (**b**) suburban sites with 'degree of urbanity' values > -0.5 and ≤ 0.5 ($n = 1828$), and (**c**) urban sites with 'degree of urbanity' values > 0.5 ($n = 652$ latrines)

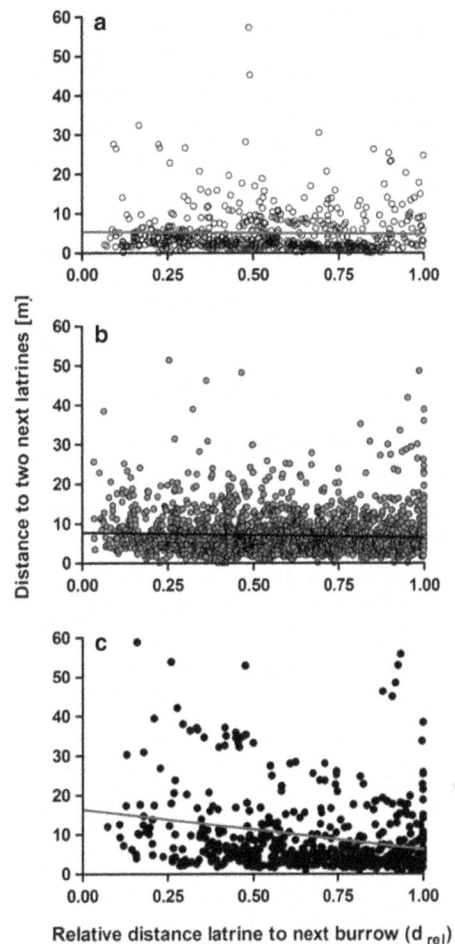

Fig. 4 Latrine density. Correlation between latrine density (expressed by the mean distance of a latrine to the nearest two neighboring latrines [m]) and the relative distance to the next burrow (d_{rel}) at (**a**) rural sites with 'degree of urbanity' values ≤ -0.5 ($n = 547$ latrines), (**b**) suburban sites with 'degree of urbanity' values > -0.5 and ≤ 0.5 ($n = 1828$), and (**c**) urban sites with 'degree of urbanity' values > 0.5 ($n = 652$ latrines)

no notable variation within the latrine-marked area (Fig. 3b).

Considering latrine densities, the 'degree of urbanity', 'd_{rel}' and the interaction term '$d_{rel} \times$ degree of urbanity' had significant effects ('Latrine density'; Table 3d). The latrine density decreased slightly with increasing distance from the next burrow system at rural study sites (Fig. 4a). By contrast, at urban sites latrine densities were considerably higher at the relative periphery of the latrine-marked area compared to latrines located close to the burrow (Fig. 4c). At suburban study sites, latrine densities did not vary throughout the latrine-marked area (Fig. 4b).

(b) Indicator of recent latrine use

As an estimate of the frequency of recent latrine use, we analyzed presence of fresh fecal pellets in each latrine. The logistic regression revealed a negative correlation between the 'degree of urbanity' and the presence of fresh fecal pellets within latrines ($B = -0.17$, $Wald = 13.96$, $SE = 0.046$, $P < 0.001$, -2log likelihood $= 2884.71$, Nagelkerke $R^2 = 0.007$; all excluded variables: $P \geq 0.29$), suggesting that the proportion of latrines that contain fresh fecal pellets decreased along the rural-to-urban gradient. Considering only the subset of latrines that contained fresh fecal pellets, our mixed model revealed no significant relations between the dependent and independent variables ('Numbers of fresh fecal pellets'; Table 3e).

(c) Indicator of territorial behavior at latrines

Regarding the presence of paw-scrapings at latrines the logistic regression uncovered a positive correlation with the 'degree of urbanity': the proportion of latrines at which paw-scrapings were present increased along the rural-to-urban gradient ($B = 0.57$, $Wald = 176.27$, $SE = 0.043$, $P < 0.001$, -2log likelihood $= 3637.78$, Nagelkerke $R^2 = 0.083$; all excluded variables: $P \geq 0.38$). In 76.6 ± 2.0 % of all latrines mapped at urban study sites paw-scrapings were present (mean percent latrines with paw-scrapings present $\pm$ SE), while this was only the case in 43.8 ± 1.8 % of all latrines at rural study sites and 70.4 ± 1.1 % of latrines at suburban sites.

(d) Effect of woody vegetation on latrine distributions

Finally, the distance of latrines to the next woody vegetation was affected by 'd_{rel}' and the 'degree of urbanity' ('Distance to next woody vegetation'; Table 3f). The distance between latrines and the next tree or shrub increased with increasing distance from the burrow, reflecting that most burrows were situated in dense vegetation. At core sections ($d_{rel} \leq 0.25$), the mean ($\pm$ SE) distance of latrines to the next woody vegetation was 8.72 ± 0.98 m ($n = 318$), while at the periphery ($d_{rel} \geq 0.75$) mean distances were 16.10 ± 0.58 m ($n = 896$). Along the rural-to-urban gradient, the mean distance of latrines to the next woody vegetation was shortest for urban areas (5.66 ± 0.69 m, $n = 652$) compared to rural (14.95 ± 0.65 m, $n = 547$) and suburban sites (16.83 ± 0.38 m, $n = 1828$).

Discussion

Our present study is the first to demonstrate gradual variation in the relative importance of different latrine marking strategies in European rabbit populations along a rural-to-urban gradient. The results comply with our prediction that higher rabbit population densities in urban regions, along with smaller group sizes (pairs and their offspring, and partly even solitary individuals [23, 31]), bring about an increased necessity for *between-group communication*, e.g., to claim territorial occupancy through peripheral marking. Not only were relatively more latrines located at the periphery of the rabbit burrow in urban populations, but those latrines were also larger in size, more densely packed and more frequently used. This trend contrasted with a strong signature of core marking in rural rabbit populations.

Fewer group members contributing to the establishment and maintenance of latrine-based communication networks in urban rabbit populations likely explain why the proportion of latrines with fresh fecal pellets was lower. Moreover, higher ambient temperatures and altered patterns of precipitation and evaporation are typical of urban regions—caused by the high proportion of sealed surfaces [48, 49]—possibly accelerating the decay of fecal pellets. Also, some fecal pellets will be regularly removed during the maintenance of green spaces, which, according to information provided by the *Frankfurter Grünflächenamt*, reaches its maximum in urban parks. Accordingly, using numbers of fecal pellets and fresh fecal pellets, respectively, as dependent variables to characterize latrine-based communication networks in urban, suburban and rural mammalian populations needs to be considered with caution. Likewise, those variables are sometimes used to estimate local rabbit population densities, which can also provide misleading information (see also [50]). Competition for space and other resources in the small and highly fragmented urban parks is probably intense, given that both the proportion of sealed surface areas and population densities were high, while home range areas marked by latrines were small. We argue that strong competition brings about an increased importance of peripheral marking behavior (see also [15–17]). This is also reflected by the fact that more paw-scrapings (which males use for territory demarcation) were found in latrines at urban study sites.

Following Domínguez-Cebrián and de Miguel [33], another important factor that likely affects latrine-based communication networks in rabbits is predation risk [33, 51]. Common predators of European rabbits in Germany can also reach high densities in cities (foxes [20]; mustelids like *Martes foina* and *Mustela erminea* [30]; domestic cats [52]; crows, *Corvus corone* and magpies, *Pica pica* that prey on juvenile rabbits [53]). However, the fact that those species can reach high densities in cities does not necessarily mean that they exert strong predation on urban rabbit populations ("the predation paradox" [54], reviewed in [22]). For example, several studies demonstrated that those predators can use other abundant food sources in cities [22, 55]. Moreover, both, predator and prey species can alter their activity patterns in urban regions, again leading to an altered predator exposure [56]. Unfortunately, we were not able to systematically quantify predation risk at our study sites. Still, decreased flight initiation distances in suburban and urban rabbits [31] and less time spent exhibiting anti-predator behavior [32] suggest that predation of urban and suburban rabbits may indeed be lower compared to rural populations. At rural sites, rabbits that use latrines at the periphery of their home ranges may be more exposed to predators, while reduced predation risk in urban populations leaves more time to establish and maintain complex communication networks involving latrines afar from the burrow.

When considering distances between latrines and the nearest woody vegetation, shorter distances in urban areas likely reflect more heterogeneous landscapes in cities [54, 56]. In contrast, rural study sites were mostly agriculturally used and are characterized by open and homogeneous landscapes with scarce woody vegetation. In line with the interpretation that sufficient shelter (shrubs and trees) eases burrow formation, a previous study found burrows to become more uniformly distributed along the rural-to-urban gradient considered here [23]. Rabbits prefer to establish latrines on bare soil, clearings, or elevated areas, often close to conspicuous landscape elements such as bushes, trees or anthropogenic objects, while avoiding densely vegetated areas [33, 36]. Not only does this increase the visibility and accessibility of latrines, but it could also reduce the risk of falling victim to avian and terrestrial predators during latrine visits [35]. At our rural study sites, most latrines were found on meadows with short grass, especially close to pathways, while crop fields were largely avoided. By contrast, landscape elements appear to not have such a strong effect on latrine distribution patterns at suburban and urban study sites, where meadows with short grass prevailed.

In contrast to European rabbits, groups of European badgers showed no peripheral marking behavior in urban regions—even at the few sites where the home ranges of different groups overlapped [25, 26]. Davison et al. [26] argued that urban badger groups were rather isolated even where population densities were high, reducing the need for territory demarcation (see also [25]). This was clearly not the case in our study, in which distinct social groups of rabbits occupied territories in close proximity to one another at urban and suburban study sites. Furthermore, crepuscular, timid species like badger are less likely to habituate to permanent anthropogenic disturbance compared to European rabbits (see above). Badgers are probably more distracted from latrine marking by human disturbance than rabbits (see also [4]). Moreover, badger home ranges are considerably larger than those of European rabbits (mean 95 % kernel group home range sizes of urban badgers: 4.71 [26] vs. 0.62 ha for suburban and urban European rabbit populations, unpubl. data). This renders peripheral marking in badgers even more challenging under intense anthropogenic disturbance.

Conclusions

Human activities affect urban wildlife populations, e.g., through anthropogenic nuisance, habitat fragmentation, and altered food availability (reviewed in [22, 29]). Behavioral changes in urban populations compared to populations inhabiting rural areas (like altered flight- or ranging behavior [22, 29]) are often interpreted as a *direct* consequence of animals having to cope with those novel ecological conditions. Our present study demonstrates behavioral changes in European rabbits, namely altered distribution patterns of latrines relative to the corresponding burrow. Based on previous studies on this and other mammalian species, we argue that increased peripheral marking in urban populations reflects an increased importance of *between-group communication* (rather than *within-group communication*), and this seems to be a consequence of higher population densities, smaller group sizes, and altered predation risk. Our study adds to our knowledge about the function of mammalian latrines as centers for information exchange between individuals, and—more generally—points towards *indirect* effects of anthropogenic landscape alteration and human nuisance on the behavior of urban wildlife populations. If our interpretations are correct, our results have implications for the conservation and management of rabbit populations: while rural rabbit populations suffer from a loss of suitable habitat [23, 31, 41–44], rabbit populations in urban areas might show higher intrinsic mortality rates arising from high intraspecific competition, while suburban habitats may currently provide an advantageous combination of structural heterogeneity and comparatively low levels of competition. Ongoing studies are trying to assess the potential role of cities in the future conservation of this species, e.g., by providing population genetic information on potential source-sink dynamics in population development. Another aspect to be considered in future studies is that urban and suburban rabbit populations may serve as ecosystem engineers; e.g., nutrients accumulate at latrines, which could have implications for local plant communities and possibly seed dispersal [56, 57]. As "fertile islands", latrines likely further increase habitat heterogeneity in urban and suburban landscapes [57].

Abbreviations
a.M.: am Main; d_{max}: maximum distance of a latrine to the center of the next burrow; d_{rel}: relative distance of a latrine to the center of the next burrow; d_{abs}: absolute distance of a latrine to the center of the next burrow.

Authors' contributions
Conceived and designed the experiments: MZ, SM, TW and MP. Collected data: MZ, SB, A-LB, MB, SW. Analyzed the data: MZ, DB, BG. Contributed analysis tools: BG. Wrote the paper: MZ, TW and MP. All authors read and approved the final manuscript.

Author details
[1] Department of Ecology and Evolution, Goethe University Frankfurt, Max-von-Laue-Str. 13, 60439 Frankfurt am Main, Germany. [2] Department of Biology and Ecology of Fishes, Leibniz-Institute of Freshwater Ecology and Inland Fisheries, Müggelseedamm 310, 12587 Berlin, Germany. [3] Department for Applied Bioinformatics, Goethe University Frankfurt, Max-von-Laue-Str. 13, 60439 Frankfurt am Main, Germany. [4] Department of Zoology, State Museum of Natural History Stuttgart, Rosenstein 1, 70191 Stuttgart, Germany. [5] Bristol Zoological Society, Conservation Science, Clifton, Bristol BS8 3HA, UK. [6] College of Animal Science and Technology, Northwest A&F University, Yangling 712100, Shaanxi, China.

Acknowledgements

T. Dieckmann from the *Frankfurter Grünflächenamt*, Dr. M. Wolfsteiner from the *Einwohnermeldeamt Frankfurt*, and P. Winkemann from the *Stadtvermessungsamt Frankfurt* kindly provided information and map material. Thanks are rendered to B. G. Atak, D. Babitsch, S. Hornung, S. Kriesten, A. Schieß, M.-L. Schrödl, S. Straskraba and M. Weinhardt for their help during field work. We would like to express our gratitude to the handling editor and both reviewers for their thoughtful comments that greatly helped improve previous manuscript versions.

Competing interests

The authors declare that they have no competing interests.

References

1. MacDonald DW. Patterns of scent marking with urine and faeces amongst carnivore communities. Symp Zool Soc Lond. 1980;45:107–39.
2. Gorman ML, Trowbridge BJ. The role of odor in the social lives of carnivores. In: Gittleman JL, editor. Carnivore behavior, ecology, and evolution. Berlin: Springer; 1989. p. 57–88.
3. Gorman ML. Scent marking strategies in mammals. Rev Suisse Zool. 1990;97:3–29.
4. Dröscher I, Kappeler PM. Maintenance of familiarity and social bonding via communal latrine use in a solitary primate (*Lepilemur leucopus*). Behav Ecol Sociobiol. 2014;68:2043–58.
5. Roper TJ, Shepherdson DJ, Davies JM. Scent marking with faeces and anal secretion in the European badger (*Meles meles*): seasonal and spatial characteristics of latrine use in relation to territoriality. Behaviour. 1986;97:94–117.
6. Roper TJ, Conradt L, Butler J, Christian SE, Ostler J, Schmid TK. Territorial marking with faeces in badgers (*Meles meles*): a comparison of boundary and hinterland latrine use. Behaviour. 1993;127:289–307.
7. Irwin MT, Samonds KE, Raharison JL, Wright PC. Lemur latrines: observations of latrine behavior in wild primates and possible ecological significance. J Mammal. 2004;85:420–7.
8. Jordan NR, Cherry MI, Manser MB. Latrine distribution and patterns of use by wild meerkats: implications for territory and mate defence. Anim Behav. 2007;73:613–22.
9. Wronski T, Apio A, Plath M. The communicatory significance of localised defecation sites in bushbuck (*Tragelaphus scriptus*). Behav Ecol Sociobiol. 2006;60:368–78.
10. Kruuk H. Spatial organization and territorial behavior of the European badger *Meles meles*. J Zool. 1978;184:1–19.
11. Wronski T, Plath M. Characterization of the spatial distribution of latrines in reintroduced mountain gazelles (*Gazella gazella*): do latrines demarcate female group home ranges? J Zool. 2010;280:92–101.
12. Wronski T, Apio A, Plath M, Ziege M. Sex difference in the communicatory significance of localized defecation sites in Arabian gazelles (*Gazella arabica*). J Ethol. 2013;31:129–40.
13. Mykytowycz R, Gambale S. The distribution of dung-hills and the behavior of free living wild rabbits, *Oryctolagus cuniculus* (L.), on them. Forma Funct. 1969;1:333–49.
14. Mykytowycz R, Hesterman ER, Gambale S, Dudziński ML. A comparison of the effectiveness of the odors of rabbits, *Oryctolagus cuniculus*, in enhancing territorial confidence. J Chem Ecol. 1976;2:13–24.
15. Schley L, Schaul M, Roper TJ. Distribution and population density of badgers *Meles meles* in Luxembourg. Mamm Rev. 2004;34:233–40.
16. Hutchings MR, Service KM, Harris S. Is population density correlated with faecal and urine scent marking in European badgers (*Meles meles*) in the UK? Mamm Biol. 2002;67:286–93.
17. Myers K, Poole WE. A study of the biology of the wild rabbit, *Oryctolagus cuniculus* (L.), in confined populations. I. The effects of density on home range and the formation of breeding groups. CSIRO Wildl Res. 1959;4:14–26.
18. Gosling LM, Roberts SC. Testing ideas about the function of scent marks in territories from spatial patterns. Anim Behav. 2001;62:F7–10.
19. Francis RA, Chadwick MA. What makes a species synurbic? Appl Geogr. 2012;32:514–21.
20. Gloor S, Bontadina F, Hegglin D, Deplazes P, Breitenmoser U. The rise of urban fox populations in Switzerland. Mamm Biol. 2001;66:155–64.
21. Prange S, Gehrt SD, Wiggers EP. Demographic factors contributing to high raccoon densities in urban landscapes. J Wildl Manag. 2003;67:324–33.
22. Rodewald AD, Gehrt SD. Wildlife Population Dynamics in Urban Landscapes. In: McCleery RA, Moorman CE, Peterson MN, editors. Urban wildlife conservation—theory and praxis. Berlin: Springer; 2014. p. 117–47.
23. Ziege M, Brix M, Schulze M, Seidemann A, Straskraba S, Wenninger S, Streit B, Wronski T, Plath M. From multifamily residences to studio apartments—shifts in burrow structures of European rabbits along a rural-to-urban gradient. J Zool. 2015;295:286–93.
24. Harris S. Activity patterns and habitat utilization of badgers (*Meles meles*) in suburban Bristol: a radio tracking study. In: symposia of the zoological society of London. Vol. 49. Published for the Zoological Society by Academic Press; 1982. p. 301–23.
25. Cresswell WJ, Harris S. Foraging behavior and home-range utilization in a suburban badger (*Meles meles*) population. Mamm Rev. 1988;18:37–49.
26. Davison J, Huck M, Delahay RJ, Roper TJ. Restricted ranging behavior in a high-density population of urban badgers. J Zool. 2009;277:45–53.
27. Bozek CK, Prange S, Gehrt DS. The influence of anthropogenic resources on multi-scale habitat selection by raccoons. Urban Ecosyst. 2007;10:413–25.
28. Adkins CA, Stott P. Home ranges, movements and habitat associations of red foxes *Vulpes vulpes* in suburban Toronto, Ontario, Canada. J Zool. 1998;244:335–46.
29. Ryan AM, Partan SR. Urban wildlife behavior. In: McCleery RA, Moorman CE, Peterson MN, editors. Urban wildlife conservation—theory and praxis. Berlin: Springer; 2014. p. 149–73.
30. Duduś L, Zalewski A, Kozioł O, Jakubiec Z, Król N. Habitat selection by two predators in an urban area: the stone marten and red fox in Wrocław (SW Poland). Mamm Biol. 2014;79:71–6.
31. Ziege M, Babitsch D, Brix M, Kriesten S, Seidemann A, Wenninger S, Plath M. Anpassungsfähigkeit des Europäischen Wildkaninchens entlang eines rural-urbanen Gradienten. Beiträge zur Jagd- und Wildtierforsch. 2013;38:189–99.
32. Ziege M, Babitsch D, Brix M, Kriesten S, Straskraba S, Wenninger S, Wronski, T, Plath M. Extended diurnal activity patterns of European rabbits along a rural–to–urban gradient. Submitted to Mamm Biol—Zeitschrift für Säugetierkd. 2016.
33. Domínguez-Cebrían I, de Miguel FJ. Selected factors influencing the spatial relationship between latrines and burrows in rabbits *Oryctolagus cuniculus* (L.) in a suburban area of Madrid (Spain). Pol J Ecol. 2013;61:819–23.
34. Mykytowycz R. Territorial marking by rabbits. Sci Am. 1968;218:116–26.
35. Monclús R, de Miguel FJ. Distribución espacial de las letrinas de conejo (*Oryctolagus cuniculus*) en el Monte de Valdelatas (Madrid). Galemys. 2003;15:157–65.
36. Sneddon IA. Aspects of olfaction, social behavior and ecology of an island population of the European rabbit (*Oryctolagus cuniculus*). Ph.D. Thesis, Scotland: St. Andrews University; 1984.
37. Sneddon IA. Latrine use by the European rabbit (*Oryctolagus cuniculus*). J Mammal. 1991;72:769–75.
38. Hesterman ER, Mykytowycz R. Some observations on the odours of anal gland secretions from the rabbit, *Oryctolagus cuniculus* (L.). CSIRO Wildl Res. 1968;13:71–81.
39. Mykytowycz R. Territorial function of chin gland secretion in the rabbit, *Oryctolagus cuniculus* (L.). Nature. 1962;193:799.
40. Mykytowycz R, Hesterman ER. The behavior of captive wild rabbits, *Oryctolagus cuniculus* (L.) in response to strange dung-hills. Forma Funct. 1970;2:1–12.
41. Lees AC, Bell DJ. A conservation paradox for the 21st century: the European wild rabbit *Oryctolagus cuniculus*, an invasive alien and an endangered native species. Mamm Rev. 2008;38:304–20.
42. Ferreira C, Touza J, Rouco C, Díaz-Ruiz F, Fernandez-de-Simon J, Ríos-Saldaña CA, Ferreras P, Villafuerte R, Delibes-Mateos M. Habitat management as a generalized tool to boost European rabbit *Oryctolagus cuniculus* populations in the Iberian Peninsula: a cost-effectiveness analysis. Mamm Rev. 2014;44:30–43.
43. Arnold JM, Greiser G, Keuling O, Martin I, Straus E. Status und Entwicklung ausgewählter Wildtierarten in Deutschland. Jahresbericht 2012. Wildtier-Informationssystem der Länder Deutschlands (WILD), Berlin.

44. Virgós E, Cabezas-Díaz S, Malo A, Lozano J, López-Huertas D. Factors shaping European rabbit abundance in continuous and fragmented populations of central Spain. Acta Theriol (Warsz). 2003;48:113–22.

45. Ruiz-Aizpurua L, Planillo A, Carpio AJ, Guerrero-Casado J, Tortosa FS. The use of faecal markers for the delimitation of the European rabbit's social territories (Oryctolagus cuniculus L.). Acta Ethol. 2013;16:157–62.

46. Eisermann K. Long-term heart rate responses to social stress in wild European rabbits: predominant effect of rank position. Physiol Behav. 1992;52:33–6.

47. Bell DJ. Aspects of the social behavior of wild and domesticated rabbits (Oryctolagus cuniculus). Cardiff: University of Wales; 1977.

48. Pickett STA, Cadenasso ML, Grove JM, Nilon CJ, Pouyat RV, Zipperer WC, Costanza R. Urban ecological systems: linking terrestrial, ecological, physical, and socioeconomic components of metropolitan areas. Annu Rev Ecol Syst. 2001;32:127–57.

49. Knapp S, Kühn I, Schweiger O, Klotz S. Challenging urban species diversity: contrasting phylogenetic patterns across plant functional groups in Germany. Ecol Lett. 2008;11:1054–64.

50. Barrio IC, Acevedo P, Tortosa FS. Assessment of methods for estimating wild rabbit population abundance in agricultural landscapes. Eur J Wildl Res. 2010;56:335–40.

51. Villafuerte R, Moreno S. Predation risk, cover type, and group size in European rabbits in Doñana (SW Spain). Acta Theriol (Warsz). 1997;42:225–30.

52. Baker PJ, Molony SE, Stone E, Cuthill IC, Harris S. Cats about town: is predation by free-ranging pet cats Felis catus likely to affect urban bird populations? Ibis (Lond 1859). 2008;150:86–99.

53. von Holst D, Hutzelmeyer H, Kaetzke P, Khaschei M, Rödel HG, Schrutka H. Social rank, fecundity and lifetime reproductive success in wild European rabbits (Oryctolagus cuniculus). Behav Ecol Sociobiol. 2002;51:245–54.

54. Shochat E, Warren PS, Faeth SH, McIntyre NE, Hope D. From patterns to emerging processes in mechanistic urban ecology. Trends Ecol Evol. 2006;21:186–91.

55. Contesse P, Hegglin D, Gloor S, Bontadina F, Deplazes P. The diet of urban foxes (Vulpes vulpes) and the availability of anthropogenic food in the city of Zurich, Switzerland. Mamm Biol. 2004;69:81–95.

56. Delibes-Mateos M, Delibes M, Ferreras P, Villafuerte R. Key role of European rabbits in the conservation of the Western Mediterranean Basin Hotspot. Conserv Biol. 2008;22:1106–17.

57. Willott SJ, Miller AJ, Incoll LD, Compton SG. The contribution of rabbits (Oryctolagus cuniculus L.) to soil fertility in semi-arid Spain. Biol Fert Soils. 2000;31:379–84.

58. Ziege M, Bierbach D, Bischoff S, Brandt AL, Brix M, Greshake B, Merker S, Wenninger S, Wronski T, Plath M. Field data from latrine mapping of rabbit populations located within urban, suburban and rural study sites in and around Frankfurt am Main. BMC Ecol. 2016. http://www.dx.doi.org/. 10.5061/dryad.8s3p0.

12

Vole abundance and reindeer carcasses determine breeding activity of Arctic foxes in low Arctic Yamal, Russia

Dorothee Ehrich[1]* ⓘ, Maite Cerezo[1], Anna Y. Rodnikova[2], Natalya A. Sokolova[3,4], Eva Fuglei[5], Victor G. Shtro[3] and Aleksandr A. Sokolov[3,4]

Abstract

Background: High latitude ecosystems are at present changing rapidly under the influence of climate warming, and specialized Arctic species at the southern margin of the Arctic may be particularly affected. The Arctic fox (*Vulpes lagopus*), a small mammalian predator endemic to northern tundra areas, is able to exploit different resources in the context of varying tundra ecosystems. Although generally widespread, it is critically endangered in subarctic Fennoscandia, where a fading out of the characteristic lemming cycles and competition with abundant red foxes have been identified as main threats. We studied an Arctic fox population at the Erkuta Tundra Monitoring site in low Arctic Yamal (Russia) during 10 years in order to determine which resources support the breeding activity in this population. In the study area, lemmings have been rare during the last 15 years and red foxes are nearly absent, creating an interesting contrast to the situation in Fennoscandia.

Results: Arctic fox was breeding in nine of the 10 years of the study. The number of active dens was on average 2.6 (range 0–6) per 100 km² and increased with small rodent abundance. It was also higher after winters with many reindeer carcasses, which occurred when mortality was unusually high due to icy pastures following rain-on-snow events. Average litter size was 5.2 (SD = 2.1). Scat dissection suggested that small rodents (mostly *Microtus* spp.) were the most important prey category. Prey remains observed at dens show that birds, notably waterfowl, were also an important resource in summer.

Conclusions: The Arctic fox in southern Yamal, which is part of a species-rich low Arctic food web, seems at present able to cope with a state shift of the small rodent community from high amplitude cyclicity with lemming dominated peaks, to a vole community with low amplitude fluctuations. The estimated breeding parameters characterized the population as intermediate between the lemming fox and the coastal fox ecotype. Only continued ecosystem-based monitoring will reveal their fate in a changing tundra ecosystem.

Keywords: Food web, Numerical response, Reindeer carcasses, Small rodent community, Vole cycle, Diet, *Vulpes lagopus*

Background

Arctic ecosystems are at present changing rapidly under the influence of climate warming [1]. At the southern margin of the Arctic, temperatures now often exceed those characteristic for the Arctic [2], rain falling in winter hardens the snow pack [3], tall shrubs and boreal species expand, whereas typical Arctic species are impacted negatively [4, 5]. The Arctic fox (*Vulpes lagopus*) is a widespread and common Arctic predator, which is endemic to the circumpolar tundra areas. It has been chosen as one of ten flagship species for the impact of climate change highlighted by the International Union for Conservation of Nature [6], and the species is a good candidate to become the object of coordinated circumpolar

*Correspondence: dorothee.ehrich@uit.no
[1] Department of Arctic and Marine Biology, University of Tromsø-The Arctic University of Norway, 9037 Tromsø, Norway
Full list of author information is available at the end of the article

monitoring [7] as asked for by the Arctic Terrestrial Bio-diversity Monitoring Plan [8]. Given the potential vulner-ability of the species to changes in the prey base and the availability of competitors [5], it is important to under-stand what drives the dynamics of Arctic fox populations in different ecological contexts, in particular in the low Arctic.

Adapted to survive scarcity and extreme cold in lit-tle productive ecosystems [9], Arctic foxes are able to exploit many different resources including lemmings (*Lemmus* sp. and *Dicrostonyx* sp.), birds, as well as marine resources or ungulate carcasses [10]. Depend-ing on the prevailing prey base, two main ecotypes have been described [11, 12]: lemming foxes and coastal foxes. Lemming or inland foxes feed preferably on lemmings, but switch to alternative prey such as birds in low lem-ming years [13]. They maximize reproductive effort in peak years with very large litters, but may skip breeding or breed poorly in years with low lemming abundance [14, 15]. This results in clear population fluctuations, which follow the lemming cycle [16]. A recent survey of Arctic fox monitoring initiatives revealed that strong multiannual fluctuations were prevalent in the majority of populations, and that most of these populations feed on lemmings [7]. This was also the case on Yamal Pen-insula, Russia, during the 1970s and 1980s, when large-scale surveys showed that Arctic fox den occupancy and the percentage of pregnant females among hunted foxes were related to lemming abundance [17].

Coastal foxes, on the contrary, rely mostly on more sta-ble marine resources such as sea bird colonies. In coastal populations, for instance on Iceland or on Mednyi Island in the Bering Sea, breeding occurs nearly every year, but litters are smaller [11, 18]. In the high Arctic Sval-bard archipelago, Norway, where native populations of small rodents are absent, Arctic foxes exploit sea birds, reindeer (*Rangifer tarandus*) carcasses and geese, with highest reproductive output in dens close to seabird colonies [19]. Year to year variation in den occupancy is mainly driven by the availability of reindeer carcasses in late winter [20, 21]. Substantial variation in the den-sity of breeding pairs in the absence of small rodents was also observed on Kolguev Island, Russia, where Arctic foxes feed mainly on geese in summer [22]. Consider-ing the varied resources used by non-lemming foxes and the diverse dynamics observed in such populations, Eide et al. [19] suggested moving beyond the distinc-tion between lemming and coastal foxes when studying resource dependency of Arctic foxes, and rather investi-gating which main resources drive the fox dynamics in a specific ecosystem context (see also [23]).

The status of Arctic fox populations close to the south-ern margin of the Arctic tundra biome varies between areas depending on specific ecological processes. The coastal fox populations in Iceland, where red foxes (*Vulpes vulpes*) are absent, have been growing until recently due to an increase in carrying capacity attrib-uted to growing populations of marine birds and geese [24, 25]. Mainland populations, on the contrary, are particularly exposed to increased pressure from com-petitively superior red foxes expanding northwards. This is the case in Fennoscandia, where Arctic foxes are critically endangered [26]. On Varanger Peninsula, northeastern Norway, two main drivers have been iden-tified for the decline of the Arctic fox: (1) an increase in the population of red foxes, which is subsidized by car-rion of semi-domestic reindeer [27]; and (2) a scarcity of lemmings due to increasing irregularity of their con-spicuous peak years [5]. A fading out of the characteristic lemming cycles leading to low density populations with detrimental consequences for specialized Arctic preda-tors has indeed been observed in several regions of the Arctic and attributed to changing winter climate [4, 28, 29]. Voles (*Myodes rutilus* and *Microtus oeconomus*), on the contrary, are abundant on Varanger Peninsula and exhibit population cycles with a period of 4–5 years [29]. However, they seem not able to replace lemmings as a resource, as Arctic fox reproduction did not respond to a vole peak without lemmings in 2015 [5]. Presently, little is known about the drivers of Arctic fox populations at the southern margin of the vast Eurasian tundra in Russia.

Here we present data from a 10-year study of Arc-tic fox at the Erkuta Tundra Monitoring site in Yamal, Russia. The aim of the study was to identify the main resources driving the dynamics of this low Arctic popula-tion, experiencing at present little competition from red foxes [30]. To answer this question, we first determined factors explaining variation in breeding activity, and sec-ond assembled available data about the diet of foxes dur-ing the breeding season and assessed whether it varied according to resource availability. We hypothesized that breeding productivity in this inland population would be related to the small rodent dynamics, in particular lem-ming abundance, as described for Fennoscandia [5, 31] and for southern Yamal during the 1980s [17]. However, we also assessed the importance of other resources avail-able in the study area in late winter, at the time when Arctic fox females initiate breeding [19], such as willow ptarmigan (*Lagopus lagopus*), mountain hare (*Lepus tim-idus*) and reindeer carcasses.

Methods
Study area and components of the vertebrate food web
The Erkuta Tundra Monitoring site is situated in the southern part of Yamal Peninsula (Russia) close to the confluence of the Payutayakha and Erkutayakha rivers

(68.2°N, 69.2°E; Fig. 1). This low Arctic area is characterized by a tundra landscape with gently rolling hills (ca 30 m high), including some steep slopes and sandy cliffs along riverbanks and lakes. Mean temperature in the area is −24.1 °C in January and 11.4 °C in July, and mean annual precipitation is about 335 mm (averages for the period 1960–1990; from [32]). The substrate consists of sandy and clayey sediments that provide good opportunities for den excavation by Arctic foxes. Permafrost is continuous [33]. Numerous water bodies sustain extensive wetlands, and dense thickets of tall shrubs more than 2 m high (willows *Salix sp.* and some alder *Alnus fruticosa*) occur along rivers and lakes. The main vegetation consists of low shrub tundra and erect dwarf shrub tundra [34].

Arctic fox is the most common mammalian predator, but least weasel (*Mustela nivalis*), stoat (*M. erminea*), wolverine (*Gulo gulo*) and wolf (*Canis lupus*) are also present. Red foxes were rare in the beginning of the study in 2007 [35], but the first two breeding events were recorded in 2014 [30]. The most common birds of prey in the area are rough-legged buzzards (*Buteo lagopus*), peregrine falcons (*Falco peregrinus*), and long-tailed and Arctic skuas (*Stercorarius longicaudus* and *S. parasiticus*). Raven (*Corvus corax*) have been breeding in the study area since 2009 and we recorded hooded crow (*Corvus cornix*) breeding for the first time in 2014 [30]. In total, 40 species of migratory birds, including numerous passerines, waders and waterfowl, are breeding in the study area [36]. Several species of geese are present

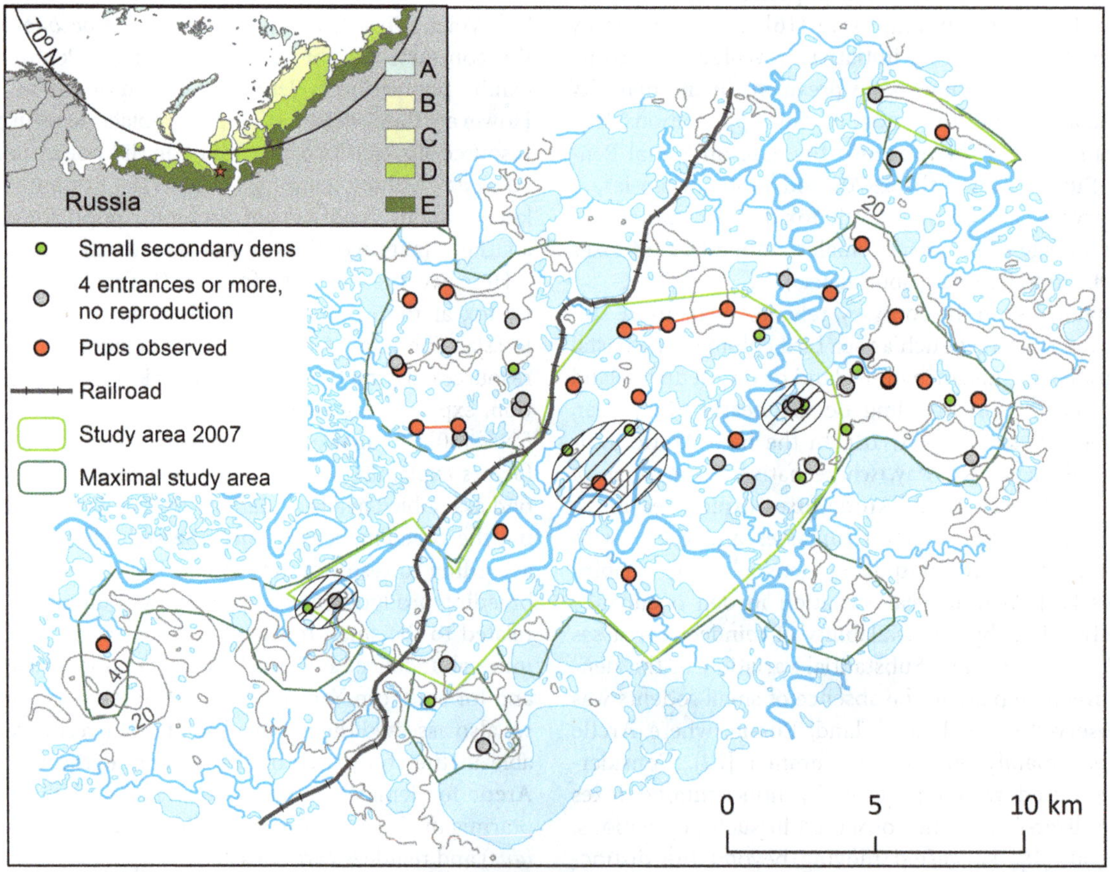

Fig. 1 Map of the Erkuta tundra monitoring site in southern Yamal, Russia. The inset indicates the location of the area in the western Eurasian Arctic (represented by the red star) and shows the five bioclimatic subzones of the Arctic according to [2]. Subzones A–C represent the high Arctic whereas subzones D and E represent the low Arctic. The hatched ellipses show the three replicate areas (units), where herbivore faeces counts and small rodent trapping were carried out. All fox dens are shown. Red dots represent dens where pups have been observed during the study period, and red lines link dens between which fox families have moved. The green lines show the extent of the study area in the first year of the study (2007) and the maximal extent of the study area. Note, however, that in most years the area actually surveyed was somewhat less than this maximal area

in rather low numbers, but ducks are numerous both as breeders and as non-breeders on rivers and lakes.

The vertebrate food web comprises many species of herbivores. The small rodent community consists of five species: Narrow-sculled voles (*Microtus gregalis*) and Middendorff's voles (*M. middendorffii*) are most abundant. Other species are collared lemming (*Dicrostonyx torquatus*), Siberian lemming (*Lemmus sibiricus*), and red-backed vole (*Myodes rutilus*) [37]. During the last decade, small rodent fluctuations were of low amplitude and rather low densities [37]. The last high amplitude small rodent peak was recorded in 1999 [38], but earlier high amplitude population cycles with a period of 3–5 years were occurring in the area and lemmings were more common (Shtro [17]). The most abundant resident medium-sized herbivores are willow ptarmigan and mountain hare. The muskrat (*Ondatra zibethica*) is also present. Domestic reindeer, herded traditionally by Nenets people, are the only large herbivore. According to official statistics there were approximately 300,000 reindeer on Yamal Peninsula in 2013, and our study area is used by herds in all seasons. During the last decade, extensive ground icing in winter resulting from heavy rain-on-snow events (ROS [3]) caused unusually high reindeer mortality during two winters: 2006–2007 [39] and 2013–2014 [30]. The most recent event caused the death of 40,000 reindeer in Yamalskyi district according to official statistics and had dramatic consequences for local herders. The high availability of reindeer carrion in that year was confirmed by numerous observations of reindeer carcasses in our study area.

Monitoring of herbivore populations

Small rodent dynamics were monitored from 2007 to 2016 by snap trapping, which was carried out according to the small quadrat method [40]. Three traps baited with raisins and rolled oats were placed at each corner of a 15×15 m quadrat for two nights in the second part of June and in the beginning of August ($2 \times 12 = 24$ trap nights per quadrat). We placed quadrats in three habitats (willow thicket edge, wet tundra and dry tundra; see [37] for a description of the habitats) and replicated the design in two spatial units from 2007 to 2011, resulting in $2 \times 3 \times 6 = 36$ quadrats in total (864 trap nights) per session. Since 2012 we surveyed three spatial units (54 quadrats = 1296 trap nights per session), and in 2007 and 2016 we carried out only one trapping session for logistic reasons. The data were summarized as total number of individuals of lemmings and voles respectively, trapped per 24 trap nights and averaged over units and habitats. We assumed that these indices reflected the relative changes in abundance over time of the respective small

rodents in our study area, although they are not estimates of density.

We carried out faeces counts on permanent removal plots according to the same design as above to obtain an estimate of the relative presence and activity of ptarmigan and hare. The 15×15 m trapping quadrats were surrounded by eight small plots of 50×50 cm where faeces were counted [41]. The data were summarized as faeces occurrence, i.e. the number of small plots with presence of faeces among the eight small plots surrounding one quadrat, and averaged over habitats and units. This resulted in an index of overall relative presence of these species in the study area, similar to the small rodent index described above. In 2007, 2009 and 2016, because of logistic reasons, faeces were counted only once.

Arctic fox den survey

From 2007 to 2016, we carried out systematic fox den surveys each summer. We started with a core area of ca 130 km^2 in 2007, comprising most breeding dens that were known at that time [35]. In subsequent years, we progressively enlarged the study area and searched for more dens. By 2014, the area covered was 230 km^2. To the extent possible given logistic constraints, we visited all known dens annually. Dens were first observed from a distance with binoculars to check for the presence of Arctic foxes. Subsequently, we walked to the den and recorded the number of entrances, whether these were cleaned and showed traces of recent digging, footprints, prey remains, and the presence of pups in the den (sounds). A den was considered active (i.e. inhabited by a breeding fox family) if pups were seen or clear sounds of pups were heard from the den. The minimum number of pups was determined for most breeding dens either by observing the den from a distance over a period of at least 5 h (or until an adult brought food and the pups emerged), or by using an automatic camera on the den. We fixed an automatic camera (Reconyx PC85/PC800; Reconyx Inc., Holmen, WI, USA) on wooden poles and placed it for a period of between 1 and 5 weeks at approximately 2–8 m from den entrances in a position providing a good overview of the den. The cameras used a motion sensor programmed to high sensitivity, and were taking ten pictures for each trigger. We visited active dens between one and five times during the summer.

Arctic fox diet

We investigated the diet of Arctic foxes using three complementary methods: scat dissection, description of prey remains on active dens and analysis of stable isotopes. Scat dissection is likely to lead to an overrepresentation of small prey such as small rodents, which are consumed with identifiable bones and teeth, whereas larger prey such as hare or reindeer carcasses, from which mostly meat is consumed, are likely to be underrepresented. Larger prey, on the contrary, are better represented in the prey remains than small rodents, which are often consumed whole or taken into the den [22, 42]. Stable isotopes reflect the mixture of resources consumed over a certain period, and have a lower resolution than the two previous methods, because they can only distinguish between resources with distinct isotopic signatures [43].

Scat dissection

We collected fresh scats on breeding dens in the summers of 2007, 2013 and 2014, several times during the summer when possible. All scats were stored in a freezer at −80 °C before analysis to prevent human exposure to eggs from the tapeworm *Echinococcus multilocularis*. In 2007, scats from seven dens were obtained, and the total material collected on a den at a particular date was analyzed as one bulk sample. Scats were soaked in water, fragmented by hand and remains of rodents (fur, bones, teeth; including muskrat), birds (feathers and bones), reindeer (fur), hare (fur, identified by comparing with reference samples), fish, insects and plants were visually sorted, using a magnifying glass when necessary. Subsequently, remains were dried at 60 °C for 2 days, and weighed. Results were summarized as percent dry weight. In 2013 and 2014, a different protocol was used, because the data resulted from a different student project: 21 scats per den and collection date were analyzed individually [44]. After soaking in water with laundry soap, solid parts were separated by washing through a sieve (mesh size 0.5 mm). As above, the scat material was visually sorted into the categories small or medium sized mammal fur (small rodents, muskrat and hare), feathers, small rodent bones and teeth, bird bones, eggs, fish, insects and plants. We estimated percentage volume for the different remain categories according to the whole faeces equivalent approach of [15]. For all years, small rodent species were identified by examining the first molar of the lower jaw [45].

Prey remains

From 2010 to 2016, we recorded prey remains during visits to dens with clear signs of recent fox presence (but not only on breeding dens). On the first visit, we described all fresh remains, and removed them from the den. Items, which still contained food, were not removed, but registered in order to avoid counting them again on the next visit. Visits during which the observer did not pay attention to prey remains, but for instance only collected an automatic camera, were excluded from the analysis.

Stable isotopes

Winter fur was collected each summer at the entrance of dens or when encountered otherwise, and analyzed for stable isotopes of carbon ($\delta^{13}C$) and nitrogen ($\delta^{15}N$; see [10] for details about the methods). The fur shed in spring reflects the diet during the period when the foxes were molting to winter fur in the previous fall. Stable isotope signatures of the main prey species were available from the International Polar Year project "Arctic Predators" [10]. These consisted of muscle samples collected in summer or in fall primarily from 2007 to 2009 (see [10] for details about collecting tissue of the different species, and choosing and aggregating of prey signatures). A few additional prey signatures, notably from muskrat, were obtained more recently from samples collected opportunistically, for instance from remains on Arctic fox dens.

Statistical analysis

Data analysis was carried out in R version 3.2.2 [46]. The probability of a den to be active in a particular year was analyzed with GLMMs with a logit link and a binomial error distribution using the function glmer of the package lme4 in R [47]. Den identity (hereafter "Den ID") was included as a random effect in all models to account for differences in the quality of dens and the surrounding territories, and potentially important resources were considered as fixed effects. As lemmings are known to be a main driver of Arctic fox breeding dynamics (e.g. [14, 15], the average number of lemmings trapped in June was used as a proxy of lemming abundance in late winter when Arctic fox females initiate breeding. Because voles were much more abundant in the study area than lemmings, and the total amount of small rodents may be important, we also included the log of the average number of all small rodents trapped in June. Other resources present in late winter and considered possible determinants of breeding activity, were reindeer carcasses, ptarmigan and hare. Since no quantitative estimates of the number of reindeer carcasses were available for our study area, we used a factor with two levels: "high" abundance for the breeding seasons 2007 and 2014 (the 2 years where unusually high reindeer mortality was documented in the media and scientific literature [30, 39]) and "Usual" abundance of reindeer carcasses for the other years. For hare and ptarmigan, we used the average occurrence indices obtained from faeces counts in June, as the faeces, which accumulated since the previous August, reflect the presence of these herbivores in winter. We assembled a set of candidate models consisting of a model with an intercept only, and one or two additive fixed effects. We started with models including lemming or total small rodent abundance as the most likely driver of breeding activity, and then included other potential resources (reindeer, hare or ptarmigan). The most suitable model was chosen according to Akaike's information criterion corrected for small sample sizes (AICc). Models with a difference in AICc (ΔAICc) of <2 were considered equally adequate. The selected model was graphically checked for constant variance of residuals, presence of outliers and normality of the random effects.

We used a similar modelling approach for the minimum number of pups at the dens. Here a GLMM with a log link and a Poisson error distribution was used. We included only data about the number of pups that had been estimated either by thorough observation or with automatic cameras. As above, lemmings, total small rodent abundance, reindeer, hare and ptarmigan were used as additive fixed effects in candidate models. Den ID was included as a random effect in all models.

We summarized the scat dissection data as mean proportions (either dry weight or volume) of different prey categories per year, and as frequencies of occurrence of the main prey types. The observations of prey remains were presented as the proportion of den visits carried out each year, at which a certain category of prey was observed. We carried out a correspondence analysis using the function dudi.coa from the R package ade4 [48] to assess differences between years in the occurrences of prey remains.

The stable isotope data of Arctic fox fur were presented graphically by plotting the fox values together with the mean signatures of main prey groups. The fox signatures were corrected for isotopic discrimination using the factors determined by [49]. Linear mixed effects models (LMM) were used to investigate whether the stable isotope signatures of winter fur varied with small rodent abundance in late summer. For each isotope, we implemented a model with the total trapping index of small rodents in August of the year preceding the fur sample collection as a fixed factor, and the collection place of the sample as a random factor. Analyses were carried out using the function lmer in lme4, and models were graphically checked for constant variance of residuals, presence of outliers and normality of the random effects.

Results
Herbivore dynamics
The two dominant small rodent species, the narrow-sculled vole and Middendorff's vole, exhibited low amplitude multiannual density fluctuations and reached relative abundance peaks in 2010 and 2013 with 1.67 and 1.38 animals per 24 trap nights in August (Fig. 2a). Abundance always increased over the summer. The three other species occurred at lower densities. Collared lemmings were present in most years, whereas siberian lemmings

were nearly absent during the study period. Red-backed voles were rare and possibly increasing [37]. In total, the abundance index fluctuated with a factor 8 in spring and 9 in fall.

The average occurrence of ptarmigan faeces decreased slightly in the beginning of the period, and was highest in the years 2012–2014 (Fig. 2b). Hare faeces occurrence was in general higher and rather stable, with a suggested relative maximum in 2013 (Fig. 2b). The overall high abundance of hare and ptarmigan was corroborated by frequent observations of these species in the study area both in winter and in summer [41].

Den survey
A total of 59 fox dens were present in the study area (Fig. 1). The smallest den, in which reproduction was observed, had four entrances, thus 47 dens with four or more entrances were considered large enough for breeding. This resulted in an overall density of 17 dens per 100 km^2 (dens with four entrances or more). Most dens were located on the slopes of hills and often close to lakes. Over the 10 years of the study, 42 breeding events of Arctic foxes were recorded in 23 different dens. In at least four cases, the Arctic foxes moved their pups to a new den in the course of the summer (Fig. 1). The number of active breeding dens was on average 2.6/100 km^2 and varied between 0 in 2008 and 6 in 2007 (Fig. 3). The proportion of active dens, taking into account only the 25 dens where pups have been observed (23 dens with breeding and two dens to which pups have been moved later in the season), varied between 0 and 0.7. The number of pups observed per den was on average 5.2 (SD = 2.1, maximum 10 pups). Considering each year, the mean number of pups was lowest in 2010 and highest in 2015, when nine pups were observed on the single den where an estimate was obtained (Fig. 3).

The probability of a den being active in a particular year was best explained by a model with the log of small rodent abundance in June and the availability of reindeer carcasses as additive fixed effects. This model was notably better supported by AICc than models with lemming abundance in June, or lemming abundance in June and the availability of reindeer carcasses as fixed effects (ΔAICc = 12.9 and ΔAICc = 3.1 respectively; Additional file 1: Table S1). Thus, more foxes were breeding after winters with high reindeer mortality (odds ratio = 5.2, 95% confidence interval CI = 2.1–13.7; Table 1; Fig. 4) and in years with higher small rodent abundance in June (odds ratio = 2.8, CI = 1.4–6.6 for an increase of the log of the trapping index by 1).

Models including an effect of ptarmigan or hare in addition to small rodents received lower support (Additional file 1: Table S1). For the minimum number of

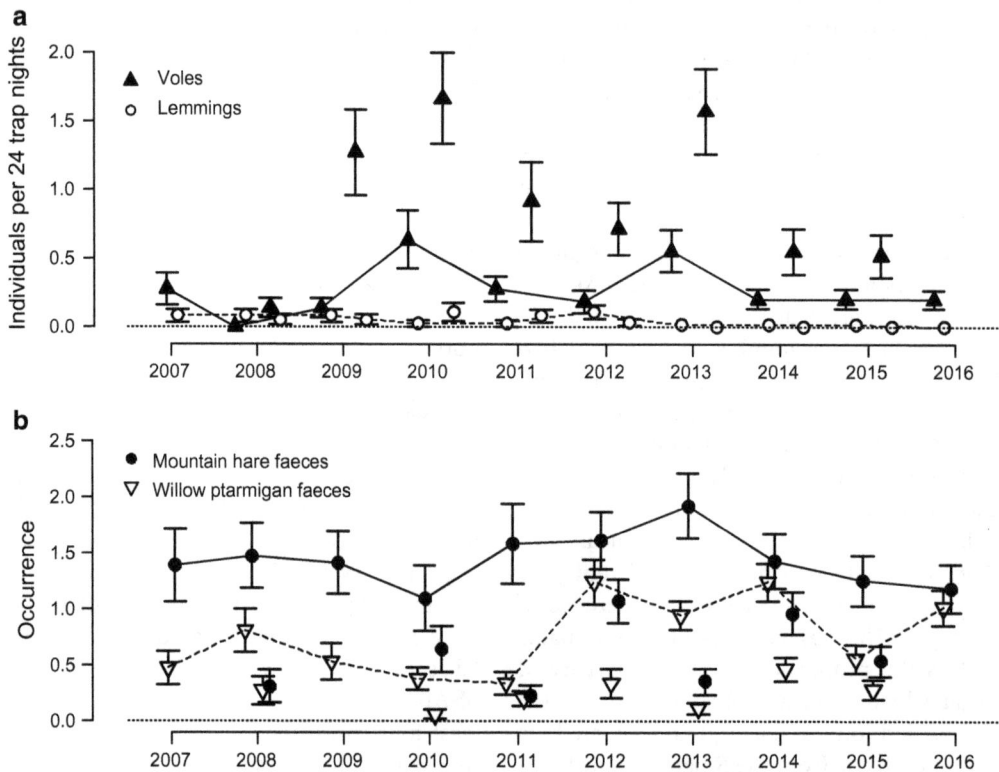

Fig. 2 Dynamics of main herbivores from 2007 to 2016 at Erkuta, southern Yamal. **a** Number of voles and lemmings caught on trapping quadrats per 24 trap nights. **b** Occurrence of mountain hare and willow ptarmigan faeces estimated as the number of small plots surrounding a quadrat where faeces were recorded (max eight small plots). For each trapping/faeces counting session (second part of June and August each year except in 2007, where one observation was carried out in July, in 2009 for faeces counts and in 2016, when only the June trapping and counts were carried out), mean results for all habitats are shown with standard error. Lines join the estimates from June

pups per den, the model with an intercept only received most support from AICc, but ΔAICc to the models with the log of small rodent abundance in June or the availability of reindeer carcasses as fixed effects was small (ΔAICc = 0.72 and 0.45 respectively; Additional file 1: Table S2). All three models had substantial Akaike weights (between 0.19 and 0.27), but none of them indicated that the explanatory variables had a strong effect on the number of pups (Additional file 1: Table S3).

Arctic fox diet

Rodent remains represented the most important component in Arctic fox scats. This was the case when considering proportions of dry weight (2007), proportions of volume (2013–2014; Fig. 5a), and frequencies of occurrence: Rodent remains were found in the bulk samples from all dens in 2007, in 75% of the scats in 2013 and in 78% of the scats in 2014. As these estimates were based mainly on the amount of fur recovered from the scats, they represented all species of rodents together including fur of muskrat in 2007. In 2013 and 2014 they also included fur of hare. Recovered bone fragments

indicated, however, that most fur belonged to small rodents (voles and lemmings). Among recovered small rodent teeth, which could be identified to species, the proportion of lemmings was 48% in 2007 (n = 64; total for all dens), but only 30% in 2013 (n = 43) and 10% in 2014 (n = 57). In addition to small rodents, remains of birds and eggs, plants, insects, hare, reindeer and fish were identified (Fig. 5a). Remains of birds occurred in the bulk samples from all dens in 2007, in 34% of the scats in 2013 and in 18% of the scats in 2014. Except for the proportion of different small rodent species, the proportion of diet components determined in 2007 could not be directly compared to the other years because of differences in methods.

Prey remains were recorded at 16 different dens during up to five visits per den in the course of the same summer. Seven records were from dens visited by foxes but without documented reproduction (not more than one record per den per summer). During 12 visits involving 8 different dens and 9 different breeding events, the observer noted that no prey remains were found. Contrary to the results from scat analyses, birds represented the most common

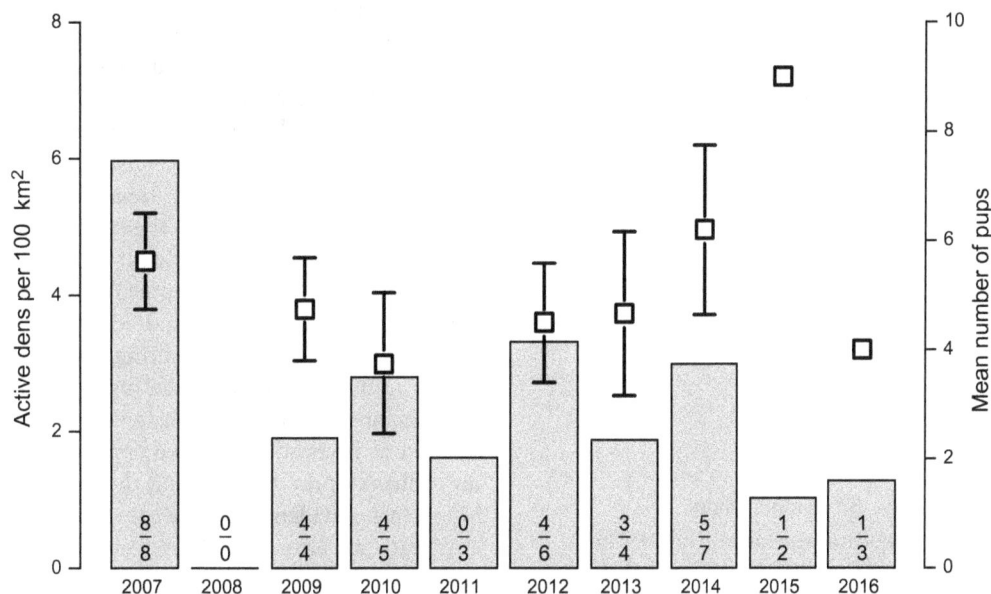

Fig. 3 Number of active dens and litter size of Arctic foxes at Erkuta, southern Yamal. Number of active Arctic fox dens (breeding) per 100 km² and mean minimum number of pups with standard errors observed between 2007 and 2016 at Erkuta. The numbers on the bars show the number of active dens per year (lower) and the number of dens where a minimum number of pups could be determined either by proper observation or with automatic cameras (upper)

Table 1 Coefficients of a generalized linear mixed effects model for the probability of a den to be active

Coefficient	Estimate	SE	CI
Intercept	−0.39	0.52	
Log (rodents index)	1.03	0.41	0.32; 1.89
Reindeer	1.64	0.46	0.73; 2.62

Random effect of den ID: var = 0.42. Estimated coefficients of a generalized linear mixed effects model (binomial error distribution) for the probability of an Arctic fox den to be active in a certain year. Fixed effects were the log of the index of small rodent abundance in June and high availability of reindeer carcasses in the previous winter. Den ID was included as random effect on the intercept. Estimates are given on the logit scale with standard error (SE) and 95% bootstrap confidence intervals (CI)

prey remains on dens and were observed during most visits (Fig. 5b). Many of the bird remains belonged to waterfowl (observed at ca 50% of den visits between 2010 and 2013; Additional file 1: Figure S1). Passerines and ptarmigan were also present in all years, although at lower frequencies, whereas remains of waders were rarely identified. Egg shells were observed in 3 of 7 years (Additional file 1: Figure S1). We observed remains of small rodents and hare in 5 years out of seven (Fig. 5b; Additional file 1: Figure S2), and recorded muskrat, reindeer and freshwater fish (northern pike *Esox lucius*) occasionally. The correspondence analysis revealed that there were no consistent differences in prey remain assemblages between the years (Additional file 1: Figure S3). The only year, that

was somewhat distinct, was 2015, when only little data were available, but remains of ptarmigan and voles were recorded at two out of three den visits.

The stable isotope signatures of Arctic fox winter fur varied little and there was considerable overlap between years (Fig. 5c). Arctic fox values (corrected for isotopic discrimination) were overlapping with the signatures of voles and small birds (waders and passerines). Moreover, they were located between collared lemmings and hare, and waterfowl and muskrat along the $\delta^{15}N$ axis, implying that a shift in diet from voles to equal proportions of waterfowl and hare could for instance remain undetected. Overall, the pattern was compatible with a heavy reliance on voles, in particular narrow-sculled voles, and/or with a mixed diet for the population in late summer. The LMM analysis revealed a small but significant decrease in $\delta^{15}N$ with increasing small rodent abundance (−0.28 ‰ for an increase in one individual per 24 trap nights, CI = −0.51; −0.05), but no effect of small rodent abundance on $\delta^{13}C$ (Additional file 1: Table S4). Such a shift would be compatible with increased consumption of collared lemmings and a lower proportion of waterfowl in the diet in years with high small rodent densities.

Discussion

Our analyses showed that small rodents, despite the low abundance of lemmings and the low amplitude fluctuations of the total abundance, were a major resource for

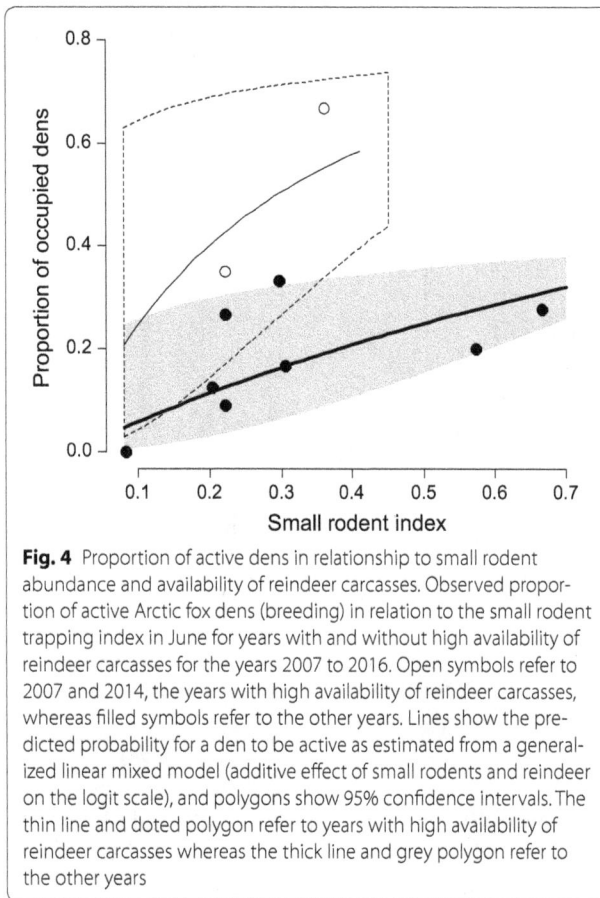

Fig. 4 Proportion of active dens in relationship to small rodent abundance and availability of reindeer carcasses. Observed proportion of active Arctic fox dens (breeding) in relation to the small rodent trapping index in June for years with and without high availability of reindeer carcasses for the years 2007 to 2016. Open symbols refer to 2007 and 2014, the years with high availability of reindeer carcasses, whereas filled symbols refer to the other years. Lines show the predicted probability for a den to be active as estimated from a generalized linear mixed model (additive effect of small rodents and reindeer on the logit scale), and polygons show 95% confidence intervals. The thin line and doted polygon refer to years with high availability of reindeer carcasses whereas the thick line and grey polygon refer to the other years

Arctic foxes in southern Yamal. They were an important determinant of breeding activity in addition to the availability of reindeer carcasses in winter. Moreover, during the breeding season they were an important component of the diet together with birds.

The discrepancy of the diet composition resulting from scat dissection and prey remains is in agreement with what has been shown previously, notably in studies of the diet of raptors [42]. Larger prey are overrepresented in prey remains, whereas raptor pellets show an exaggerated proportion of small mammal prey [50]. For Arctic foxes, it is likely that the importance of small rodents is underestimated from prey remains, as these are usually consumed rapidly and eaten whole. Birds, on the contrary, may be underrepresented in scat dissection data, as small bone fragments are difficult to identify, and feathers are not consumed in large amounts, but often left outside the fox dens.

The proportion of small rodents inferred from the scat analysis (up to 60% based on volume estimates) was lower than what has been reported with a similar approach for lemming fox populations for instance along the Siberian Arctic coast (76–87%; [15] or in northern

Sweden (more than 80%; [14]). The scat analysis data did not allow to address a functional response of the foxes to the small rodent population fluctuations. Stable isotopes, however, revealed a slight decrease in $\delta^{15}N$ with increasing small rodent abundance, which could be compatible with a small increase in the consumption of lemmings, but could also reflect other diet changes (Fig. 3c). The stable isotope data showed that the signatures of Arctic fox fur at Erkuta varied little compared to other regions in the Arctic [10, 51]. A high degree of similarity in isotopic composition among prey limited, however, the inference of diet we could make using stable isotope data at Erkuta [10]. Together with the high proportion of birds and remains of other prey near the dens, this indicates that Arctic foxes at Erkuta feed opportunistically and rely on a diverse prey base during the summer. There was, however, no evidence for the use of marine resources in any year (see also [10]). Moreover, the analysis of prey remains also showed that the diet of this fox population varied little from year to year, a situation that is rather untypical for inland foxes [23].

At Erkuta, Arctic foxes were breeding in nine out of the 10 years of our study. The number of active dens varied more than in Svalbard, where foxes of the coastal ecotype were breeding every year (11-year study; [19], but less than what is typical for lemming fox populations, such as on Bylot Island [52] or in Scandinavia [53]. The density of active dens was lower than observed on Kolguev Island in a non-rodent population, where breeding occurred every year (6-year study) and the main resource during the breeding season was colonially nesting geese [22]. The reproductive dynamics of the Arctic fox population at Erkuta thus represented an intermediate position along the gradient from highly fluctuating lemming fox populations to more stable non-rodent or coastal fox populations. Therefore, our results, describing an inland mixed-diet vole fox population, support the suggestion of Eide et al. [19] to move beyond the distinction between lemming foxes and coastal foxes.

The probability of a den being active increased with the overall abundance of small rodents, but models including only lemmings received low support (Additional file 1: Table S1). Voles were considerably more abundant than lemmings in our study area (Fig. 2a). Lemmings represented <10% of the small rodents trapped in June in 6 out of 10 years, and in the year when only lemmings were caught in June (2008), Arctic foxes did not breed. This clearly demonstrates that voles (Microtus spp.) are a driver of reproductive activity in this Arctic fox population, contrary to results from Fennoscandia, where Arctic foxes responded to lemming densities, but not to voles [5, 15]. It is thus likely that voles are to some degree accessible to Arctic foxes also in winter, contrary to what seems

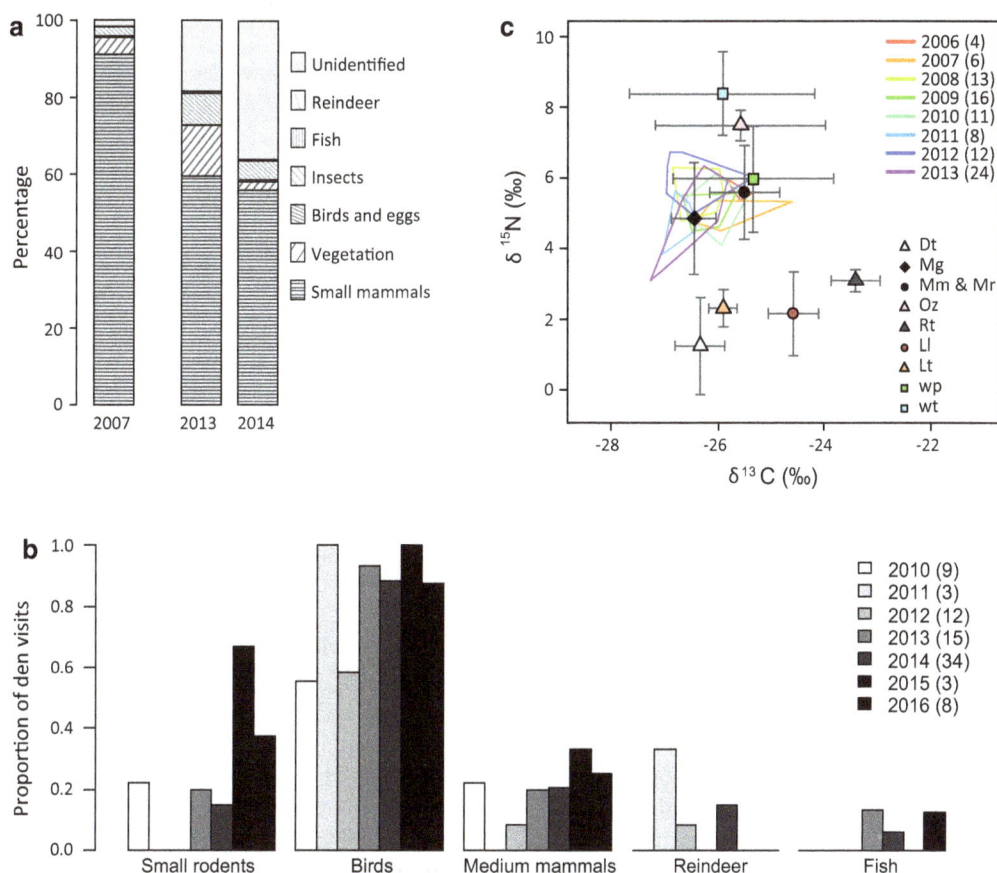

Fig. 5 Diet of Arctic foxes at Erkuta, southern Yamal. **a** Percentage of each prey category identified from scat dissection. In 2007 percentage of dry weight was estimated, whereas in 2013 and 2014 percentage of volume was estimated. These estimates can thus not be directly compared. **b** Proportion of den visits in the years 2010–2016 during which remains of different categories of prey were recorded. Number of visits per year are given in parentheses. **c** Polygons surrounding the stable isotope signatures of Arctic fox winter fur are shown in different colours for each year (i.e. the year during which the fur was growing; sample sizes in parentheses). Values have been corrected for isotopic discrimination according to Lecomte et al. [49]. Average signatures of main prey groups are shown with standard deviations: Dt, collared lemming; Mg, narrow-sculled vole; Mm and Mr, Middendorff's vole and red vole; Oz, muskrat; Rt, domestic reindeer; Ll, willow ptarmigan; Lt, mountain hare; wp, waders and passerines; wt, waterfowl

to be the case in Fennoscandia [27]. This might be due to lower amounts of snow than in Fennoscandia (total precipitation during the coldest quarter is 55 mm in Erkuta compared to 120–130 mm for instance on Varanger Peninsula in northeast Norway; http://www.worldclim.org/bioclim), and to different wintering habitats of the various vole species. Accessibility of voles to Arctic foxes in winter has previously been suggested by [54], who reported that about 1/3 of small rodent remains were voles, when considering remains identified in stomach contents from foxes shot in southern Yamal in the winters 1939–1941—a period when lemming outbreaks were occurring in the area.

High availability of reindeer carcasses after winters with icy pastures following ROS events (resulting in bad feeding conditions) also had a positive effect on breeding

activity. The resource pulses created by dramatic reindeer mortality during winters 2006–2007 and 2013–2014 benefitted all generalist predators [30], including the Arctic fox, as they do in Svalbard [20, 21]. Together with the varied alternative resources available in summer, notably waterfowl, ptarmigan, waders, hare and muskrat, which provide food for the growing pups, they probably enhanced the productivity of the foxes. For the development of the Arctic fox population on a decadal scale, these resource pulses might have replaced to a certain degree the lemming outbreaks missing since 2000 [38]. In this study, we estimated the availability of reindeer carcasses through a coarse index, nevertheless its effect was clear. In the future, however, the identified relationship should be confirmed using quantitative data on the availability of carcasses in the study area. The absence of a clear effect of our hare and ptarmigan indices on den occupancy may be due to the fact that these species were indeed abundant in the study area in all years [41]. It is thus likely that they contributed to the late winter diet of the foxes in all years, and that a strong decrease in their abundance might lead to a decrease of den occupancy.

With an average of 5.2, litter sizes were only slightly above those observed in coastal foxes in Svalbard [19], but lower than those observed in lemming foxes in Scandinavia in years with increasing or peak lemming densities [53]. Interestingly, the identified determinants of den occupancy did not have any clear effect on litter size, as we expected based on knowledge from lemming fox populations [11]. A similar pattern was however observed in Svalbard, where the number of active dens was related to the availability of reindeer carcasses, but average litter size was not [19]. Data from Bylot Island also showed that lemming abundance affected litter size much less than den occupancy [52]. Litter size estimates obtained on the dens are likely to reflect both the number of embryos, which may be related to resource availability in late winter as is the case for den occupancy, but also early survival of the pups, which is likely to be related to resource availability at the beginning of the summer [11]. They are, however, not estimates of breeding success at the end of the summer. In our study, the faeces indices for ptarmigan and hare reflect the situation in late winter, whereas small rodent abundance in June represents the early summer. The availability of reindeer carcasses is likely to be most important in late winter, but the carcasses are actively used at least until June (own observations) and may provide additional resources during lactation. We lack, however, estimates of the yearly variation in the availability of the diverse alternative resources used by the foxes in summer, notably migratory birds. The absence of correlation between determinants of breeding activity and litter size in several studies may indicate that when foxes in these populations breed, they produce a certain relatively constant number of pups, but probably only a few of them survive if resources during the summer are scarce. True estimates of reproductive success are likely to be mainly correlated with resource availability in summer.

Conclusions

Contrary to what has been observed in Fennoscandia [5], the Arctic fox population at Erkuta in southern Yamal is a lemming fox population that seems at present to be able to cope with a shift of the small rodent community from high amplitude cyclicity with lemming dominated peaks, to a vole community with low amplitude fluctuations and a consistent population increase over the summer [55]. Such changes in small rodent dynamics have been related to warmer and less stable winters [56, 57]. Due to climate change, it is likely that such winters will occur more frequently [3]. Arctic foxes at Erkuta showed a reproductive response to vole abundance, but reacted as well to the availability of reindeer carcasses in winters with high mortality among domestic reindeer induced by heavy ROS events. The observed flexibility in resource use is in agreement with the generally opportunistic trophic position of the species [23] and observations, for instance, from eastern Greenland, where the fading out of lemming cycles has affected Arctic foxes less than it has affected snowy owls or long-tailed skuas [4]. At Erkuta, such an adaptation to changing resource dynamics may have been facilitated by occasional winters with high availability of reindeer carcasses, which might have replaced lemming outbreaks in creating sporadic resource pulses stimulating high breeding activity among females, as well as by the high abundance of medium sized herbivores such as ptarmigan and hare. Ongoing climate change leading to warmer autumns, shorter winters and more frequent ROS events may, however, in the future also affect these alternative resources negatively as has been shown for instance for ptarmigan on Svalbard [21] and for mountain hare in southern Norway [58]. Thus, the present state of the Arctic fox population at Erkuta may be transient and future climate induced ecosystem changes might be more difficult to cope with.

At present red foxes are rare compared to Arctic foxes in this low Arctic area, but the resource pulses created by reindeer carcasses seem to promote their expansion [30], and an increase in their population may become detrimental to Arctic foxes [26]. Access to constant resource subsidies in winter are a determining factor for the establishment of red fox populations in tundra [59, 60]. As long as winters with high reindeer mortality remain the exception, and Nenets herders generally manage to keep most of their animals alive, red fox expansion may not reach critical levels for Arctic foxes. If winters with ROS events inducing extensive icing become more frequent due to climate warming [3], it is likely that the herders will adapt their seasonal migrations to avoid catastrophic mortality of their herds. Moreover, occasional red foxes appearing in the area are actively controlled by Nenets hunters, who consider them as much more attractive hunting targets than Arctic foxes. Only continued ecosystem-based monitoring will reveal the fate of this southern Arctic fox population in a changing tundra ecosystem.

Additional file

Additional file 1: Table S1. Candidate models evaluated for arctic fox den occupancy. Table S2. Candidate models evaluated for for the number of arctic fox pups per litter. Table S3. Estimated coefficients of two generalized linear mixed effects models for the minimum number of pups per den. Table S4. Estimated coefficients of a linear mixed effects model for the yearly variation in a) $\delta^{13}C$ and b) $\delta^{15}C$ values of winter fur of arctic foxes. Figure S1. Proportion of den visits each year during which remains of different categories of birds were recorded among prey remains. Figure S2. Proportion of den visits each year during which remains of different categories of mammalian prey remains were recorded. Figure S3. Results of a correspondence analysis of prey remains recorded at each visit on active dens or dens with clear signs of recent presence of foxes.

Authors' contributions
AAS, NAS and DE are leading the monitoring program at Erkuta, which was initiated together with EF and VGS. AAS, NAS, DE, MC, AYR and EF contributed to field work. MC and AYR carried out the feces analyses. DE and MC performed the stable isotope analyses. DE carried out the statistical analyses and wrote the first draft of the manuscript. All authors read and approved the final manuscript.

Author details
[1] Department of Arctic and Marine Biology, University of Tromsø-The Arctic University of Norway, 9037 Tromsø, Norway. [2] Faculty of Biology, Lomonosov Moscow State University, GSP-1, Leninskie Gory, Moscow 119991, Russia. [3] Arctic Research Station of Institute of Plant and Animal Ecology, Ural Branch, Russian Academy of Sciences, 629400, Zelenaya Gorka Str., 21, Labytnangi, Russia. [4] Arctic Research Center of Yamal-Nenets Autonomous District, Salekhard, Russia. [5] Norwegian Polar Institute, Fram Centre, PostBox 6606, Langnes, 9296 Tromsø, Norway.

Acknowledgements
We thank Rolf A. Ims and Nigel Yoccoz for help and inspiration to establish our monitoring program, and the Laptander family for essential support in the field and sharing of their local knowledge on ecosystem functioning. Sissel Kaino contributed to the laboratory work. This study would not have been possible without the contribution of numerous field assistants. Six anonymous reviewers contributed to improve the manuscript through Peerage of Science. The publication charges for this article have been funded by a grant from the publication fund of UiT The Arctic University of Norway.

Competing interests
The authors declare that they have no competing interests.

Funding
Financial support was received from the Research Council of Norway through the International Polar year project "Arctic Predators", from the program UD RAS 15-15-4-35, from RFBR Grant No. 16-44-890108, the Government of Yamalo-Nenetsky Autonomous Region (Department of Science and Innovation), the Norwegian Environment Agency, and Kometen (North Norwegian Research Fund; to MC). Logistic support was partly covered by the Interregional Expedition Center "Arctic".

References
1. Ims RA, Ehrich D, Forbes BC, Huntley B, Walker DA, Wookey PA, Berteaux D, Bhatt US, Bråthen KA, Edwards ME, et al. Terrestrial ecosystems. In: Meltofte H, editor. Arctic biodiversity assessment status and trends in Arctic biodiversity. Akureyri: Conservation of Arctic Flora and Fauna; 2013.
2. Walker DA, Raynolds MK, Daniels FJA, Einarsson E, Elvebakk A, Gould WA, Katenin AE, Kholod SS, Markon CJ, Melnikov ES, et al. The circumpolar Arctic vegetation map. J Veg Sci. 2005;16:267–82.
3. Hansen BB, Isaksen K, Benestad RE, Kohler J, Pedersen ÅØ, Loe LE, Coulson SJ, Larsen JO, Varpe Ø. Warmer and wetter winters: characteristics and implications of an extreme weather event in the high Arctic. Environ Res Lett. 2014;9(11):114201.
4. Schmidt NM, Ims RA, Høye TT, Gilg O, Hansen LH, Hansen J, Lund M, Fuglei E, Forchhammer MC, Sittler B. Response of an Arctic predator guild to collapsing lemming cycles. Proc R Soc B-Biol Sci. 2012;279:4417–22.
5. Ims RA, Killengreen ST, Ehrich D, Flagstad Ø, Hamel S, Henden J-A, Jensvoll I, Yoccoz NG. Ecosystem drivers of an Arctic fox population at the western fringe of the Eurasian Arctic. Polar Res. 2017;36:8.
6. IUCN. Species and climate change: more than just the polar bear. Gland: International Union for Conservation of Nature; 2009.
7. Berteaux D, Thierry A-M, Alisauskas R, Angerbjörn A, Buchel E, Doronina L, Ehrich D, Eide NE, Erlandsson R, Flagstad Ø, et al. Harmonizing circumpolar monitoring of Arctic fox: benefits, opportunities, challenges and recommendations. Polar Res. 2017;36:2.
8. Christensen TR, Payne J, Doyle M, Ibarguchi G, Taylor J, Schmidt NM, Gill M, Svoboda M, Aronsson M, Behe C, et al. The Arctic terrestrial biodiversity monitoring plan. Akureyri: CAFF International Secretariat; 2013.
9. Fuglei E, Øritsland NA. Seasonal trends in body mass, food intake and resting metabolic rate, and induction of metabolic depression in Arctic foxes (Alopex lagopus) at Svalbard. J Comp Physiol B-Biochem Syst Environ Physiol. 1999;169:361–9.
10. Ehrich D, Ims RA, Yoccoz NG, Lecomte N, Killengreen ST, Fuglei E, Rodnikova AY, Ebbinge BS, Menyushina IE, Nolet BA, et al. What can stable isotope analysis of top predator tissues contribute to monitoring of tundra ecosystems? Ecosystems. 2015;18:404–16.
11. Tannerfeldt M, Angerbjörn A. Fluctuating resources and the evolution of litter size in the Arctic fox. Oikos. 1998;83:545–59.
12. Braestrup FW. A study on the Arctic fox in Greenland: immigrants, fluctuations in numbers based mainly on trading statistics. Copenhagen: Kommissionen for videnskabelige undersøgelser i Grønland; 1941.
13. Bêty J, Gauthier G, Korpimäki E, Giroux JF. Shared predators and indirect trophic interactions: lemming cycles and Arctic-nesting geese. J Anim Ecol. 2002;71:88–98.
14. Elmhagen B, Tannerfeldt M, Verucci P, Angerbjörn A. The Arctic fox (Alopex lagopus): an opportunistic specialist. J Zool. 2000;251:139–49.
15. Angerbjörn A, Tannerfeldt M, Erlinge S. Predator-prey relationships: Arctic foxes and lemmings. J Anim Ecol. 1999;68:34–49.
16. Elton CS. Periodic fluctuations in the numbers of animals—their causes and effects. Br J Exp Biol. 1924;2:119–63.
17. Shtro VG. Pesec Yamala (the Arctic fox of Yamal). Ekaterinburg, Russia: Institute of Plant and Animal Ecology, Ural Branch of the Russian Academy of Sciences; 2009.
18. Goltsman M, Kruchenkova EP, Sergeev S, Volodin I, Macdonald DW. 'Island syndrome' in a population of Arctic foxes (Alopex lagopus) from Mednyi Island. J Zool. 2005;267:405–18.
19. Eide NE, Stien A, Prestrud P, Yoccoz NG, Fuglei E. Reproductive responses to spatial and temporal prey availability in a coastal Arctic fox population. J Anim Ecol. 2012;81:640–8.
20. Fuglei E, Øritsland NA, Prestrud P. Local variation in Arctic fox abundance on Svalbard, Norway. Polar Biol. 2003;26:93–8.
21. Hansen BB, Grøtan V, Aanes R, Saether BE, Stien A, Fuglei E, Ims RA, Yoccoz NG, Pedersen ÅØ. Climate events synchronize the dynamics of a resident vertebrate community in the high Arctic. Science. 2013;339:313–5.
22. Pokrovsky I, Ehrich D, Ims RA, Kondratyev AV, Kruckenberg H, Kulikova O, Mihnevich J, Pokrovskaya L, Shienok A. Rough-legged buzzards, Arctic foxes and red foxes in a tundra ecosystem without rodents. PLoS ONE. 2015;10:e0118740.
23. Fuglei E, Ims RA. Global warming and effects on the Arctic fox. Sci Prog. 2008;91:175–91.

24. Palsson S, Hersteinsson P, Unnsteinsdóttir ER, Nielsen OK. Population limitation in a non-cyclic Arctic fox population in a changing climate. Oecologia. 2016;180:1147–57.

25. Unnsteinsdóttir ER, Hersteinsson P, Palsson S, Angerbjörn A. The fall and rise of the Icelandic Arctic fox (*Vulpes lagopus*): a 50-year demographic study on a non-cyclic Arctic fox population. Oecologia. 2016;181:1129–38.

26. Angerbjörn A, Eide NE, Dalén L, Elmhagen B, Hellström P, Ims RA, Killengreen S, Landa A, Meijer T, Mela M, et al. Carnivore conservation in practice: replicated management actions on a large spatial scale. J Appl Ecol. 2013;50:59–67.

27. Killengreen ST, Lecomte N, Ehrich D, Schott T, Yoccoz N, Ims RA. The importance of marine vs. human-induced subsidies in the maintenance of an expanding mesocarnivore in the Arctic tundra. J Anim Ecol. 2011;80:1049–60.

28. Gilg O, Sittler B, Hanski I. Climate change and cyclic predator-prey population dynamics in the high Arctic. Glob Change Biol. 2009;15:2634–52.

29. Ims RA, Yoccoz NG, Killengreen ST. Determinants of lemming outbreaks. Proc Natl Acad Sci USA. 2011;108:1970–4.

30. Sokolov AA, Sokolova NA, Ims RA, Brucker L, Ehrich D. Emergent rainy winter warm spells may promote boreal predator expansion into the Arctic. Arctic. 2016;69:121–9.

31. Angerbjörn A, Tannerfeldt M, Bjarvall A, Ericson M, From J, Norén E. Dynamics of the Arctic fox population in Sweden. Ann Zool Fenn. 1995;32:55–68.

32. Worldclim version 1.0. http://www.worldclim.org. Accessed 15 Jan 2016.

33. Pavlov AV, Moskalenko NG. The thermal regime of soils in the north of western Siberia. Permafr Periglac Process. 2002;13:43–51.

34. Magomedova MA, Morozova LM, Ektova SN, Rebristaya OV, Chernyadeva IV, Potemkin AD, Knyazev MC. Poluostrov Yamal: Rastitel'nyi pokrov (Yamal Peninsula: vegetation cover). Tyumen: City Press; 2006.

35. Rodnikova A, Ims RA, Sokolov A, Skogstad G, Sokolov V, Shtro V, Fuglei E. Red fox takeover of Arctic fox breeding den: an observation from Yamal Peninsula, Russia. Polar Biol. 2011;34:1609–14.

36. Sokolov V, Ehrich D, Yoccoz N, Sokolov A, Lecomte N. Bird communities of the Arctic shrub tundra of Yamal: habitat specialists and generalists. PLoS ONE. 2012;7(12):e50335.

37. Sokolova NA, Sokolov AA, Ims RA, Skogstad G, Lecomte N, Sokolov VA, Yoccoz NG, Ehrich D. Small rodents in the shrub tundra of Yamal (Russia): density dependence in habitat use? Mamm Biol. 2014;79:306–12.

38. Sokolov AA. Functional'nye svyazi mohnonogogo kaniuka i melkih gryzunov yuzhnyh kustarnikovyh tundr Yamala (functional response of the rough-legged buzzard (*Buteo lagopus*) to small rodents of the southern shrub tundra of Yamal). Russia: Perm State University; 2002.

39. Bartsch A, Kumpula T, Forbes BC, Stammler F. Detection of snow surface thawing and refreezing in the Eurasian Arctic with QuikSCAT: implications for reindeer herding. Ecol Appl. 2010;20:2346–58.

40. Myllymäki A, Paasikallio A, Pankakoski E, Kanervo V. Removal experiments on small quadrats as a means of rapid assessment of the abundance of small mammals. Ann Zool Fenn. 1971;8:177–85.

41. Ehrich D, Henden JA, Ims RA, Doronina LO, Killengren ST, Lecomte N, Pokrovsky IG, Skogstad G, Sokolov AA, Sokolov VA, et al. The importance of willow thickets for ptarmigan and hares in shrub tundra: the more the better? Oecologia. 2012;168:141–51.

42. Redpath SM, Clarke R, Madders M, Thirgood SJ. Assessing raptor diet: comparing pellets, prey remains, and observational data at hen harrier nests. Condor. 2001;103:184–8.

43. Layman CA, Araujo MS, Boucek R, Hammerschlag-Peyer CM, Harrison E, Jud ZR, Matich P, Rosenblatt AE, Vaudo JJ, Yeager LA, et al. Applying stable isotopes to examine food-web structure: an overview of analytical tools. Biol Rev. 2012;87:545–62.

44. Cerezo M. Arctic fox diet in Yamal Peninsula. Norway: University of Tromsø; 2015.

45. Borodin AW. Opredelitel´ zubov polevok Yrala i zapadnoy Sibiri (guide of the teeth of voles in the Urals and western Siberia). Ekaterinburg, Russia: Institute of Ecology of Plants and Animals, Ural Branch of the Russian Academy of Sciences; 2009.

46. Team RC. R: a language and environment for statistical computing. Vienna: R Foundation for Statistical Computing; 2016.

47. Bates D, Mächler M, Bolker B, Walker S. Fitting linear mixed-effects models using lme4. J Stat Softw. 2015;67:1–48.

48. Dray S, Dufour AB. The ade4 package: implementing the duality diagram for ecologists. J Stat Softw. 2007;22:1–20.

49. Lecomte N, Ahlstrøm Ø, Ehrich D, Fuglei E, Ims RA, Yoccoz NG. Intrapopulation variability shaping isotope discrimination and turnover: experimental evidence in Arctic foxes. PLoS ONE. 2011;6(6):e21357.

50. Francksen RM, Whittingham MJ, Baines D. Assessing prey provisioned to common buzzard buteo buteo chicks: a comparison of methods. Bird Study. 2016;63:303–10.

51. Tarroux A, Bety J, Gauthier G, Berteaux D. The marine side of a terrestrial carnivore: intra-population variation in use of allochthonous resources by Arctic foxes. PLoS ONE. 2012;7(8):e42427.

52. Giroux MA, Berteaux D, Lecomte N, Gauthier G, Szor G, Bety J. Benefiting from a migratory prey: spatio-temporal patterns in allochthonous subsidization of an Arctic predator. J Anim Ecol. 2012;81:533–42.

53. Meijer T, Elmhagen B, Eide NE, Angerbjörn A. Life history traits in a cyclic ecosystem: a field experiment on the Arctic fox. Oecologia. 2013;173:439–47.

54. Dunaeva TN. Comparative review of the ecology of tundra voles of Yamal Peninsula. In: Formozov AN, editor. Materials of the institute of geography of academy of sciences of USSR. Moscow: Academy of Sciences of USSR; 1948. p. 78–143 (**in Russian**).

55. Ims RA, Fuglei E. Trophic interaction cycles in tundra ecosystems and the impact of climate change. Bioscience. 2005;55:311–22.

56. Kausrud KL, Mysterud A, Steen H, Vik JO, Østbye E, Cazelles B, Framstad E, Eikeset AM, Mysterud I, Solhøy T, et al. Linking climate change to lemming cycles. Nature. 2008;456:93.

57. Ims RA, Henden JA, Killengreen ST. Collapsing population cycles. Trends Ecol Evol. 2008;23:79–86.

58. Pedersen S, Odden M, Pedersen HC. Climate change induced molting mismatch? Mountain hare abundance reduced by duration of snow cover and predator abundance. Ecosphere 2017;8(3):e01722.

59. Henden JA, Ims RA, Yoccoz NG, Hellström P, Angerbjörn A. Strength of asymmetric competition between predators in food webs ruled by fluctuating prey: the case of foxes in tundra. Oikos. 2010;119:27–34.

60. Elmhagen B, Berteaux D, Burgess RM, Ehrich D, Gallant D, Henttonen H, Ims RA, Killengreen ST, Niemimaa J, Norén K, et al. Homage to Hersteinsson and Macdonald: climate warming and resource subsidies cause red fox range expansion and Arctic fox decline. Polar Res. 2017;36:3.

13

BioVeL: a virtual laboratory for data analysis and modelling in biodiversity science and ecology

Alex R. Hardisty[1][*][iD], Finn Bacall[2], Niall Beard[2], Maria-Paula Balcázar-Vargas[3], Bachir Balech[4], Zoltán Barcza[5], Sarah J. Bourlat[6], Renato De Giovanni[7], Yde de Jong[3,8], Francesca De Leo[4], Laura Dobor[5], Giacinto Donvito[9], Donal Fellows[2], Antonio Fernandez Guerra[10,11], Nuno Ferreira[12], Yuliya Fetyukova[8], Bruno Fosso[4], Jonathan Giddy[1], Carole Goble[2], Anton Güntsch[13], Robert Haines[14], Vera Hernández Ernst[15], Hannes Hettling[16], Dóra Hidy[17], Ferenc Horváth[18], Dóra Ittzés[18], Péter Ittzés[18], Andrew Jones[1], Renzo Kottmann[10], Robert Kulawik[15], Sonja Leidenberger[19], Päivi Lyytikäinen-Saarenmaa[20], Cherian Mathew[13], Norman Morrison[2], Aleksandra Nenadic[2], Abraham Nieva de la Hidalga[1], Matthias Obst[6], Gerard Oostermeijer[3], Elisabeth Paymal[21], Graziano Pesole[4,22], Salvatore Pinto[12], Axel Poigné[15], Francisco Quevedo Fernandez[1], Monica Santamaria[4], Hannu Saarenmaa[8], Gergely Sipos[12], Karl-Heinz Sylla[15], Marko Tähtinen[23], Saverio Vicario[24], Rutger Aldo Vos[3,16], Alan R. Williams[2] and Pelin Yilmaz[10]

Abstract

Background: Making forecasts about biodiversity and giving support to policy relies increasingly on large collections of data held electronically, and on substantial computational capability and capacity to analyse, model, simulate and predict using such data. However, the physically distributed nature of data resources and of expertise in advanced analytical tools creates many challenges for the modern scientist. Across the wider biological sciences, presenting such capabilities on the Internet (as "Web services") and using scientific workflow systems to compose them for particular tasks is a practical way to carry out robust "in silico" science. However, use of this approach in biodiversity science and ecology has thus far been quite limited.

Results: BioVeL is a virtual laboratory for data analysis and modelling in biodiversity science and ecology, freely accessible via the Internet. BioVeL includes functions for accessing and analysing data through curated Web services; for performing complex in silico analysis through exposure of R programs, workflows, and batch processing functions; for on-line collaboration through sharing of workflows and workflow runs; for experiment documentation through reproducibility and repeatability; and for computational support via seamless connections to supporting computing infrastructures. We developed and improved more than 60 Web services with significant potential in many different kinds of data analysis and modelling tasks. We composed reusable workflows using these Web services, also incorporating R programs. Deploying these tools into an easy-to-use and accessible 'virtual laboratory', free via the Internet, we applied the workflows in several diverse case studies. We opened the virtual laboratory for public use and through a programme of external engagement we actively encouraged scientists and third party application and tool developers to try out the services and contribute to the activity.

*Correspondence: hardistyar@cardiff.ac.uk
[1] School of Computer Science and Informatics, Cardiff University, Queens Buildings, 5 The Parade, Cardiff CF24 3AA, UK
Full list of author information is available at the end of the article

Conclusions: Our work shows we can deliver an operational, scalable and flexible Internet-based virtual laboratory to meet new demands for data processing and analysis in biodiversity science and ecology. In particular, we have successfully integrated existing and popular tools and practices from different scientific disciplines to be used in biodiversity and ecological research.

Keywords: Biodiversity science, Ecology, Computing software, Informatics, Workflows, Virtual laboratory, Biodiversity virtual e-laboratory, Data processing, Analysis, Automation

Background

Environmental scientists, biologists and ecologists are pressed to provide convincing evidence of contemporary changes to biodiversity, to identify factors causing biodiversity decline, to predict the impact of, and suggest ways of combating biodiversity loss. Altered species distributions, the changing nature of ecosystems and increased risks of extinction, many of which arise from anthropogenic activities all have an impact in important areas of societal concern (human health and well-being, food security, ecosystem services, bioeconomy, etc.). Thus, scientists are asked to provide decision support for managing biodiversity and land-use at multiple scales, from genomes to species and ecosystems, to prevent or at least to mitigate such losses. Generating enough evidence and providing decision support increasingly relies on large collections of data held in digital formats, and the application of substantial computational capability and capacity to analyse, model, simulate and predict using such data [1–3]. Achieving the aims of the recently established Intergovernmental Science-Policy Platform on Biodiversity and Ecosystem Services (IPBES) [4] requires progressive developments in approach and method.

The complexity and scope of analyses in biodiversity science and ecology is growing very fast. It is becoming more common to carry out complex analysis using hundreds of data files with different structures and data types (e.g., genetic, species, geographical, environmental) combined with a variety of algorithms; producing results that need to be visualized in innovative ways. The requirement for scientists to work together, with collaborations that integrate datasets across many different parties and synthesize answers computationally to address larger scientific questions are becoming the norm. Biodiversity science and ecology are now in the era of data-intensive science [5, 6]. New research practices that productively exploit data pipelines and data-driven analytics need infrastructure that enables reliability, robustness, repeatability, provenance and reproducibility for large and complex scientific investigations. Methods evolve, exploiting tendencies to base on variants of previous processes, composed of common steps. However, usage statistics from developed science-wide e-Infrastructures show that biodiversity, conservation, and ecology scientists do not

carry out large-scale experiments to the same extent as scientists in the physical sciences [7].

Scientific workflow systems, such as Kepler [8], Pegasus [9], Apache Taverna [10], VisTrails [11], KNIME [12], Galaxy [13] and RapidMiner [14] are mature technology for practical ways to carry out computer-based experimentation and analysis of relevant data in disciplines as diverse as medical 'omics'/life sciences, heliophysics and toxicology [15–17]. Scientific workflow systems can be broadly organised into three categories. First, those developed for specialist domains, often with capabilities to be extended to other disciplines (e.g., LONI pipeline for neuro-imaging [18]; Galaxy for omics data processing; KNIME for pharmacological drug discovery). Secondly, there are workflow systems developed to be independent of any particular science discipline, with features for incorporating specialised customisations (e.g., Apache Taverna). Thirdly, there are those that cut across disciplines and focus on specific tasks (e.g., RapidMiner for data mining).

Workflows support both automation of routine tasks (data transformation, mining, integration, processing, modelling) and pursuit of novel, large-scale analysis across distributed data resources. Today, such activities are typically done in the R environment and here the integration with workflow systems can add value to current practice in ecological research. Most importantly, the record of what was done (the provenance) and the reproducibility of complex analyses can be enhanced when migrating ecological analysis into workflow environments, while workflow systems are able to handle the procedures for scaling computation on cloud infrastructure, for example. For this purpose, scientific workflow systems are starting to become used in biodiversity science and ecology for example: in ecology [19, 20], niche and distribution modelling [21–23], and digitisation of biological collections [24, 25].

Workflows can be deployed, executed, and shared in virtual laboratories. A modern virtual laboratory (sometimes also known as a virtual research environment, VRE) is a web-enabled software application that brings the digital, Internet-based data resources (which may include data collections, databases, sensors and/or other instruments) together with processing and analytical tools needed to carry out "in silico" or "e-science" work. As in a real laboratory, the essence of a virtual laboratory is providing the capability to carry out experimental work as a sequence of interconnected work processes i.e., a workflow. Data and tools are combined harmoniously to present a consistent joined-up computer-based work environment to the scientist user. The laboratory keeps track of the details of experiments designed and executed, as well as creating relevant provenance information about the data and tools used; to assist repeatability and replication of results. A virtual laboratory often also incorporates elements to provide assistance and to support collaborations between persons and across teams. These can include sharing and publishing mechanisms for data, experiments and results, as well as supplemental communications capabilities (either built-in or external) for Web-based audio/video conferencing, email and instant messaging, technical training and support.

The aim of our work has been to explore use of the workflow approach in ecology and to encourage wider adoption and use by developing, deploying and operating this Biodiversity Virtual e-Laboratory as a showcase for what is possible and as an operational service [26]. We have demonstrated this with results from a number of scientific studies that have used the BioVeL platform (Additional file 4).

Implementation

The biodiversity virtual e-laboratory (BioVeL) provides a flexible general-purpose approach to processing and analysing biodiversity and ecological data (such as species distribution, taxonomic, ecological and genetic data). It is based on distributed computerised services accessible via the Internet that can be combined into sequences of steps (workflows) to perform different tasks.

The main components of the platform are illustrated in Fig. 1 and described following, with cross-references (A)–(F) between the text and the figure. Additional file 1 to the present article provides 'how-to' guidelines on how to make use of the various components.

Web services for biodiversity science and ecology: (A) in Fig. 1

In computing terms, Web services are pieces of computing functionality (analytical software tools and data resources) deployed at different locations on the Internet (Worldwide Web) [27]. The idea of presenting data resources and analytical tools as Web services is an essential principle of the notion of the Worldwide Web as a platform for higher value "Software as a Service" applications, meaning users have to install less and less specialised software on their local desktop computers. Web services are central to the concept of workflow

composition and execution; increasingly so with proliferation of third-party data resources and analytical tools, and trends towards open data and open science. Wrapping data resources and analytical tools to present the description of their interfaces and capabilities in a standard way aids the process of matching the outputs of one element in a workflow sequence to the inputs of the next. Where such matches are inexact, specialised services can be called upon to perform a translation function. Another benefit of describing resources and functions in a standardised way is the ability to register and advertise details in a catalogue akin to a 'Yellow Pages' directory, such that the resources and tools can be more easily discovered by software applications.

Many candidate Web services, representing useful biodiversity data resources and analytical tool capabilities can be identified from the different thematic subdomains of biodiversity science. These include services coming from domains of enquiry such as: taxonomy, phylogenetics, metagenomics, ecological niche and population modelling, and ecosystem functioning and valuation; as well as more generally useful services relating to statistics, data retrieval and transformations, geospatial processing, and visualization. Working with domain experts via a series of workshops during 2012–2013 and other community networking mechanisms, we considered and prioritised more than 60 candidate services in seven groups (Table 1) many of which went on to be further developed, tested and deployed by their owning "Service Providers". A full list of services is available in the Additional information.

We have catalogued these capabilities (Web services) in a new, publicly available, curated electronic directory called the Biodiversity Catalogue (http://www.biodiversitycatalogue.org) [29]. This is an openly available online registry of Web services targeted towards the biodiversity science and ecology domain. It is an instance of software developed originally by the BioCatalogue project for the life sciences community [30], branded and configured for use in ecology. Our intention is that this catalogue should be well-founded through careful curation, and should become well-known and used, as is the case for the BioCatalogue in life sciences. The catalogue uses specialised service categories and tag vocabularies for describing services specific to biodiversity science and ecology and it has been operational since October 2012. Currently (date of writing), it catalogues 70+ services (some of which are aggregates of multiple capabilities) from 50+ service providers, including Global Biodiversity Information Facility (GBIF), European Bioinformatics Institute (EBI), EUBrazilOpenBio, PenSoft Publishers, Royal Botanic Gardens Kew, Species2000/ITIS Catalogue of Life (CoL), Pangaea, World Register of Marines Species (WoRMS), Naturalis

Biodiversity Virtual e-Laboratory

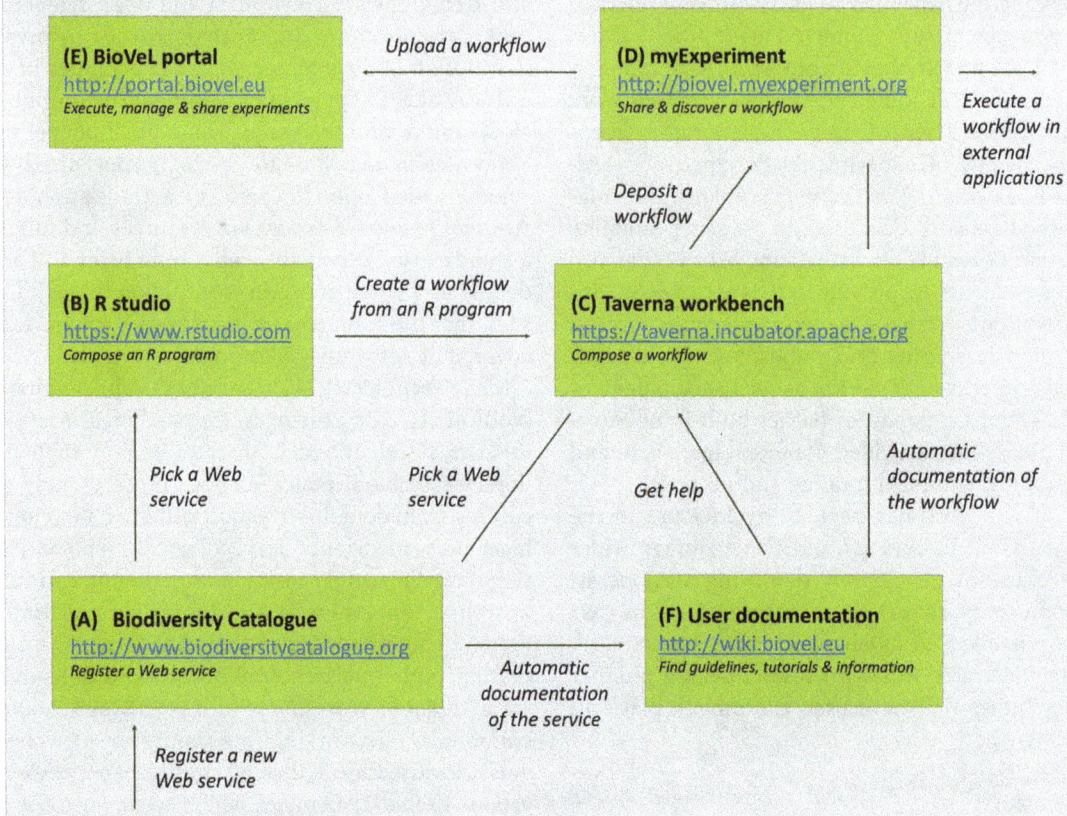

Fig. 1 Biodiversity virtual laboratory (BioVeL) is a software environment that assists scientists in collecting, organising, and sharing data processing and analysis tasks in biodiversity and ecological research. The main components of the platform are: *A* the Biodiversity Catalogue (a library with well-annotated data and analysis services); *B* the environment, such as RStudio for creating R programs; *C* the workbench for assembling data access and analysis pipelines; *D* the myExperiment workflow library that stores existing workflows; *E* the BioVeL Portal that allows researchers and collaborators to execute and share workflows; and *F* the documentation wiki. Infrastructure is indicated in bold, while processes related to research activities are indicated in *italics*. Components *A–F* are referred to from the text, where they are described in detail. See also 'how-to' guidelines in the Additional information

Biodiversity Center and Canadensys). It has 130+ contributing members and is open for any provider of similar kinds of capabilities to register their Web services there.

The catalogue supports registration, discovery, curation and monitoring of Web services. Catalogue entries are contributed by the community and also curated by the community. Experts oversee the curation process to ensure that descriptions are high quality and that the services entries are properly annotated. We developed a 4-level service maturity model to measure the quality of service descriptions and annotations. Biodiversity Catalogue supports service 'badging' using this model. In this way, users can distinguish between services that are poorly described and perhaps unlikely to perform reliably, and those with higher quality descriptions and annotations. This encourages service providers to invest more

time and effort in annotating their services and improving their documentation. It eases discovery and use of services by end users and scientists.

Within the catalogue we have provided an automated framework for service availability monitoring. Monitoring is performed on a daily basis. Service providers and curators are notified of potential availability problems when these are detected. The statistics collected over time are compiled into service reliability reports to give end users some indication of longer-term reliability of services and to help them choose the most reliable services for their scientific workflows and applications. This public portrayal of service performance information encourages Service Providers to invest time and effort in maintaining and improving the availability and reliability of their offering.

Table 1 Services for data processing and analysis (Additional file 2)

Service group	Capabilities (web services)
General purpose, including mapping and visualization	General-purpose capabilities needed in many situations, such as for: Interactive visualization of spatio-temporal data (BioSTIF) e.g., occurrence data; Execution of R programs embedded as steps in workflows; Temporary workspace for data file movements between services
Ecological niche modelling	Built up from the existing openModeller web service [28] to offer a wide range of algorithms and modelling procedures integrated with geospatial management of environmental data, enabling researchers to create, test, and project ecological niche models (ENM)
Ecosystem modelling	A basic toolbox for studies of carbon sequestration and ecosystem function. It includes data-model integration and calibration services, model testing and Monte Carlo Experiment services, ecosystem valuation services, and bioclimatic services
Metagenomics	A basic set of services for studying community structure and function from metagenomic ecological datasets. It includes services for geo-referenced annotation, metadata services, taxonomic binning and classification services, metagenomic traits services, and services for multivariate analysis
Phylogenetics	Services to enable DNA sequence mining and alignment, core phylogenetic inference, tree visualization, and phylogenetic community structure, for broad use in evolutionary and ecological studies
Population modelling	Services for demographic data and their integration into matrix projection models and integral projection models (MPM, IPM)
Taxonomy	Services for taxonomic name resolution, checklists and classification, and species occurrence data retrieval

Composing custom programs and workflows with Web services: (A), (B) and (C) in Fig. 1

Today, it is not only reliable and open Web services that are still scarce in ecology, but also easy-to-use applications and orchestration mechanisms that connect such services in a sequence of analytical steps. It takes significant effort to compose and prove an efficient workflow when the sequence of steps is complex—from tens to perhaps many hundreds of individual detailed steps. The inter-relations and transformations between components have to be properly understood to generate confidence in the output result.

In the R language [31] for example, interaction with servers via the HTTP protocol is built-in, so that a program client for a RESTful service only needs to compose the right request URL and decode the response. Both URL parameters and response formats are exhaustively documented in the Biodiversity Catalogue and off-the-shelf parsers exist for the syntax formats that our services return (e.g., JSON, CSV, domain-specific formats). For SOAP services, the R open source community uses the 'S SOAP' package to build more complex, stateful client–server interaction workflows, where an analysis is built up in multiple, small steps rather than a single request/response cycle.

The general-purpose Apache Taverna Workflow tool suite [10] is a widely used and popular approach to creating, managing and running workflows. With an established community of more than 7500 persons, organised into more than 300 specialised groups, having publicly shared more than 3700 workflows (information correct at March 2016) it represents a rich resource for scientists developing new analysis methods. We have chosen Apache Taverna as the basis for workflows that we developed to build on this already extensive platform, gaining advantage in expertise, familiarity, opportunities for cross-fertilisation and interdisciplinarity that increasingly characterises the science of biodiversity and ecology. With comprehensive capabilities to mix distributed Web Services, local programs/command line tools and other service types (e.g., BioMart queries or R programs) into a single workflow that can be executed locally, on specialised community or institutional computing facilities or "in the cloud", Taverna was a suitable candidate for the task. We adapted the Taverna tools to meet new requirements we anticipated would arise during the course of this work.

We have developed appropriate interfaces in the Biodiversity Catalogue to invoke Web services directly from the R environment [32], or from within the Apache Taverna workflow management system [10]. We developed 20 interactive workflows to explore and showcase the utility of the Web services in ecological research. The workflows can be executed through the BioVeL Portal (E) described below. They are summarised in Table 2, with references to scientific studies using these workflows. A more detailed version of Table 2 is available as Additional file 3; also further Additional information describing different scientific studies that have made use of them.

Table 2 Workflows for biodiversity science (Additional file 3)

Workflow (family)	*Capability/purpose (i.e., what is it for?)* *incl. persistent identifier (purl) to locate the workflow and references to scientific studies that have exploited it*
Data refinement	The data refinement workflow (DRW) is for preparing taxonomically accurate species lists and observational data sets for use in scientific analyses such as: species distribution analysis, species richness and diversity studies, and analyses of community structure purl: http://purl.ox.ac.uk/workflow/myexp-2874.13 Portal: https://portal.biovel.eu/workflows/641 Scientific studies: [33, 34]
Ecological niche modelling (ENM)	The generic ENM workflow creates, tests, and projects ecological niche models (ENM), choosing from a wide range of algorithms, environmental layers and geographical masks purl: http://purl.ox.ac.uk/workflow/myexp-3355.20 Portal: https://portal.biovel.eu/workflows/440 The BioClim workflow retrieves environmentally unique points from a species occurrence file under a given set of environmental layers, and calculates the range of the environmental variables (min–max) for a given species purl: http://purl.ox.ac.uk/workflow/myexp-3725.2 Portal: https://portal.biovel.eu/workflows/443 Scientific studies: [33, 35, 36]
ENM statistical difference (ESW)	Statistical post-processing of results from ecological niche modelling ESW DIFF workflow computes extent, direction and intensity of change in species potential distribution through computation of the differences between two models, including change in the centre point of the distribution purl: http://purl.ox.ac.uk/workflow/myexp-3959.2 Portal: https://portal.biovel.eu/workflows/442 ESW STACK workflow computes extent, intensity, and accumulated potential species distribution by computing the average sum from multiple models purl: http://purl.ox.ac.uk/workflow/myexp-3856.3 Portal: https://portal.biovel.eu/workflows/70 Scientific studies: [33, 36]
Population modelling	Matrix population model construction and analysis workflows provide a complete environment for creating a stage-matrix with no density dependence, and then to perform several analyses on it. Each of the workflows in the collection is also available separately. The expanded version of this table, available as Additional information contains a link purl: http://purl.ox.ac.uk/researchobj/myexp-483 Portal: https://portal.biovel.eu/workflows/596 Integral projection models workflow provides an environment to create and test an integral projection model and to perform several analyses on that purl: http://purl.ox.ac.uk/researchobj/myexp-482 Portal: https://portal.biovel.eu/workflows/599 Scientific studies: no publication yet
Ecosystem modelling	Based around the Biome-BGC biogeochemical model, a collection of five workflows for calibrating and using Biome-BGC for modelling ecosystems and calculating a range of ecosystem service indicators. The Biome-BGC projects database and management system provides a user interface for setting of model parameters, for support sharing and reusing of datasets and parameter settings purl: http://purl.ox.ac.uk/researchobj/myexp-687 Portal: https://portal.biovel.eu/workflows/81 https://portal.biovel.eu/workflows/289 https://portal.biovel.eu/workflows/300 https://portal.biovel.eu/workflows/48 https://portal.biovel.eu/workflows/507 Scientific studies: [37–40]
Metagenomics	Microbial metagenomic trait calculation and statistical analysis (MMT) workflow calculates key ecological traits of bacterial communities as observed by high throughput metagenomic DNA sequencing. Typical use is in the analysis of environmental sequencing information from natural and disturbed habitats as a routine part of monitoring programs purl: http://purl.ox.ac.uk/workflow/myexp-4489.3 Portal: access on request (BioMaS) Bioinformatic analysis of Metagenomic ampliconS is a bioinformatic pipeline supporting biomolecular researchers to carry out taxonomic studies of environmental microbial communities by a completely automated workflow, comprehensive of all the fundamental steps, from raw sequence data arrangement to final taxonomic identification. This workflow is typically used in meta-barcoding high-throughput-sequencing experiments url: https://www.biodiversitycatalogue.org/services/71 Scientific studies: [41–43]

Table 2 continued

Phylogenetics	Bayesian phylogenetic inference workflows are for performing phylogenetic inference for systematics and diversity research. Bayesian methods guide selection of the evolutionary model and a post hoc validation of the inference is also made. Phylogenetic partitioning of the diversity across samples allows study of mutual information between phylogeny and environmental variables purl: http://purl.ox.ac.uk/researchobj/myexp-370 Portal: https://portal.biovel.eu/workflows/466 https://portal.biovel.eu/workflows/549 https://portal.biovel.eu/workflows/550 https://portal.biovel.eu/workflows/525 PDP workflow, using PhyloH for partitioning environmental sequencing data using both categorical and phylogenetic information purl: http://purl.ox.ac.uk/workflow/myexp-3570.5 Portal: https://portal.biovel.eu/workflows/434 https://portal.biovel.eu/workflows/71 MSA-PAD workflow performs a multiple DNA sequence alignment coding for multiple/single protein domains invoking two alignment modes: gene and genome Gene mode purl: http://purl.ox.ac.uk/workflow/myexp-4549.1 Portal: https://portal.biovel.eu/workflows/712 (access on request) Genome mode purl: http://purl.ox.ac.uk/workflow/myexp-4551.1 Portal: https://portal.biovel.eu/workflows/713 (access on request) SUPERSMART (self-updating platform for estimating rates of speciation and migration, ages and relationships of taxa) is a pipeline analytical environment for large-scale phylogenetic data mining, taxonomic name resolution, tree inference and fossil-based tree calibration url: https://www.biodiversitycatalogue.org/services/78 Scientific studies: [44–47]

Creating R programs that use Web services: (B) in Fig. 1

Users can interact with the Biodiversity Catalogue and its services in a variety of ways, one of which is by developing their own analysis programs that invoke services in the catalogue. Both the catalogue itself and the services that are advertised in it are exposed through Applications Programming Interfaces (API) that are accessible using standard Internet protocols (HTTP, with RESTful or SOAP functionality). Hence, writing custom analysis code is relatively straightforward in commonly-used programming languages, such as R [31]. The advantage of this way of interacting with the Biodiversity Catalogue services is that users can do this within a development environment (such as RStudio [48]). This enables them to go through their analysis one step at a time (in a "read-eval-print loop") visually probing their data as it accumulates. Users can include additional functionalities accessible through Web services [49] as well as from relevant third-party R packages for biodiversity and ecological analysis; many of which have been developed in recent years. These latter are available, for example via CRAN [50]. The Additional file 1 'how-to' guidelines points to an example of how to create an R program that calls a Web service. Given the popularity of the R programming language in biodiversity and ecology, we expect to see not just ad hoc analysis programs but also published, re-usable analysis libraries written against Web services APIs. The Biodiversity Catalogue provides a single place where such Web services can be found.

Several Web services can be linked together in sequence in an R program to create a 'work flow'. However, this can rapidly become quite complex. Outputs of one Web service may not match the inputs of the next. Conversions and other needs (such as conditional branching, nesting of sub-flows, parallel execution of multiple similar steps, or waiting asynchronously for a long-running step to complete) all add to the complexity, which has to be managed. Here, workflow management systems, like Apache Taverna [10] can hide some of the complexity and make workflows easier to create, test and manage. Such systems often offer graphical 'what you see is what you get' user interfaces to compose workflows from Web and other kinds of services, such as embedded R programs. Reasonably complex custom workflows can be created (see below) without writing a single line of programming code, which can be attractive for scientists with little or no programming background.

Combined with other capabilities of the BioVeL platform, including transparent access to greater levels of computing capability and capacity for processing large amounts of data, managing the complexity of multiple workflow runs, sharing workflows and provenance, offering data services, etc.) Web services can be applied (i) consistently and (ii) in combination. We have given further examples of potential areas of application where these functions can be combined to support and accelerate new research in the "Discussion" section below (under 'Towards more comprehensive and global investigations').

Creating a workflow from an R program: (B) and (C) in Fig. 1
It is possible to convert pre-existing R programs for inclusion into Taverna workflows as discrete 'R service' steps. We have developed some recommendations [51] to make this as easy as possible. We have, for example taken an existing R program that uses data from a local directory and incorporated this into a workflow that generates graphical plots from Ocean Sampling Day (OSD) data [41] to visualise the metagenomic sequence diversity in ocean water samples. We exposed the inputs and outputs of the R program as 'ports' of the corresponding R service, such that the program can be easily re-run using different data. A user could re-use this or another R program, wrapped as a service into their own workflow. Because the workflow is executed on the BioVeL platform, including execution of the R service, there is no need to run R locally on their own computer. This approach gives the possibility to combine R programs and workflows in complementary fashion, the full power of which becomes evident when workflows are embedded as executable objects in 3rd party web sites and web applications (see "Execute a workflow in external applications" section, below).

Building a workflow from Web services: (C) in Fig. 1
Using the Apache Taverna Workbench [10], we devised workflows meeting scientists' own needs or fulfilling common needs for routine tasks performed by many scientists in a community. The design and creation of a workflow from Web services requires some programming skills and has often been done by service curators at institutes that also provide Web services.

The Taverna Workbench is a 'what you see is what you get' graphical tool, locally installed on the user's desktop computer that can be used to create and test workflows using a 'drag and drop' approach. In the Workbench, users select processing steps from a wide-ranging list of built-in local processing steps and on-line Web services to create a workflow. They do this by dragging and dropping the step into a workflow and linking to its other steps. Each step is in reality an encapsulation of a software tool (an R program, for example) with its own inputs and outputs. Workbench users link the inputs of a step to the outputs from a preceding step and the outputs to the inputs of the next step. Links can be edited when steps are inserted, removed or re-organised. The user can test the workflow by running it locally on their desktop computer or by uploading it to BioVeL Portal (described below).

We have provided a customised version of the Workbench, Taverna Workbench for Biodiversity [52], configured with a selectable palette of services especially relevant to biodiversity science and ecology. This version provides a direct link to the Biodiversity Catalogue, allowing users to search for the most recent and useful external Web services provided by the community as a whole.

Customising existing workflows: (D) and (C) in Fig. 1
Scientists with programming skills can inspect and modify existing workflows available in the myExperiment workflow library (http://biovel.myexperiment.org), again using the Taverna Workbench tool. There is a direct link to the public myExperiment workflow library (described below), allowing to search for and download existing workflows.

As an example, a scientist used an existing workflow for statistical calculations of differences between ENM output raster files (Table 2, ESW DIFF) to create a new variant that additionally calculates the magnitude and direction of shift in distribution between two model projections. Enhancing the underlying logic (R program in this case) with additional code to compute the weighted centre point of each model projection, and the geographic distance between them was all that was needed. The required data management and visualization resources were already in place, provided by other elements of the existing workflow and the BioSTIF service (Table 1). The ESW DIFF workflow was modified to include the functionality of the new variant.

In a further example: Aphia, the database underlying the World Register of Marine Species (WoRMS) [53] is a consolidated database for marine species information, containing valid species names, synonyms and vernacular names, higher taxon classification information and extra information such as literature and biogeographic data. Its principal aim is to aid data management, rather than to suggest any taxonomic or phylogenetic opinion on species relationships. As such it represents a resource that is complementary to those already programmed as part of the Data Refinement Workflow (Table 2). After working with the Service Provider to register the service in the Biodiversity Catalogue, we easily modified the Data Refinement Workflow to present the AphiaName lookup service as a choice alongside the Catalogue of Life and GBIF Checklist Bank lookup services when carrying out the taxonomic name resolution stage of the workflow.

Building and using workflow components (not illustrated in Fig. 1)
Packaging a series of related processing steps into a reusable component eases the complexity of building workflows. For example, the task of dynamically defining a geographically bounded area (known as a mask)

within which something of interest should be modelled (or selecting from a list of pre-defined masks) involves a lengthy sequence of steps and interactions between a user and a Web service that is used to do the mask creation and selection. To create this from scratch every time it is needed in a workflow would be time-consuming and error-prone. A "create_or_select_mask" component makes it easier to do.

Such components serve as basic building blocks in larger or more complex workflows, making workflows quicker and easier to assemble. We have developed a series of ecological niche modelling (ENM) components that have been mixed and matched for investigating the effects of mixing different spatial resolutions in ENM experiments [35]; as well as to assemble a jack-knife resampling workflow to study the influence of individual environmental parameters as part of our study on species distribution responses to scenarios of predicted climate change [54]. The packages of population modelling workflows (Table 2) are also based on component families, allowing mix-and-match configuration of population modelling analyses. Well-designed and well-documented sets of workflow components can effectively allow a larger number of scientists without in-depth programming skills to more easily assemble new analytical pipelines.

Discovering workflows: (D) in Fig. 1

As with making Web services available in a directory to encourage discovery and re-use, sharing workflows publicly encourages re-use and adoption of new methods. It makes those methods available to users having less skills or time and effort to create such methods. More importantly, sharing enables more open science, repeatability and reproducibility of science, as well as favouring peer-review of both the methods themselves and results arising from their use.

One mechanism for sharing nurtures a distinctive community of biodiversity workflow practitioners within the well-established myExperiment online workflows repository [55]. This social repository provides workflow publishing, sharing and searching facilities. Within myExperiment we have established a discrete group with its own distinctive branding, where our workflows are shared. The BioVeL group [56] allows scientists from the biodiversity community to upload their workflows, in silico experiments, results and other published materials. Currently (at the time of writing) the BioVeL domain of myExperiment features almost 40 workflows. Through active participation and collaboration, users can contribute to and benefit from a pool of scientific methods in the biodiversity domain, and be acknowledged when their workflows have been re-used or adapted.

Executing workflows: (E) in Fig. 1

As a part of the BioVeL virtual laboratory, we designed and deployed the BioVeL Portal (http://portal.biovel.eu) [26], an Internet Web browser based execution environment for workflows. The Portal does not require any local software installations and scientists can use a Web browser interface to upload and execute workflows from myExperiment (or they can choose one already uploaded). Once initiated, users are able to follow the progress of the analysis and interact with it to adjust parameters or to view intermediate results. When satisfied with the final results, a user can share these with others or download them to their local computer. Results can be used as inputs to subsequent work or incorporated into publications, with citation to the workflow and parameters that produced them.

We adopted and adapted SEEK, the systems biology and data model platform [57] to meet the needs of the biodiversity science and ecology community. We re-branded SEEK for BioVeL and gave it a user interface suited to typical tasks associated with uploading and executing workflows and managing the results of workflow runs. We equipped it to execute workflows on the users' behalf, for multiple users and multiple workflows simultaneously.

The BioVeL Portal offers functions for discovering, organising and sharing both blueprints for analyses (i.e., workflows) as well as results of analyses (i.e., workflow runs) among collaborators and groups. The Portal provides users with their own personal workspace in which to execute workflows using their own data and to keep their results. Users can manage how their results are shared. At any time, they can share workflows and results publicly, within and between projects, or in groups of individuals. Users can return to their work at any time and pick up where they left off. This ability to create 'pop-up' collaborations by inviting individuals into a shared workspace to explore an emerging topic, and to keep track of work offers an immediate way to establish exciting new collaborations with little administrative overhead.

Presently (at the time of writing) there are 50 workflows publicly available within the BioVeL Portal. To support them we have provided a Support Centre, including training materials, documentation (http://wiki.biovel.eu) and helpdesk (mailto:support@biovel.eu) where users can obtain assistance. The expanded version of Table 2 in the Additional file 3 gives full details.

As an example, workflows created for invasive alien species studies [54] have been frequently re-used in other scientific analyses; for example, to predict potential range shifts of commercially important species under scenarios of climatic change [36], and to describe the biogeographic range of Asian horseshoe crabs [58]. Here especially, seamless linkage of data access to species occurrence records and environmental data layers, as well as the partly automated cleaning and processing procedures are useful functions when running niche modelling experiments for several species across a large number of parameter settings. The Data Refinement Workflow [59] has likewise been used in both preparation of niche modelling experiments as well as in analysis of historical changes in benthic community structure [34]. Here especially, the Taxonomic Name Service and data cleaning functions were helpful in resolving synonyms, correcting misspellings, and dealing with other inconsistencies in datasets compiled from different sources.

The Portal also offers functions for data and parameter sweeping. This includes batch processing of large quantities of separate input data using the same parameters (data sweeping) and batch processing the same data using different parameters (parameter sweeping). As example, the niche modelling workflow (Table 2, ENM) has 15 user interaction steps where parameters or files have to be supplied. When repeated manually multiple times this is error-prone. The sweep functions can be used to automate systematic exploration of how data and parameters affect the results in a larger analysis. In such cases the Portal can automatically initiate multiple workflow runs in parallel, significantly reducing the time needed to complete all the planned experiments. It is possible to delegate computing intensive operations to 3rd party computing facilities such as a high performance computing (HPC) centre or a cloud computing service.

Scientists have used the batch processing capability of BioVeL to explore parameter space in models and to generate comparable results for a large number of species. For example, in investigations of present and future distributions of shellfish (Asian Horseshoe Crabs) under predicted climate changes, the technique has been used to generate consensus outputs based on several different, individually executed niche modelling algorithms (for example: MaxEnt, Support Vector Machine and Environmental Distance) to build and evaluate a wide range of models with different combinations of environmental data layers (parameter sweep with 12 different combinations of environmental layers); and to build models for multiple ecologically similar species (data sweep for six intertidal shellfish species). Such calculations, running three modelling algorithms with 12 different environmental datasets for six species (i.e., 216 models) can be concluded in a single day via the Portal.

Execute a workflow in external applications

Finally, BioVeL supports executable workflows to be embedded in other web sites and applications; just like YouTube™ videos can be embedded in web sites. Such embedding would allow, for example a web site giving statistical information about fluctuations in a species population to be rapidly updated as soon as the most recent survey data is entered. Or, it could allow members of the public (e.g., school students) to explore 'what-if' scenarios by varying the data and parameter detail without specific knowledge of the workflow executing behind the website. In Scratchpads [60], 6000+ users have the possibility now to embed workflows into their personal and collaborative Scratchpad websites to repeatedly process their data; as in BioAcoustica [61] for example. BioAcoustica is an online repository and analysis platform for scientific recordings of wildlife sounds. It embeds a workflow based on an R package that allows scientists contributing data to the site to analyse the sounds.

Distributed computing infrastructure and high performance computing

Although not reported in detail in the present paper, we configured and deployed the underlying information and communications technology (ICT) infrastructure needed to support a multi-party distributed heterogeneous network of biodiversity and ecology Web services (the Biodiversity Service Network), and the execution of workflows simultaneously by multiple users. We offered a pilot operational service. In doing so we utilised different kinds of distributed computing infrastructure, including: Amazon web services (AWS), EGI.eu Federated Cloud/INFN ReCaS Network Computing, SZTAKI Desktop Grid, as well as various localised computer servers under the administration of the partner and contributing organisations. This demonstrates the ability of the BioVeL Web services Network to cope with heterogeneity of underlying infrastructures by adopting a service-oriented computing approach.

Discussion
Principal findings

We wanted to kick-start familiarisation and application of the workflow approach in biodiversity science and ecology. Our work shows that the Biodiversity Virtual e-Laboratory (BioVeL) is a viable operational and flexible general-purpose approach to collaboratively processing and analysing biodiversity and ecological data. It integrates existing and popular tools and practices from different scientific disciplines to be used in biodiversity and ecological research. This includes functions for: accessing data through curated Web services; performing complex in silico analysis through exposure of R programs, workflows, and batch processing functions; on-line collaboration through sharing of workflows and workflow runs; experiment documentation through reproducibility and repeatability; and computational support via seamless connections to supporting computing infrastructures.

Most of these functions do exist today individually and are frequently used by biodiversity scientists and ecologists. However, our platform unites them as key components of large-scale biodiversity and ecological research in a single *virtual research environment*.

We developed scientifically useful workflows in thematic sub-domains (taxonomy, phylogenetics, metagenomics, niche and population modelling, biogeochemical modelling) useful to address topical questions related to ecosystem functioning and valuation, biospheric carbon sequestration and invasive species management. These topical science areas have real unanswered scientific questions, with a potentially high societal impact arising from new knowledge generated. We applied our workflows to case studies in two of these areas, as well as to case studies more generally in niche modelling and phylogenetics. Our scientific results (Additional file 4) demonstrate that the combination of functions in BioVeL have potential to support biodiversity and ecological research involving large amounts of data and distributed data, tools and researchers in the future.

Strengths and weaknesses
Productivity gains
The key criterion for success of the infrastructure and the associated use of Web services is delivering the ability to perform biodiversity and ecology research faster, and/or cheaper, and/or with a higher quality. From the scientists' perspective, we have seen increased ease of use and improved ability to manage complexity when faced with manipulation and analysis of large amounts of data. The upfront investment to design new workflows pays off not only in the multiple applications of it to different scientific questions and re-uses of it across data and parameter sweeps; but also in terms of time to accomplish work, especially when large analysis can be easily delegated to appropriate computing infrastructures.

Exploiting distributed data resources and processing tools via the Internet opens access to vastly greater computing capacity and analytical capability than is normally available in a desktop or local cluster computer. Our work with the Biome-BGC workflows (see Additional file 4) model and supporting database reused 1100 datasets and 84 parameter sets 84 times, achieving a performance of about 92,000 model runs during 22 days (three simulations per minute on average).

Meeting conditions for reproducibility of work
Wrapping R programs and Web service interactions in workflows removes the repetitiveness, inconsistency and lack of traceability of manual work, while permitting consistent repetition of an experiment. The BioVeL system keeps track of how the analysis was done, documents

the research steps and retains the provenance of how the workflow executed. This provenance information helps in recording and tracing back to decisions, reducing time for error discovery and remedy; as well as for malization for reporting. It is these consistent processing and tracking features (rather than speed of execution per se) that are a principal advantage when dealing with large amounts of data, and when running many algorithms and different parameter settings across that data. They give an investigator the ability to document, overview, share and collaboratively evaluate the results from a complex large-scale study.

A progressive drive towards more open research, including with greater reproducibility [62, 63] and stronger emphasis on '*elevating the status of code in ecology*' [64] is leading journal publishers (including those of the present article, BioMed Central) to make it a condition of publication that data (and increasingly, software) should be accessible and easy to scrutinise. As noted in a BMC Ecology editorial [65] the idea that the data underlying a study should be available for validation of conclusions is not unreasonable. By implication, "*...readily reproducible materials... freely available...*" includes the workflows and software that have been used for preparation and analysis of that data. Using the BioVeL ecosystem is an easy way of meeting such conditions.

Increased levels of inter-disciplinary working
The infrastructure enables increased levels of inter-disciplinary working and more scalable scientific investigations. The first generation of publications resulting from the e-laboratory is encouraging and shows that BioVeL services start providing these features. The majority of the users of ecological niche modelling workflows (for example) may not be experts in this field. They can be scientists with backgrounds in ecology, systematics, and environmental sciences that use the workflows to become familiar with new analytical methods [33, 36, 54]. Similarly, the taxonomic, phylogenetic and metagenomic services have been used by scientists to complement their existing analytical expertise with that from another field [36, 44]. A further example: Amplicon-based metagenomics approaches have been widely used to investigate both environmental and host associated microbial communities. The BioMaS (Bioinformatic analysis of Metagenomic ampliconS) Web service (Table 2; [43]) offers a way to simply and accurately characterize fungal and prokaryotic communities, overcoming the necessity of computer-science skills to set up bioinformatics workflows. This is opening the field to a wide range of researchers, such as molecular biologists [66] and ecologists [42].

Towards more comprehensive and global investigations

The principal BioVeL functionalities support more comprehensive and global investigations of biodiversity patterns and ecological processes. Such investigations are not impossible today but they are expensive and often can only be addressed with large and resourceful scientific networks. Exploiting such scalability is particularly attractive, for example to prepare and verify large-scale data products relating to the essential biodiversity variables (EBV) [67]; for phyloclimatic investigations [68]; and for characterisation of biogeographic regions [69]. In addition, complex predictive approaches that couple mechanistic with statistical models may benefit from the use of the BioVeL environment [70]. All these kinds of processing usually require integration of distributed biological, climate and environmental data, drawn from public databases as well as personal sources. They depend on a wide range of analytical capabilities, computational power and, most importantly the combined knowledge of a large number of experts. The BioVeL platform can connect these critical resources on the fly. In conjunction with an easy-to-use interface (the Portal) they can be used to dynamically create ad hoc scientific networks and cross-disciplinary collaborations fast. In the absence of dedicated funding it is a mechanism that can help scientists to react more quickly to newly emerging socio-environmental problems. The infrastructure is increasingly used for this purpose of 'next-generation action ecology' [71].

Dependency on supporting infrastructure and robust Web services

One apparent drawback of the approach we describe is dependency on the ready availability of robust infrastructure to provide access to data and to processing capabilities. This is out of the control of the end-user scientists but it is a matter for service providers. It is the same issue we face as consumer users of the Internet, whereby we rely on a well-developed portfolio of robust related services; for example, for making our travel arrangements with airlines, rental cars and hotels. In the biodiversity and ecology domain this is not the case. The portfolio of services is not yet well developed. There are only a limited number of robust large-scale service providers thus far (GBIF, EMBL-EBI, OBIS, PANGAEA to name just four examples) and not many smaller ones. Compare this with the life sciences community, where more than 1000 Web services from more than 250 service providers are listed in the BioCatalogue [30]. By promoting the Biodiversity Catalogue [29] as the well-founded one-stop shop to keep track of high-quality Web services as they appear; and annotating entries in the catalogue to

document their capabilities we are hoping to encourage steps towards greater maturity. As with all software, the services and workflows, and the platforms on which they run have to be maintained. There is a cost associated with that. Projects like Wf4ever ("Workflow for Ever") [72] have examined some of the challenges associated with preserving scientific experiments in data-intensive science but long-term it is a community responsibility that still has to be addressed.

Results in context

Prototypes to operational service

Historical projects such the UK's BBSRC-funded Biodiversity World project, 2003–2006 [73] and the USA's NSF-funded SEEK project 2002–2007 [74] (not to be confused with the SEEK platform for systems biology) successfully explored the potential of automated workflow systems for large-scale biodiversity studies. Moving from concept-proving studies towards a reliable infrastructure supporting collaboration is a substantial challenge. In the long-run such infrastructure has to robustly serve many thousands of users simultaneously.

With BioVeL we offer a pilot-scale operational service, delivered continuously and collaboratively by multiple partner organisations. This "Biodiversity Commons" of workflows, services and technology products can be used by anyone. Embedding elements of it within third party applications and contexts such as Scratchpads [75], Jupyter/iPython Notebooks [76], data analysis for Ocean Sampling Day collection events [41], national level biodiversity information infrastructures [77] and biodiversity observation networks has a multiplier effect, making it possible for all users of those wider communities and others to execute and exploit the power of workflows.

The underlying SEEK platform [57] on which BioVeL is based (not to be confused with the SEEK project mentioned above) is designed fundamentally to assist scientists to organise their digital data analysis work. As well as supporting execution of workflows, it allows them to describe, manage and execute their projects. These normally consist of experiments, datasets, models, and results. It helps scientists by gathering and organising pieces of information related to these different artefacts into different categories and making links between them; namely: yellow pages (programmes, people, projects, institutions); experiments (investigations, studies, assays); assets (datasets, models, standard operating procedures, publications); and activities (presentations, events). Not all the functionality of SEEK is presently enabled in the BioVeL Portal variant but in future it can be enabled as the needs of the community grow.

Global research infrastructures

Globally, organisations with data and processing facilities across the world are working to deliver research infrastructure services to their respective scientific user communities. Initiatives in Europe (LifeWatch), Australia (Atlas of Living Australia), Brazil (speciesLink network, SiBBr Brazilian Biodiversity Information System), China (Academy of Sciences National Specimen Information Infrastructure and the World Federation of Culture Collections), South Africa (SANBI Integrated Biodiversity Information System), USA (DataONE and NEON) as well as GBIF, Catalogue of Life, Encyclopedia of Life, Biodiversity Heritage Library, and others are all mutually interdependent. They are driven not only by the direct needs of curiosity science but also more and more by the science needs of global policy initiatives. All research infrastructure operators recognise the need to remove barriers to global interoperability through common approaches based on interoperable Web services and promoting the development, sharing and use of workflows [78]. Our work is relevant to and supports this goal.

IPBES, GEO BON, and essential biodiversity variables

The Intergovernmental Science-Policy Platform on Biodiversity and Ecosystem Services (IPBES) has to provide assessments of the state of the environment [4]. Guidelines for authors of assessments focus on several areas highly relevant in the context of the present paper: (i) improving access to data, information and knowledge of all types; (ii) managing data uncertainty and quality; and (iii) performing various model simulations and scenario-based analysis of future developments [79]. Additionally, some key principles and practices are given to ensure respect for and to consistently apply transparency at all steps of data collection, selection, analysis and archiving. This is so that IPBES can enable replication of results and informed feedback on assessments; comparability across scales and time; and use of systematic methodology and shared approach in all steps of the assessment process. The workflow approach, applied via BioVeL tools and infrastructure with specific additional developments to support Essential Biodiversity Variables in conjunction with other partners from the Group on Earth Observations Biodiversity Observation Network (GEO BON) would be a very progressive move to fulfil these requirements [80].

Towards wider use of workflows

Tools for creating, executing and sharing workflows to process and analyse scientific data (see third paragraph of the introduction) have been around for 15 years. Most of these started life as desktop tools. Indeed, Kepler was a product of the previously mentioned SEEK project [74], with origins in ecological science. Despite variable usage across disciplines the cumulative experience is that the general approach of configurable, flexible workflows to assist the process of transforming, analysing and modelling with large amounts of data is well accepted. Workflows as a paradigm for orchestrating disparate capabilities to pursue large-scale data intensive ecological science are an important next step for the community. They represent *"primacy of method"* for a community evolving towards a new research culture that is increasingly dependent on working collaboratively, exchanging and aggregating data and automating analyses [63, 81]. They balance shareability, repeatability and flexibility with simplicity.

Conclusions

In conclusion, we have presented a virtual laboratory that unites critical functions necessary for supporting complex and data intensive biodiversity science and ecological research in the future. We have created and deployed multiple Web services and 'off-the-shelf' packs of pre-defined workflows that meet the specific needs for several types of scientific study in biodiversity science and ecology. We have made these available respectively through a catalogue of services, the Biodiversity Catalogue and via a public repository of workflows, myExperiment. Each part can be used independently of the others or as an integrated part of the platform as a whole. BioVeL is operational and we have provided guidelines for its use (Additional file 1). We can refer (via Additional file 4) to many scientific studies that have used and are using the platform. We have raised awareness of what is possible and have laid foundations for further adoption and convergence activities as more ecologists encounter the worlds of big data and open science.

We foresee two main directions of future development

Firstly, building complete, flexible, independent virtual laboratories will become more commonplace. Scientists want to be in control of their own real physical laboratories and there is no reason to assume they will not want to be in control of their own virtual laboratories for data processing and analysis. As with their physical laboratories, scientists will not want to build all elements from scratch. They will wish to take advantage of proven ready-built workflows and workflow components built and tested by trusted suppliers. Such workflows and components are part of an emerging Biodiversity Commons those labs can draw upon. We already have the first cases where scientists use BioVeL to expose and share their own analytical assets, and begin to pool and aggregate tools developed by the community rather than for the community. Capabilities for data management have been built into the core of BioVeL because the users requested it. Combined, such functionality provides a comprehensive collaborative platform that supports the needs of modern-day reproducible digital science.

The second direction extends towards building the sufficient base of robust data and computational services in biodiversity and ecological sciences (Web services, R programs and command line tools) that can be combined to automate multi-stage processing and analysis tasks. This will give scientists the freedom to compose sequences of processing steps to perform the scientific tasks they know and are familiar with today. At the same time the approach retains a dual flexibility. It permits the addition of new capabilities as those develop and evolve. It also allows for composing capabilities in ways that cannot presently be foreseen; thus meeting scientists' needs of the future.

Additional files

Additional file 1. 'How-to' guidelines for the biodiversity virtual e-laboratory.

Additional file 2: Table S1. Service groups and capabilities for processing and analysis in biodiversity science.

Additional file 3: Table S2. Workflows for biodiversity science.

Additional file 4. Scientific studies using the BioVeL platform.

Abbreviations

API: application programming interface; BioMaS: bioinformatic analysis of metagenomic ampliconS; BioVeL: biodiversity virtual e-laboratory; BPI: Bayesian phylogenetic inference; CoL: Species 2000/ITIS catalogue of life; CRAN: comprehensive R archive network; CRIA: Centro de Referência em Informação Ambiental; DataONE: data observation network for earth; DRW: data refinement workflow; EBI: European Bioinformatics Institute; EBV: essential biodiversity variable; EMBL: European Molecular Biology Laboratory; EMBL-EBI: EMBL European Bioinformatics Institute; ENM: ecological niche modelling; ESI: ecosystem service indicator; ESW: ENM statistical difference workflow; GBIF: Global Biodiversity Information Facility; GEO BON: Group on Earth Observations Biodiversity Observation Network; HPC: high performance computing; IPBES: Intergovernmental Science-Policy Platform on Biodiversity and Ecosystem Services; IPM: integral projection model; LONI: Laboratory of Neuro Imaging; MMT: microbial metagenomic trait calculation and statistical analysis workflow; MPM: matrix projection model; MSA-PAD: multiple sequence alignment based on PFAM accessed domain information; NEON: National Ecological Observatory Network; OBIS: Ocean Biogeographic Information System; OSD: Ocean Sampling Day; OTU: operational taxonomic unit; PURL: persistent uniform resource locators; REST: representational state transfer (a mechanism for communicating between application programs and some web services cf. SOAP); SANBI: South African National Biodiversity Institute; SEEK: science environment for ecological knowledge (historical project); SEEK: systems biology and data model platform (unrelated to SEEK, the project); SOAP: simple object access protocol (a mechanism for communicating between application programs and some web services cf. REST); VRE: virtual research environment; WoRMS: World Register of Marine Species.

Authors' contributions

The study was conceived and designed by AH, MO and CG with assistance from FH, AP, HS, YdJ and SV. It was directed by AH, MO, CG, EP and HS. All authors contributed to the detailed implementation of the work, to its reporting and to writing parts of the present article and Additional information. All authors read and approved the final manuscript.

Author details
[1] School of Computer Science and Informatics, Cardiff University, Queens Buildings, 5 The Parade, Cardiff CF24 3AA, UK. [2] School of Computer Science, University of Manchester, Kilburn Building, Oxford Road, Manchester M13 9PL, UK. [3] Institute for Biodiversity and Ecosystem Dynamics (IBED), University of Amsterdam, PO Box 94248, 1090 Amsterdam, The Netherlands. [4] Institute of Biomembranes and Bioenergetics (IBBE), National Research Council (CNR), via Amendola 165/A, 70126 Bari, Italy. [5] Department of Meteorology, Eötvös Loránd University, Pázmány sétány 1/A, Budapest 1117, Hungary. [6] Department of Marine Sciences, University of Gothenburg, Box 463, 405 30 Gothenburg, Sweden. [7] Centro de Referência em Informação Ambiental, Avenida Dr. Romeu Tórtima, 388, Campinas, SP 13084-791, Brazil. [8] SIB Labs, Joensuu Science Park, University of Eastern Finland, P.O. Box 111, 80101 Joensuu, Finland. [9] Institute of Nuclear Physics (INFN), Via E. Orabona 4, 70125 Bari, Italy. [10] Max Planck Institute for Marine Microbiology, Celsiusstrasse 1, 28359 Bremen, Germany. [11] Jacobs University Bremen GmbH, Campus Ring 1, 28359 Bremen, Germany. [12] Stichting EGI (EGI.eu), Science Park 140, 1098 Amsterdam, The Netherlands. [13] Botanic Garden and Botanical Museum Berlin, Freie Universität Berlin, Königin-Luise-Strasse 6-8, 14195 Berlin, Germany. [14] IT Services, University of Manchester, Kilburn Building, Oxford Road, Manchester M13 9PL, UK. [15] Fraunhofer Institute for Intelligent Analysis and Information Systems (IAIS), Schloss Birlinghoven, 53757 Sankt Augustin, Germany. [16] Naturalis Biodiversity Center, Postbus 9517, 2300 Leiden, The Netherlands. [17] MTA-SZIE Plant Ecology Research Group, Szent István University, Páter K. u.1., Gödöllő 2103, Hungary. [18] Institute of Ecology and Botany, Centre for Ecological Research, Hungarian Academy of Sciences, Alkotmány u. 2-4., Vácrátót 2163, Hungary. [19] Swedish Species Information Centre/ArtDatabanken, Swedish University of Agricultural Sciences, Bäcklösavägen 10, 750 07 Uppsala, Sweden. [20] Department of Forest Sciences, University of Helsinki, P.O. Box 27, 00014 Helsinki, Finland. [21] Fondation pour la Recherche sur la Biodiversité (FRB), 195, rue Saint-Jacques, 75005 Paris, France. [22] Department of Biosciences, Biotechnology and Biopharmaceutics, University of Bari "A. Moro", via Orabona, 1514, 70126 Bari, Italy. [23] Finnish Museum of Natural History, University of Helsinki, P.O. Box 17, 00014 Helsinki, Finland. [24] Institute of Biomedical Technology (ITB), National Research Council (CNR), via Amendola 122/D, 70126 Bari, Italy.

Acknowledgements
More than 80 persons across 15 partner organisations have been involved in the work over 4 years as part of the BioVeL consortium. Without the contributions of each of them we would not have arrived at the position of being able to offer a pilot service to the scientific community for in silico digital biodiversity and ecosystems research. We thank you for your contribution. We acknowledge the contribution of Yannick Legré, who contributed to the conception and design of the work but ultimately was unable to participate in the project.

We acknowledge the helpful support and constructive criticism we received during the validation stage of our work from A. Townsend Peterson and Jorge Soberón (both of Biodiversity Institute and Natural History Museum, University of Kansas), Robert Guralnick (Florida Museum of Natural History, University of Florida) and Mark P. Schildhauer (National Center for Ecological Analysis and Synthesis, University of California). They gave us the benefit of their expertise and experience.

Finally, we are grateful for the continuing support for our ideas from among our friends in the wider community of biodiversity and ecological informatics specialists around the world, and from those scientists who "get" the idea of e-Science. They number more than 400 persons. Together we are on a journey.

Competing interests
The authors declare that they have no competing interests.

Funding
The work reported in this article has been performed by the Biodiversity Virtual e-Laboratory (BioVeL) project (http://www.biovel.eu) funded by the European Union 7th Framework Programme, Grant no. 283359 (2011–2014). The work was furthermore supported by the Swedish LifeWatch project funded by the Swedish Research Council (Grant no. 829-2009-6278). The funders had no role in study design, data collection and analysis, decision to publish, or preparation of the present article.

References

1. Evans MR. Modelling ecological systems in a changing world. Philos Trans R Soc Lond B Biol Sci. 2012;367:181–90. doi:10.1098/rstb.2011.0172.
2. Evans MR, Bithell M, Cornell SJ, Dall SRX, Díaz S, Emmott S, et al. Predictive systems ecology. Proc Biol Sci. 2013;280:20131452. doi:10.1098/rspb.2013.1452.
3. Purves D, Scharlemann J, Harfoot M, Newbold T, Tittensor DP, Hutton J, et al. Ecosystems: time to model all life on Earth. Nature. 2013;493:295–7. doi:10.1038/493295a.
4. Díaz S, Demissew S, Carabias J, Joly C, Lonsdale M, Ash N, et al. The IPBES conceptual framework—connecting nature and people. Curr Opin Environ Sustain. 2015;14:1–16. doi:10.1016/j.cosust.2014.11.002.
5. Hampton SE, Strasser CA, Tewksbury JJ, Gram WK, Budden AE, Batcheller AL, et al. Big data and the future of ecology. Front Ecol Environ. 2013;11:156–62. doi:10.1890/120103.
6. Michener WK, Jones MB. Ecoinformatics: supporting ecology as a data-intensive science. Trends Ecol Evol. 2012;27:85–93. doi:10.1016/j.tree.2011.11.016.
7. Koureas D, Arvanitidis C, Belbin L, Berendsohn W, Damgaard C, Groom Q, et al. Community engagement: the "last mile" challenge for European research e-infrastructures. Res Ideas Outcomes. 2016;2:e9933.
8. Altintas I, Berkley C, Jaeger E, Jones M, Ludascher B, Mock S. Kepler: an extensible system for design and execution of scientific workflows. In: Proceedings 16th international conference on scientific and statistical database management. IEEE. 2004:423–424. doi:10.1109/SSDM.2004.1311241.
9. Deelman E, Vahi K, Juve G, Rynge M, Callaghan S, Maechling PJ, et al. Pegasus, a workflow management system for science automation. Futur Gener Comput Syst. 2015;46:17–35. doi:10.1016/j.future.2014.10.008.
10. Wolstencroft K, Haines R, Fellows D, Williams A, Withers D, Owen S, et al. The Taverna workflow suite: designing and executing workflows of web services on the desktop, web or in the cloud. Nucleic Acids Res. 2013;41:W557–61. doi:10.1093/nar/gkt328.
11. Callahan SP, Freire J, Santos E, Scheidegger CE, Silva CT. Managing the evolution of dataflows with VisTrails. In: 22nd international conference on data engineering workshops (ICDEW'06). IEEE. 2006:71. doi:10.1109/ICDEW.2006.75.
12. Berthold MR, Cebron N, Dill F, Gabriel TR, Kötter T, Meinl T, Ohl P, Thiel K, Wiswedel B. KNIME-the Konstanz information miner: version 2.0 and beyond. In: ACM SIGKDD explorations Newsletter. vol. 16. New York: ACM. 2009. p. 26–31.
13. Goecks J, Nekrutenko A, Taylor J. Galaxy: a comprehensive approach for supporting accessible, reproducible, and transparent computational research in the life sciences. Genome Biol. 2010;11:R86. doi:10.1186/gb-2010-11-8-r86.
14. Hofmann M, Klinkenberg R. RapidMiner: data mining use cases and business analytics applications. Boca Raton: CRC Press, Taylor & Francis Group; 2013.
15. Fisher P, Hedeler C. A systematic strategy for large-scale analysis of genotype–phenotype correlations: identification of candidate genes involved in African trypanosomiasis. Nucleic Acids Res. 2007;35:5625–33. doi:10.1093/nar/gkm623.
16. Bentley RD, Csillaghy A, Aboudarham J, Jacquey C, Hapgood MA, Bocchialini K, et al. HELIO: the heliophysics integrated observatory. Adv Space Res. 2011;47:2235–9. doi:10.1016/j.asr.2010.02.006.
17. Hardy B, Douglas N, Helma C, Rautenberg M, Jeliazkova N, Jeliazkov V, et al. Collaborative development of predictive toxicology applications. J Cheminform. 2010;2:7. doi:10.1186/1758-2946-2-7.
18. Rex DE, Ma JQ, Toga AW. The LONI pipeline processing environment. Neuroimage. 2003;19:1033–48. doi:10.1016/S1053-8119(03)00185-X.
19. Lu Y, Yue T, Wang C, Wang Q. Workflow-based spatial modeling environment and its application in food provisioning services of grassland ecosystem. In: 2010 18th international conference on geoinformatics. IEEE. 2010:1–6. doi:10.1109/GEOINFORMATICS.2010.5567853.
20. Krüger F, Clare EL, Greif S, Siemers BM, Symondson WOC, Sommer RS. An integrative approach to detect subtle trophic niche differentiation in the sympatric trawling bat species Myotis dasycneme and Myotis daubentonii. Mol Ecol. 2014;23:3657–71. doi:10.1111/mec.12512.
21. Michener W, Beach J, Bowers S, Downey L, Jones M, Ludäscher B, et al. Data integration and workflow solutions for ecology. In: Proceedings, 2nd International Workshop on Data Integration in the Life Sciences July 20-22, 2005 Univ Calif, San Diego, San Diego, USA. Lecture Notes in Computer Science. Vol. 3615. Berlin: Springer; 2005. p. 321–324.
22. Pennington D, Higgins D, Peterson A, Jones M, Ludäscher B, Bowers S. Ecological niche modeling using the Kepler workflow system. In: Taylor I, Deelman E, Gannon D, Shields M, editors. Workflows for e-Science scientific workflows for grids. London: Springer; 2007. p. 91–108. doi:10.1007/978-1-84628-757-2_7.
23. Jarnevich CS, Holcombe TR, Bella EM, Carlson ML, Graziano G, Lamb M, et al. Cross-scale assessment of potential habitat shifts in a rapidly changing climate. Invasive Plant Sci Manag. 2014;7:491–502. doi:10.1614/IPSM-D-13-00071.1.
24. Dou L, Zinn D, McPhillips T, Köhler S, Riddle S, Bowers S, et al. Scientific workflow design 2.0: demonstrating streaming data collections in Kepler. In: IEEE 27th international conference on Data engineering (ICDE). IEEE. 2011. doi:10.1109/ICDE.2011.5767938.
25. Dou L, Cao G, Morris P, Morris R, Ludäscher B, Macklin J, et al. Kurator: a Kepler package for data curation workflows. Proc Comput Sci. 2012;9:1614–9. doi:10.1016/j.procs.2012.04.177.
26. BioVeL portal. http://portal.biovel.eu/. Accessed 21 Mar 2016.
27. Papazoglou MP, Georgakopoulos D. Introduction: service-oriented computing. Commun ACM. 2003;46:24. doi:10.1145/944217.944233.
28. De Giovanni R, Torres E, Amaral R, Blanquer I, Rebello V, Canhos V. OMWS: a web service interface for ecological niche modelling. Biodivers Inform. 2015;10:35–44. doi:10.17161/bi.v10i0.4853.
29. Biodiversity catalogue. http://www.biodiversitycatalogue.org/. Accessed 4 Apr 2015.
30. Bhagat J, Tanoh F, Nzuobontane E, Laurent T, Orlowski J, Roos M, et al. BioCatalogue: a universal catalogue of web services for the life sciences. Nucleic Acids Res. 2010;38((Web Server issue)):W689–94. doi:10.1093/nar/gkq394.
31. Ihaka R, Gentleman R. R: a language for data analysis and graphics. J Comput Graph Stat. 1996;5:299–314. doi:10.1080/10618600.1996.10474713.
32. Racine JS. RStudio: a platform-independent IDE for R and sweave. J Appl Econom. 2012;27:167–72. doi:10.1002/jae.1278.
33. Leidenberger S, Obst M, Kulawik R, Stelzer K, Heyer K, Hardisty A, et al. Evaluating the potential of ecological niche modelling as a component in marine non-indigenous species risk assessments. Mar Pollut Bull. 2015;97:470–87. doi:10.1016/j.marpolbul.2015.04.033.
34. Obst M, Vicario S, Berggren M, Karlsson A, Lundin K, Haines R, et al. Marine long-term biodiversity assessment indicates loss of species richness in the Skagerrak and Kattegat region. Mar Biodivers (**In review**).
35. De Giovanni R, Williams AR, Vera Hernández E, Kulawik R, Fernandez FQ, Hardisty AR. ENM components: a new set of web service-based workflow components for ecological niche modelling. Ecography (Cop). 2015;. doi:10.1111/ecog.01552.
36. Laugen AT, Hollander J, Obst M, Strand Å. The Pacific Oyster (Crassostrea gigas) invasion in Scandinavian coastal waters in a changing climate: impact on local ecosystem services. In: Canning-Clode J, editor. Biological invasions in aquatic and terrestrial systems: biogeography, ecological impacts, predictions, and management. Berlin: De Gruyter Open; 2015. p. 230–52.
37. Hidy D, Barcza Z, Haszpra L, Churkina G, Pintér K, Nagy Z. Development of the Biome-BGC model for simulation of managed herbaceous ecosystems. Ecol Model. 2012;226:99–119. doi:10.1016/j.ecolmodel.2011.11.008.
38. Sándor R, Ma S, Acutis M, Barcza Z, Ben Touhami H, Doro L, et al. Uncertainty in simulating biomass yield and carbon-water fluxes from grasslands under climate change. Adv Anim Biosci. 2015;6:49–51.
39. Sándor R, Barcza Z, Hidy D, Lellei-Kovács E, Ma S, Bellocchi G. Modelling of grassland fluxes in Europe: evaluation of two biogeochemical models. Agric Ecosyst Environ. 2016;215:1–19. doi:10.1016/j.agee.2015.09.001.
40. Sándor R, Acutis M, Barcza Z, Doro L, Hidy D, Köchy M, et al. Multi-model simulation of soil temperature, soil water content and biomass in Euro-Mediterranean grasslands: uncertainties and ensemble performance. European. Eur J Agron. 2016; (**In Press**). doi:10.1016/j.eja.2016.06.006.
41. Kopf A, Bicak M, Kottmann R, Schnetzer J, Kostadinov I, Lehmann K, et al. The ocean sampling day consortium. Gigascience. 2015;4:27. doi:10.1186/s13742-015-0066-5.
42. Manzari C, Fosso B, Marzano M, Annese A, Caprioli R, D'Erchia AM, et al. The influence of invasive jellyfish blooms on the aquatic microbiome in a coastal lagoon (Varano, SE Italy) detected by an Illumina-based deep sequencing strategy. Biol Invasions. 2014;17:923–40. doi:10.1007/s10530-014-0810-2.

43. Fosso B, Santamaria M, Marzano M, Alonso-Alemany D, Valiente G, Donvito G, et al. BioMaS: a modular pipeline for Bioinformatic analysis of Metagenomic AmpliconS. BMC Bioinformatics. 2015;16:203. doi:10.1186/s12859-015-0595-z.

44. Sandionigi A, Vicario S, Prosdocimi EM, Galimberti A, Ferri E, Bruno A, et al. Towards a better understanding of Apis mellifera and Varroa destructor microbiomes: introducing "phyloh" as a novel phylogenetic diversity analysis tool. Mol Ecol Resour. 2014;15:697–710. doi:10.1111/1755-0998.12341.

45. Antonelli A, Hettling H, Condamine FL, Vos K, Nilsson RH, Sanderson MJ, et al. Towards a self-updating platform for estimating rates of speciation and migration, ages, and relationships of taxa (SUPERSMART). Syst Biol. 2016. doi:10.1093/sysbio/syw066.

46. Balech B, Vicario S, Donvito G, Monaco A, Notarangelo P, Pesole G. MSA-PAD: DNA multiple sequence alignment framework based on PFAM accessed domain information. Bioinformatics. 2015;31:2571–3. doi:10.1093/bioinformatics/btv141.

47. Delić D, Balech B, Radulović M, Lolić B, Karačić A, Vukosavljević V, et al. Vmp1 and stamp genes variability of "Candidatus phytoplasma solani" in Bosnian and Herzegovinian grapevine. Eur J Plant Pathol. 2016;45:221–5. doi:10.1007/s10658-015-0828-z.

48. RStudio integrated development environment. https://www.rstudio.com/. Accessed 15 Jul 2016.

49. CRAN task view: web technologies and services. https://cran.r-project.org/web/views/WebTechnologies.html. Accessed 16 Oct 2015.

50. CRAN task view: analysis of ecological and environmental data. https://cran.r-project.org/web/views/Environmetrics.html. Accessed 22 Oct 2015.

51. Creating a workflow from an R script. https://wiki.biovel.eu/x/iYSz. Accessed 28 Aug2015.

52. Taverna workbench for biodiversity. http://www.taverna.org.uk/download/workbench/2-5/biodiversity/. Accessed 16 Jul 2016.

53. World register of marine species (WoRMS). http://www.marinespecies.org/. Accessed 21 Mar 2016.

54. Leidenberger S, De Giovanni R, Kulawik R, Williams AR, Bourlat SJ. Mapping present and future potential distribution patterns for a meso-grazer guild in the Baltic Sea. J Biogeogr. 2015;42:241–54. doi:10.1111/jbi.12395.

55. De Roure D, Goble C, Stevens R. The design and realisation of the myExperiment virtual research environment for social sharing of workflows. Futur Gener Comput Syst. 2009;25:561–7. doi:10.1016/j.future.2008.06.010.

56. myExperiment BioVeL group. http://biovel.myexperiment.org/. Accessed 21 Mar 2016.

57. Wolstencroft K, Owen S, Krebs O, Nguyen Q, Stanford NJ, Golebiewski M, et al. SEEK: a systems biology data and model management platform. BMC Syst Biol. 2015;9:33. doi:10.1186/s12918-015-0174-y.

58. Funch P, Obst M, Quevedo F. et al. Present and future distributions of horseshoe crabs under predicted climate changes. The third international workshop on the science and conservation of horseshoe crabs, June 15–19, 2015, Saikai national park kujukushima, Sasebo-City, Nagasaki, Japan. http://forskningsdatabasen.dk/en/catalog/2297666567. Accessed 18 Oct 2016.

59. Mathew C, Güntsch A, Obst M, Vicario S, Haines R, Williams AR, et al. A semi-automated workflow for biodiversity data retrieval, cleaning, and quality control. Biodivers Data J. 2014;2:e4221. doi:10.3897/BDJ.2.e4221.

60. Haines R, Goble C, Rycroft S, Smith V. Integrating taverna player into scratchpads. Manchester: University of Manchester; 2014. http://zenodo.org/record/10871. Accessed 20 Apr 2015.

61. Baker E, Price BW, Rycroft SD, Hill J, Smith VS. BioAcoustica: a free and open repository and analysis platform for bioacoustics. Database. 2015;2015:bav054. doi:10.1093/database/bav054.

62. The royal society. Science as an open enterprise. Final report june. 2012. http://royalsociety.org/uploadedFiles/Royal_Society_Content/policy/projects/. Accessed 1 Sept 2016.

63. Hampton SE, Anderson SS, Bagby SC, Gries C, Han X, Hart EM, et al. The Tao of open science for ecology. Ecosphere. 2015;6:1–13. doi:10.1890/ES14-00402.1

64. Mislan KAS, Heer JM, White EP. Elevating the status of code in ecology. Trends Ecol Evol. 2016;31:4–7. doi:10.1016/j.tree.2015.11.006.

65. Kenall A, Harold S, Foote C. An open future for ecological and evolutionary data? BMC Ecol. 2014;14:10. doi:10.1186/1472-6785-14-10.

66. Rigoni R, Fontana E, Guglielmetti S, Fosso B, D'Erchia AM, Maina V, et al. Intestinal microbiota sustains inflammation and autoimmunity induced by hypomorphic RAG defects. J Exp Med. 2016;213:355–75. doi:10.1084/jem.20151116.

67. Pereira HM, Ferrier S, Walters M, Geller GN, Jongman RHG, Scholes RJ, et al. Essential biodiversity variables. Science. 2013;339:277–8. doi:10.1126/science.1229931.

68. Verbruggen H, Tyberghein L, Pauly K, Vlaeminck C, Van Nieuwenhuyze K, Kooistra WHCF, et al. Macroecology meets macroevolution: evolutionary niche dynamics in the seaweed Halimeda. Glob Ecol Biogeogr. 2009;18:393–405. doi:10.1111/j.1466-8238.2009.00463.x.

69. Vilhena DA, Antonelli A. A network approach for identifying and delimiting biogeographical regions. Nat Commun. 2015;6:6848. doi:10.1038/ncomms7848.

70. Kearney MR, Wintle BA, Porter WP. Correlative and mechanistic models of species distribution provide congruent forecasts under climate change. Conserv Lett. 2010;3:203–13. doi:10.1111/j.1755-263X.2010.00097.x.

71. White RL, Sutton AE, Salguero-Gómez R, Bray TC, Campbell H, Cieraad E, et al. The next generation of action ecology: novel approaches towards global ecological research. Ecosphere. 2015;6:1–16. doi:10.1890/ES14-00485.1.

72. Belhajjame K, Zhao J, Garijo D, Hettne K, Palma R, Corcho Ó, et al. The research object suite of ontologies: sharing and exchanging research data and methods on the open web. 2014. arXiv:1401.4307.

73. Pahwa JS, Brewer P, Sutton T, Yesson C, Burgess M, Xu X, et al. Biodiversity World: a problem-solving environment for analysing biodiversity patterns. In: Sixth IEEE international symposium on cluster computing and the grid (CCGRID'06). IEEE. 2006:1. doi:10.1109/CCGRID.2006.23.

74. Michener W, Beach J, Jones M, Ludäscher B, Pennington D, Pereira R, et al. A knowledge environment for the biodiversity and ecological sciences. J Intell Inf Syst. 2007;29:111–26. doi:10.1007/s10844-006-0034-8.

75. Smith VS, Rycroft SD, Brake I, Scott B, Baker E, Livermore L, et al. Scratchpads 2.0: a virtual research environment supporting scholarly collaboration, communication and data publication in biodiversity science. Zookeys. 2011;150:53–70. doi:10.3897/zookeys.150.2193.

76. Pérez F, Granger B. IPython: a system for interactive scientific computing. Comput Sci Eng. 2007;9:21–9. doi:10.1109/MCSE.2007.53.

77. Gärdenfors U, Jönsson M, Obst M, Wremp AM, Kindvall O, Nilsson J. Swedish LifeWatch — a biodiversity infrastructure integrating and reusing data from citizen science, monitoring and research. Hum Comput. 2014;1:147–61. doi:10.15346/hc.v1i2.6.

78. García EA, Bellisari L, De Leo F, Hardisty A, Keuchkerian S, Konijn J, et al. Flock together with CReATIVE-B: a roadmap of global research data infrastructures supporting biodiversity and ecosystem science. 2014. http://orca.cf.ac.uk/88151/. Accessed 21 Mar 2016.

79. IPBES Secretariat. Guide on the production and integration of assessments from and across all scales. 2014. http://www.ipbes.net/work-programme/guide-production-assessments. Accessed 21 Mar 2016.

80. Kissling WD, Hardisty A, García EA, Santamaria M, De Leo F, Pesole G, et al. Towards global interoperability for supporting biodiversity research on essential biodiversity variables (EBVs). Biodiversity. 2015:1–9. doi:10.1080/14888386.2015.1068709.

81. Goble C, De Roure D. The impact of workflow tools on data-centric research. In: Hey T, Tansley S, Tolle K, editors. The fourth paradigm: data intensive scientific discovery. Redmond: Microsoft Research; 2009. p. 137–46.

14

Crop diversity loss as primary cause of grey partridge and common pheasant decline in Lower Saxony, Germany

Katrin Ronnenberg[1*], Egbert Strauß[1,2] and Ursula Siebert[1]

Abstract

Background: The grey partridge (*Perdix perdix*) and the common pheasant (*Phasianus colchicus*) are galliform birds typical of arable lands in Central Europe and exhibit a partly dramatic negative population trend. In order to understand general habitat preferences we modelled grey partridge and common pheasant densities over the entire range of Lower Saxony. Spatially explicit developments in bird densities were modelled using spatially explicit trends of crop cultivation. Pheasant and grey partridge densities counted annually by over 8000 hunting district holders over 10 years in a range of 3.7 Mio ha constitute a unique dataset (wildlife survey of Lower Saxony). Data on main landscape groups, functional groups of agricultural crops (consisting of 9.5 million fields compiled by the Integrated Administration and Control System) and landscape features were aggregated to 420 municipalities. To model linear 8 or 10 year population trends (for common pheasant and grey partridge respectively) we use rho correlation coefficients of densities, but also rho coefficients of agricultural crops.

Results: All models confirm a dramatic decline in population densities. The habitat model for the grey partridge shows avoidance of municipalities with a high proportion of woodland and water areas, but a preference for areas with a high proportion of winter grains and high crop diversity. The trend model confirms these findings with a linear positive effect of diversity on grey partridge population development. Similarly, the pheasant avoids wooded areas but showed some preference for municipalities with open water. The effect of maize was found to be positive at medium densities, but negative at very high proportions. Winter grains, landscape features and high crop diversity are favorable. The positive effect of winter grains and higher crop diversity is also supported by the trend model.

Conclusions: The results show the strong importance of diverse crop cultivation. Most incentives favor the cultivation of specific crops, which results in large areas of monocultures. The results confirm the importance of sustainable agricultural policies.

Keywords: Habitat model, Trend analysis, Grey partridge, Common pheasant, Citizen science, Diversity, Maize cultivation

Background

Agricultural intensification has led to a dramatic loss in biodiversity from the middle of the twentieth century until now in Europe [1, 2]. Farmland birds suffer especially from these changes. In general, traditional, heterogeneous small structured fields were found to be beneficial for farmland birds across Europe [3]. Agricultural policies, such as EU-directives or national regulations, play a major role in the rapid change of wildlife habitat in agricultural landscapes. A national act in Germany to increase the proportion of biogas implemented in the EEG directive led to a rapid increase in maize cultivation. This act was hold responsible for recent population developments of farmland birds [4, 5]. In 2008, the EU stopped subsidizing set-aside land which had been

*Correspondence: Katrin.ronnenberg@tiho-hannover.de
[1] Institute for Terrestrial and Aquatic Wildlife Research, University of Veterinary Medicine Hannover, Foundation , Bischofsholer Damm 15, 30173 Hannover, Germany
Full list of author information is available at the end of the article

beneficial to the grey partridge and other farmland birds [6–8]. An increase in production of winter crops, as opposed to more traditional summer grains, was found to be one cause for decreasing segetal flora and to generally less diverse habitats [9, 10].

Most habitat models use abundance data for one to a few years, or point samples over longer periods of time. The use of citizen science data on species densities can provide larger spatial and temporal scales and can be adequate for developing adaptive management programs [11]. A monitoring program is a core component to the management of endangered species; however, it often fails due to financial shortcomings. Many citizen science programs use opportunistic online based databases [12], which often suffer from biased data collection. A long term wildlife survey WTE (Wildtiererfassung Niedersachsen) was established in 1991 with annual questionnaires of hunters in Lower Saxony [13, 14]. With a participation rate of roughly 90 % of hunting districts (equals ca. 8000 districts, 3.7 Mio ha) the hunters built a highly motivated group and provided reasonable estimates of small game species at low cost from a small district to a federal state scale, with confirmed high reliability of estimates of population densities for the grey partridge [15].

The grey partridge (*Perdix perdix*) and the common pheasant (*Phasianus colchicus*) are two small game species typical of the agricultural landscapes in Central Europe. The grey partridge is listed as "endangered" in the Red List of threatened birds in Lower Saxony [16]. As a typical species of traditional small structured farmland the grey partridge was originally widespread in Lower Saxony, but less so in the densely wooded sub-mountainous areas of southern Lower Saxony and the sandy heathlands of eastern Lower Saxony. Today, we find great differences between areas of intensive agricultural use with 0.3:0.9 pairs/km² and <0.2 pairs/km² near the North Sea and in the South. In 2014 the grey partridge still occurred in 41 % of hunting districts [17]. The overall population size in Lower Saxony was estimated at 25,000 breeding pairs in 2008 [18].

The pheasant is not listed on the red list. It was introduced to Central Europe by the Romans and is now a typical species of arable land, pastures and reed edges of open waters. By limiting hunting to pheasant cocks, the actual sex ratio is at about 1:2 for cocks and hens [17]. In the main distribution areas the hen density is between 8:12 hens/km². In the wooded areas of southern and eastern Lower Saxony the density is at <5 hens/km².

Although reintroductions and hunting may confound population trends; for Lower Saxony they currently play a minor role. The grey partridge, specifically, is only released in very limited instances. Pheasant releases have

also decreased considerably. As assessed in 2008, pheasants are released in only 3 % of hunting districts amounting to roughly 5000 per year in the entire area of Lower Saxony [17]. Shooting time for the pheasant is between 1 Oct–15 Jan and 16 Sep–30 Nov for the grey partridge. Since 2012 the Hunting Association of Lower Saxony has encouraged a complete voluntary stop of grey partridge hunting in response to continuous population decline. Hunters have followed this recommendation with very few exceptions [17]. Pheasant hunting still occurs in two thirds of hunting districts; however the hen density should be unaffected. Whereas many studies use hunting bags to model trends [i.e. 19–21], their reliability for small game species is questionable. Differences in motivation biotic and abiotic, as well as legislative changes may influence hunting bags more than changes in population densities [14, 17].

Due to agricultural intensification and two extreme winters, grey partridge and common pheasant populations have experienced dramatic declines since the late 1970s. The common pheasant recovered partly due to artificial reintroductions but the grey partridge has continued to and is currently declining. Since 2006 for the grey partridge and since 2008 for the pheasant, parts of Lower Saxony have seen a rapid decline in population densities. Although their population trend is not identical, the two species show some parallelism in habitat preferences and a common cause of decline is suspected. In analyzing both trends separately we hope to validate the reliability of results.

The areas traditionally inhabiting the highest abundances, most notably the Dümmer and Osnabrücker land in western Lower Saxony, are also the areas with the steepest decline (see also Fig. 1). Resource limitation increases competition, and with scarcer resources, density increases the severity of competition. For the grey partridge, a density dependence of reproductive success was found across Europe [22–24]. After hatching, grey partridge and pheasant chicks rely on insects for survival (first 2–6 weeks for the grey partridge and 2–7 weeks for the pheasant), and beetle banks were found to be beneficial in England [25]. The historic decline of farmland birds, including grey partridges, was due to a decrease in insect diversity and abundance caused by pesticides ([26, 27] and publications therein). Thus, a likely cause may be specific agricultural practices i.e. pesticides that reduce insect and consequently bird abundance [28, 29]. Spatially and temporally explicit data on pesticide application are difficult to obtain. However, with specific crops being particularly unfavorable this may point to adverse cultivation practices. The latest decline causes may be different from the well-established causes as seen above; one example being the preference for maize cultivation

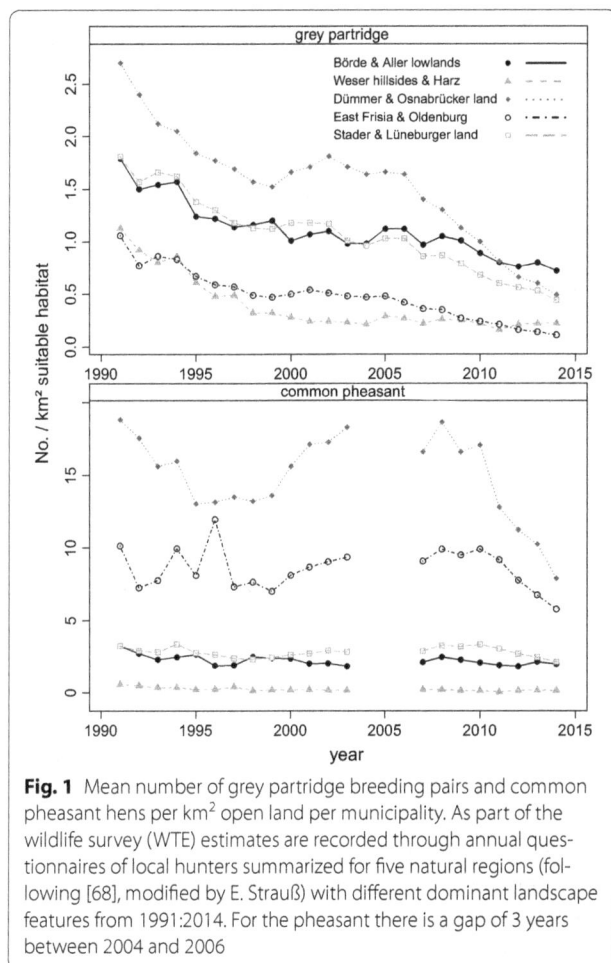

Fig. 1 Mean number of grey partridge breeding pairs and common pheasant hens per km² open land per municipality. As part of the wildlife survey (WTE) estimates are recorded through annual questionnaires of local hunters summarized for five natural regions (following [68], modified by E. Strauß) with different dominant landscape features from 1991:2014. For the pheasant there is a gap of 3 years between 2004 and 2006

for use as biofuel which was politically provoked. Over the last 8 and 10 years, trends at municipal scale were compared to try and establish causes by comparing changes in crop proportions per municipality.

In 2005, Lower Saxony implemented a directive of the European Union to establish a program to monitor and control spatially explicit data on field crops and agricultural subsidies. For research projects depersonalized data of crops can be accessed for research. These are the most detailed data on agricultural use in the European Union and give detailed information on roughly 90 % of agriculturally used lands in Lower Saxony (LEA-Portal (see below) as measure of 100 % area).

In this paper we modelled both the habitat preferences and spatially explicit trends and its causes for the grey partridge and the common pheasant. The first aim is to explain the overall density with landscape traits and functional crop groups and discuss differences in habitat preferences. The second approach is aimed at explaining the negative trends observed over the last 8 and 10 years. Specifically, we expect negative effects of (1) increasing

maize cultivation (2) decreasing proportions of set aside fields (3) increasing proportions of winter crops, (4) the general intensification as indicated by declining crop diversity and increasing field sizes.

Methods

The federal state of Lower Saxony (Germany) has a total area of 47,620 km², of which 60.2 % are in agricultural use and 21.6 % is forested land. 2.3 % of the area is open water. The remaining area is dominated by industrial, traffic and housing areas. North-west Lower Saxony adjoins the North Sea, in the south-east the Harz Mountains rise up to 1000 m elevation. Main habitat conditions differ considerably between natural regions. The north east has predominantly sandy soils and is generally dominated by coniferous forests (Additional file 1: Figure S1a) and arable lands with high proportions of potatoes. East Frisia at the North Sea has the highest proportion of open water (Additional file 1: Figure S1b). The west is mostly dominated by animal husbandry and cattle farms, especially in the north where there are high proportions of grassland (Additional file 1: Figure S1c). In the west there are industrial indoor poultry and pig farms and the crops have the highest level of diversity as indicated by the Shannon index (see below, Additional file 1: Figure S1d). The arable land of the southeast Börde and Aller lowlands has predominantly fertile soils (Chernozems) with a dominating cultivation of wheat and sugar beet. Here the land is poorly structured with large field blocks and few hedges and tree rows (Additional file 1: Figure S1e, f). The climate is oceanic in the west but increasingly continental towards the east. The lowest mean annual temperatures of 6 °C, and at the same time the wettest climate with >1000 mm annual precipitation sums, are recorded in the Harz Mountains. The warmest places are in the westernmost areas at >8.5 °C and the driest area is in the east with <600 mm annual precipitation.

Lower Saxony is politically structured in 47 administrative districts and 455 greater municipalities which are the sample units in the analyses (see below).

Data

Wildlife survey: Lower Saxony is divided into approximately 9000 hunting districts. Since 1991 holders of the hunting districts have provided estimates of wildlife in their hunting ground for a number of species including the grey partridge and common pheasant. Participation of hunting district holders was high throughout the years 1991–2014 and ranged between 80 and 90 % of hunting districts (6151–8300). Also, over 90 % of the huntable area of Lower Saxony was recorded (43,000 km²). As rigorous quality control for the grey partridge, the estimates were evaluated and directly compared to the counts of ornithologists and found to be reliable [15]. Unfortunately a comparable evaluation of pheasants was not undertaken; however even if the accurateness of estimates in absolute numbers cannot be guaranteed, the trend over the years is likely to portray the real population trend.

The counts from all hunting districts were aggregated to 420 greater municipalities (excluding 35 municipalities that were either unincorporated land as well as the islands in the North Sea). The mean huntable area of hunting districts was 500 ha (min 75 ha, max 4877 ha in 2012). The total area between years stayed almost constant as a high participation was achieved in every year. Hunting district holders report huntable area, area of wooded land and open land. For the pheasant district holders reported numbers of pheasant hens and cocks in spring for their hunting district, for the grey partridge they estimated breeding pairs. Pheasants and grey partridge densities were extrapolated to numbers/pairs/km^2 open land for the grey partridge excluding water bodies. In the years 2003:2005 the pheasant population was not assessed. In 2006 participation for the pheasant was lower and thus the year was also omitted from analyses. Tests for plausibility were run every year and obvious mistakes were removed.

IACS: In 2005 the Integrated Administration and Control System (IACS) of the European Union was introduced in Lower Saxony. Since then data on all arable fields, or of all farmers that receive subsidies in any form, have been documented. For Lower Saxony, that comprised roughly 90 % of all agricultural land as indicated by the total area of the LEA-Portal (http://sla.niedersachsen. de/landentwicklung/LEA/). IACS data were provided by the SLA ("Servicezentrum Landentwicklung und Agrarförderung" in Lower Saxony). For each field in the database we obtained data on crop type, the size of the field and the municipality it was situated in. Over 10 years this roughly amounts to 9.5 million fields. For reasons of privacy protection no further details on geographic locations or field ID were provided. Thus, no data on crop rotation or neighborhood statistics were possible. Data were summed to percentage area agricultural land per municipality. IACS data are grouped in over 164 different crops and had to be summarized into ecologically sensible groups (Additional file 2: Table S1 for details). From those groups the Shannon index (see below) was calculated. Further simplification of crop groups was used for the habitat and trend models (see below).

The Shannon Index is a standard measure for alpha diversity in vegetation science [30]. Here we use the same metric for crops within municipalities instead of species abundance per vegetation unit. Thus it is defined as:

$$H_s = -\sum_{i=1}^{S} p_i * \ln p_i$$

Hs = diversity of s municipalitiess, s = no. of crops, p_i = relative abundance of the i-th crop from 0, 0 to 1, 0. The Shannon Index was calculated in the R package vegan [31].

Mean field size was calculated per municipality. Field block size and landscape features were achieved from the LEA-Portal (http://www.sla.niedersachsen.de/landentwicklung/LEA/) for 2014 only; therefore they were only used for the habitat model. Percent land cover of main landscape features like woodland, water expanse and grassland were obtained from the LSA (Landesamt für Statistik Niedersachsen) (http://www1.nls.niedersachsen.de/statistik) these were available for the years 2005, and 2009–2014. For the missing years 2006–2008 the values for 2005 were replicated to decrease unduly reduction in sample size of the overall habitat models. This seemed like a moderate flaw as these broad categories only changed marginally over the 10 years. E.g. woodland cover changed in 85 % of municipalities to less than 1 % of total area, the maximum value was less than 6 % change of total municipal area.

Before starting statistical modeling extensive data mining was applied and all data were examined for plausibility and confounding effects. Multivariate analyses (i.e. PCA, CCA, NMDS and indicator species analyses) were undertaken to find specific crops that might be used to explain trends in grey partridge or common pheasant densities. Summer barley and triticale explained the highest proportion in the multivariate tests, which could not be supported in univariate tests. Therefore, we decided to analyze functional groups of crops instead of individual crops, as a number of explanatory variables had to be reduced to enable conversion of models.

Statistical analyses

All data preparation and analyses were conducted in R 3.1.2 [32]. The statistical models were conducted in the R package mgcv [33, 34]. Model selection on fixed effects was accomplished by AIC comparisons using maximum likelihood estimations (see Additional files 2, 3: Tables S2–S5 for an overview on all tested candidate models and the process of model selection).

Habitat modeling

The relations between response and explanatory variables were partly non-linear, thus additive mixed models were applied to model density data. Municipality was integrated as random effect. An autoregressive correlation structure was found to improve the model fit as measured by AIC comparisons. Latitude and longitude as two-dimensional tensor product smoothers were incorporated to account for spatial autocorrelation.

For the fixed effects, the following parameters were included: year as factor, winter grain, summer grain, maize, set aside, woodland (including deciduous woodland and coniferous forest), water expanse (including all water courses, rivers, canals, lakes, the shore of the North Sea and swamps). Grasslands are highly negatively collinear with the Shannon Index ($r = -0.7$). Since crop diversity resulted in better model fit than grassland we tested the Shannon Index rather and followed the $|r| > 0.7$ rule [35]. The full model had the structure (Eq. 1):

$$E\left[\frac{No.\,pheasant\,hens - No.\,grey\,partridge\,breeding\,pairs}{ha\,potential\,habitat}\right]$$
$$= f\left(winter\,grains\,\%_i\right) + f\left(summer\,grains\,\%_i\right)$$
$$+ f\left(maize\,\%_i\right) + f\left(mean\,field\,block\,size_i\right)$$
$$+ f\left(landscape\,features\,(\%)/(field\,block\,area)_i\right)$$
$$+ f\left(set\,aside\,fields\,\%_i\right) +$$
$$+ f\left(Shannon\,index_i\right) + f\left(woodland_i\right)$$
$$+ f\left(water\,courses_i\right) + factor\,(year)$$
$$+ random\left(municipality\right) + f\left(latitude_i * longitude_i\right)$$
$$+ \Phi e_{i-1} + \epsilon_i \, . \tag{1}$$

The response variables are: 1. Grey partridge breeding pair/km^2 open land (excluding water bodies). 2. Pheasant hen numbers/km^2 open land. The term $f()$ indicates a smooth term (Spline-Regression), $random()$ a random effect structure. The term $\Phi e_{i-1} + \epsilon_i$ describes an autoregressive term to control for temporal autocorrelation.

Due to some missing values in the response or explanatory variables, the total sample size of the model resulted in 3877 observations for the grey partridge model and 3101 observations in the pheasant model.

Trend analysis

In order to model the trend we calculated spearman rho correlations between all tested parameters and year for each municipality separately. The rho values of grey partridge and common pheasant were used as dependent variables and rho of all agricultural crops as explanatory variables. For the grey partridge the correlation was calculated of 10 values (10 years) for the pheasant only 8 years were available (see above). We favored the rho coefficient over Pearson r or the estimates of (generalized) linear models. These are more sensitive and require stricter assumptions in model fit. A correlation coefficient calculated from 8 or 10 values is a rather crude value and rho is heavily influenced by the 1st years of survey. A dramatic decline starting in the middle of the observation period would result in a weak rho coefficient. However, we argue that a decline only over a few years is not necessarily a real trend and an underestimation is potentially ecologically worthwhile. Rho as dependent

variable was arc sine transformed to account for upper and lower boundedness. A GAM (generalized additive model) was fitted to account for non-linear trends in the data.

Full model in Eq. 2:

$$arc\,sine[rho\,pheasant\,hens - rho\,grey\,partridge\,breeding\,pairs]$$
$$= f\left(rho\,winter\,grain_i\right) + f\left(rho\,summer\,grain_i\right) + f\left(rho\,maize_i\right)$$
$$+ f\left(rho\,mean\,field\,size_i\right) + f\left(rho\,Shannon\,Index_i\right)$$
$$+ f\left(rho\,set\,aside\,fields_i\right) + f\left(latitude_i * longitude_i\right)$$

$$\tag{2}$$

The response variables are: 1. Rho grey partridge breeding pair/km^2 open land (excluding water bodies) calculated over 10 years. 2. Rho pheasant hen numbers/km^2 open land calculated over 8 years. The term $f()$ indicates a smooth term (Spline-Regression).

Main landscape types were not tested in the trend model as changes were minor and not all years were available (see above). Also data for landscape features and field block size were only available for 2014, thus, we could not quantify changes. Latitude and longitude as two-dimensional tensor product smoothers were incorporated to account for spatial autocorrelation.

The total sample size, due to missing values resulted in 413 observations in the grey partridge model and 395 observations in the pheasant model.

Results

Both species showed a dramatic decline over the years 1991–2014 (Fig. 1). The population collapse for the grey partridge was most severe, and until 2014 its population density decreased about 60–90 % as compared to 1991 (Fig. 1a). The decline differed between geographic regions and showed that the pheasant had larger fluctuations with a phase of increasing population densities between 1995 and 2005 (Fig. 1b). But ultimately, the pheasant lost between 36 and 90 % of its original population size between 1991 and 2014.

Between 1991 and 2005 grey partridge density was highest in western and central Lower Saxony (Figs. 1a, 2a). Also, for the shorter study period (2005–2014) that was used for the habitat and trend models, the grey partridge showed a strong negative trend over the entire study area (median: rho $= -0.79$, 1st and 3rd quantiles: rho $= -0.94$ and -0.47), with the severest decline in western Lower Saxony (Fig. 2b). Since the start of data collection in 1991 the common pheasant had its highest population densities within the westernmost parts of Lower Saxony (Figs. 1, 3a), it also showed a negative trend between 2007 and 2014; however, it was somewhat less severe (median: rho $= -0.60$, 1st and 3rd quantiles: rho $= -0.81$ and -0.17). The pheasant also declined

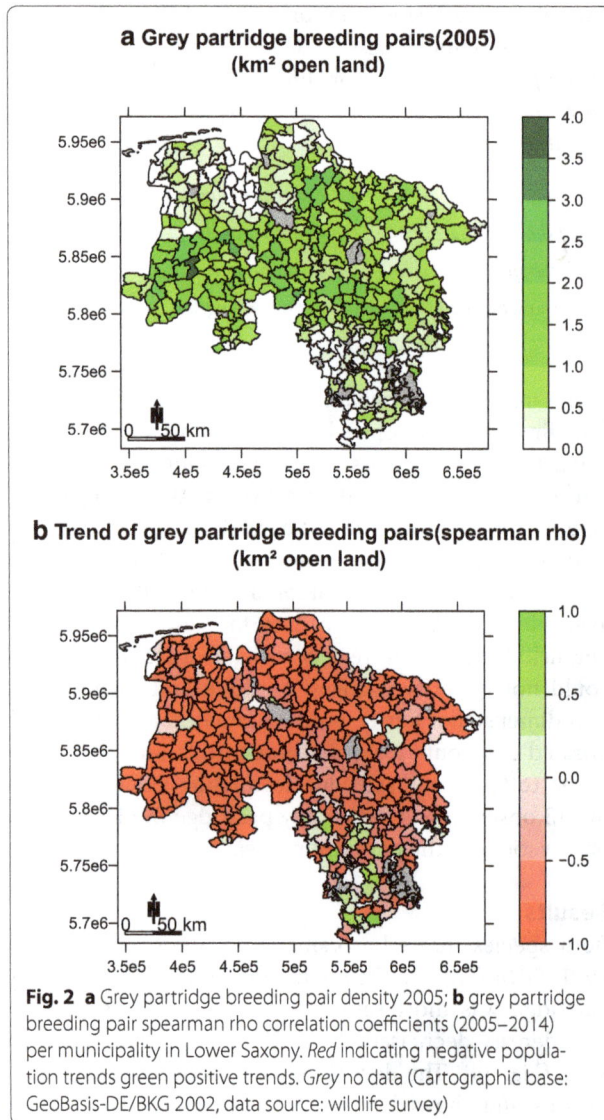

Fig. 2 a Grey partridge breeding pair density 2005; **b** grey partridge breeding pair spearman rho correlation coefficients (2005–2014) per municipality in Lower Saxony. *Red* indicating negative population trends green positive trends. *Grey* no data (Cartographic base: GeoBasis-DE/BKG 2002, data source: wildlife survey)

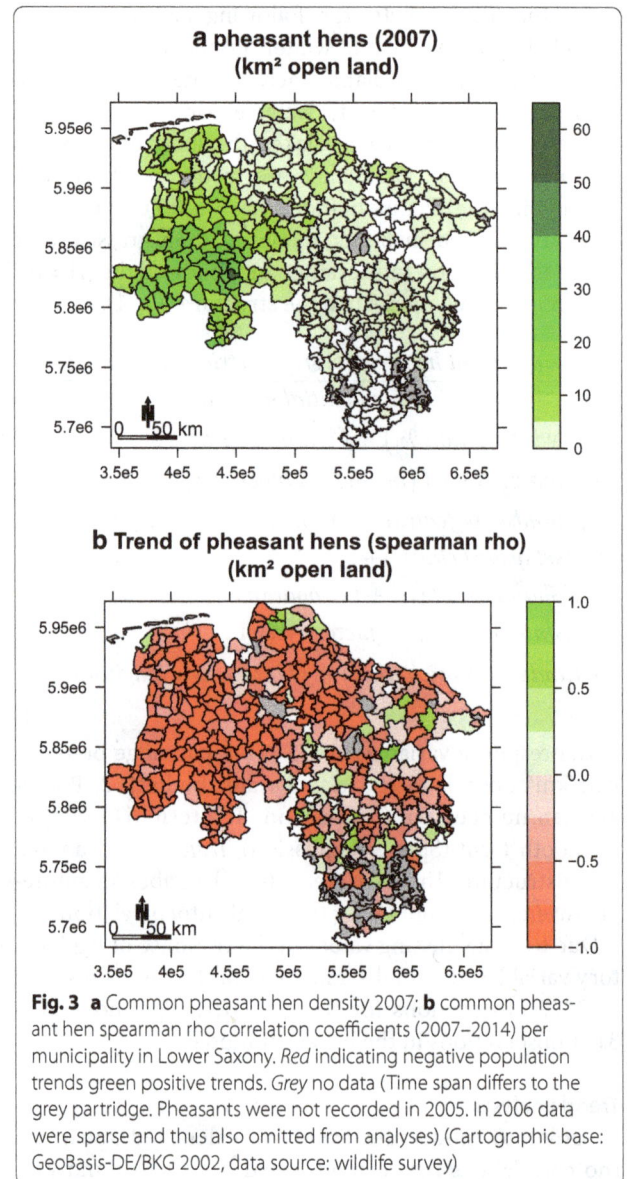

Fig. 3 a Common pheasant hen density 2007; **b** common pheasant hen spearman rho correlation coefficients (2007–2014) per municipality in Lower Saxony. *Red* indicating negative population trends green positive trends. *Grey* no data (Time span differs to the grey partridge. Pheasants were not recorded in 2005. In 2006 data were sparse and thus also omitted from analyses) (Cartographic base: GeoBasis-DE/BKG 2002, data source: wildlife survey)

most severely in the westernmost areas of Lower Saxony (Fig. 3b).

Habitat modeling

The minimum adequate model for the habitat model of grey partridge breeding pairs underlines the dramatic loss in grey partridge abundance, with all years, except for 2006, having significantly lower densities than 2005 (Table 1, p < 0.001). When percentages of winter grain are <20 %, grey partridges are less abundant. In municipalities with higher proportions the model shows overall positive responses, however when it is above 55 %, the effect is non-significant and the standard error gets larger (Fig. 4a). The non-significant smoother for field block size should not be overestimated; nonetheless it improves model fit and vaguely points to a preference

of relatively large field blocks of >6 ha (Fig. 4b). As the second most important smoother (Table 1; F = 20.7, p < 0.001) for crop diversity per municipality, the Shannon index shows that highly diverse municipalities are of benefit for the grey partridge (Fig. 4c). The grey partridge is rare in areas with a high proportion of forest/woodland, which is at the same time the most important coefficient (Table 1; F = 35.6, p < 0.001), and also a negative response to water expanse (Fig. 4 d, e). The tensor product of longitude by latitude shows the high density areas in central Lower Saxony and the lower abundances in the north and the south (Fig. 4f). The model with R^2 adjusted at 0.48 explains roughly half the variance.

Table 1 Summary of GAMM showing habitat preferences of grey partridge breeding pairs as modelled by % share of arable crop groups and other important landscape features per municipality

	Estimate	SE	t value	Pr (>|t|)	
Parametric coefficients					
(Intercept)	0.617	0.012	51.546	<0.001	***
Year = 2006	−0.010	0.009	−1.198	0.231	
Year = 2007	−0.073	0.009	−7.969	<0.001	***
Year = 2008	−0.057	0.009	−6.228	<0.001	***
Year = 2009	−0.103	0.009	−11.160	<0.001	***
Year = 2010	−0.130	0.009	−13.776	<0.001	***
Year = 2011	−0.175	0.010	−17.988	<0.001	***
Year = 2012	−0.208	0.010	−20.047	<0.001	***
Year = 2013	−0.207	0.009	−21.897	<0.001	***
Year = 2014	−0.245	0.010	−25.325	<0.001	***
	edf	Ref.df	F	p value	
Approximate significance of smooth terms					
s(Winter grain)	5.704	5.704	5.586	<0.001	***
s(Field block size)	1.000	1.000	2.245	0.134	
s(SHANNON index)	5.900	5.900	20.662	<0.001	***
s(Forest)	1.000	1.000	35.606	<0.001	***
s(Water expanse)	1.164	1.164	11.704	<0.001	***
te(Longitude, latitude)	12.905	12.905	8.844	<0.001	***

R^2 adjusted = 0.48

The minimum adequate GAMM for pheasant hens shows a unimodal relationship to percentage of maize per area. Between approximately 15 and 35 % the effect is moderately positive, whereas at the highest percentages, maize has a negative effect on pheasant hen densities (Fig. 5a). In contrast, the effect of winter grains is mostly positive. Below 20 % the effect is negative; above 40 % it is positive (Fig. 5b). The effect of landscape features shows a linear positive trend (Table 2; edf = 1, p < 0.001, Fig. 5c). Municipalities with a low Shannon index as measure for crop diversity host fewer pheasants than more diverse areas. The positive effect of highly diverse municipalities is not very pronounced, however, the negative effect of municipalities with few crop types is more evident (Fig. 5d). Municipalities with a high proportion of woodland or forests are generally unfavorable habitats for pheasants (Fig. 5e) and at the same time the most important smoother to describe pheasant hen abundance (Table 2; F = 61.8, p < 0.001), whereas the percentage of some water (approximately 1–7 %) is positive in general. At values higher than 8 percent the sample size is very low and thus also the standard error is large (Fig. 5f). Longitude and latitude generally show a west east gradient with highest density in the westernmost areas, with lower densities near the coast and lowest densities in the north and south east of Lower Saxony (Fig. 5g). R^2 adjusted is with a value of 0.87 comparably high.

Trend analysis

Modeling the rho coefficient of grey partridge trend resulted in a model underlining the importance of diverse agricultural crops (Fig. 6a). With an F value of 10 its importance is even higher than the spatial effect (F = 8, Table 3). Here again the spatial trend indicates the highest population losses in the westernmost areas and more stable conditions in southern Lower Saxony (Fig. 6b). Explained deviance of the grey partridge trend model is at 31.2 %.

The minimum adequate GAM for common pheasant trend shows an overall positive effect of increasing winter grain proportions. In municipalities with a steady decrease in winter grains the population trend of the common pheasant is negative, whereas in municipalities with generally increasing proportions of winter grain the population trend is positive (Fig. 7a). Similarly in municipalities with increasing crop diversity, pheasant hen abundance generally increases. However, the results also show that municipalities with rho of >0.6 are rare, thus confidence intervals get larger and the trend is not significant at the highest values (Fig. 7b). The two dimensional smoother of longitude and latitude indicate the highest decrease in the westernmost areas of Lower Saxony and a slight increase for the southernmost areas (Fig. 7c). The spatial trend explains most of the variance of the model (F = 6.2), followed by winter grain F = 3.5 and Shannon index F = 2.4. The model generally explains 36.2 % of deviance (Table 4).

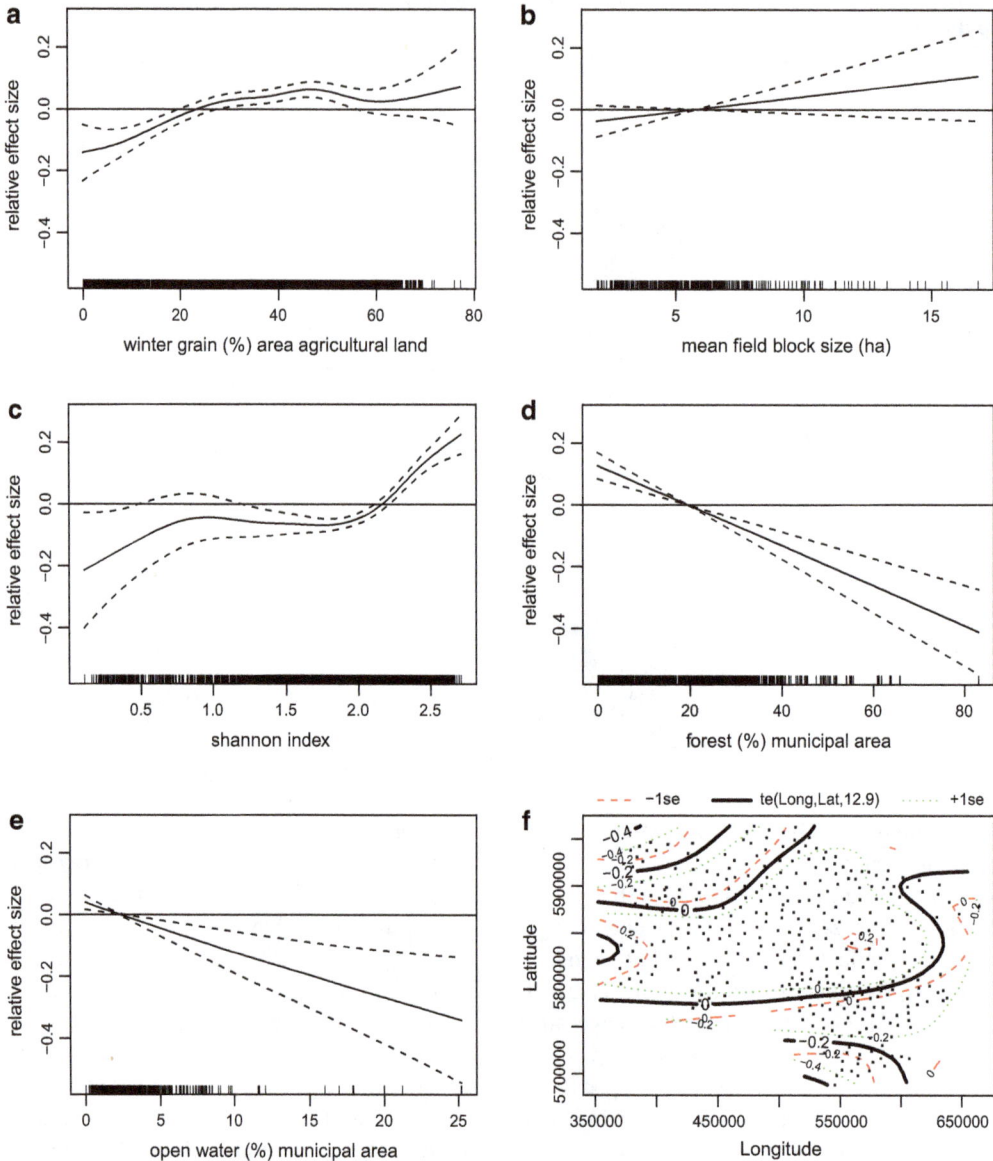

Fig. 4 Minimum adequate habitat model of grey partridge breeding pairs. Figure displays results of GAMM showing significant smoothers: **a** winter grain (%) agricultural land, **b** mean field block size, **c** Shannon Index, **d** % forest/municipal area, **e** % open water/municipal area, **f** longitude × latitude. R^2 adjusted = 0.48

Discussion

Habitat modeling

Generally, the data were found to be well suited to model common pheasant and grey partridge density. With a value of 0.87 of R^2 adjusted, the pheasant habitat model explains most of the variance. The grey partridge model explains considerably less, but still, a value of R^2 adjusted = 0.48. However, the coordinates were the second most important explanatory variables for the pheasant model and are among the more important variables

for the grey partridge model but do not explain habitat preferences. They were necessary to control for spatial autocorrelation and may indicate that some important explanatory variables are missing from the model [36, 37].

The percentage of woodland per municipal area was the most important explanatory variable for both species. As both are typical species of open landscapes, [38] a negative effect of higher proportions of woodland is straightforward. The remaining parameters of landscape

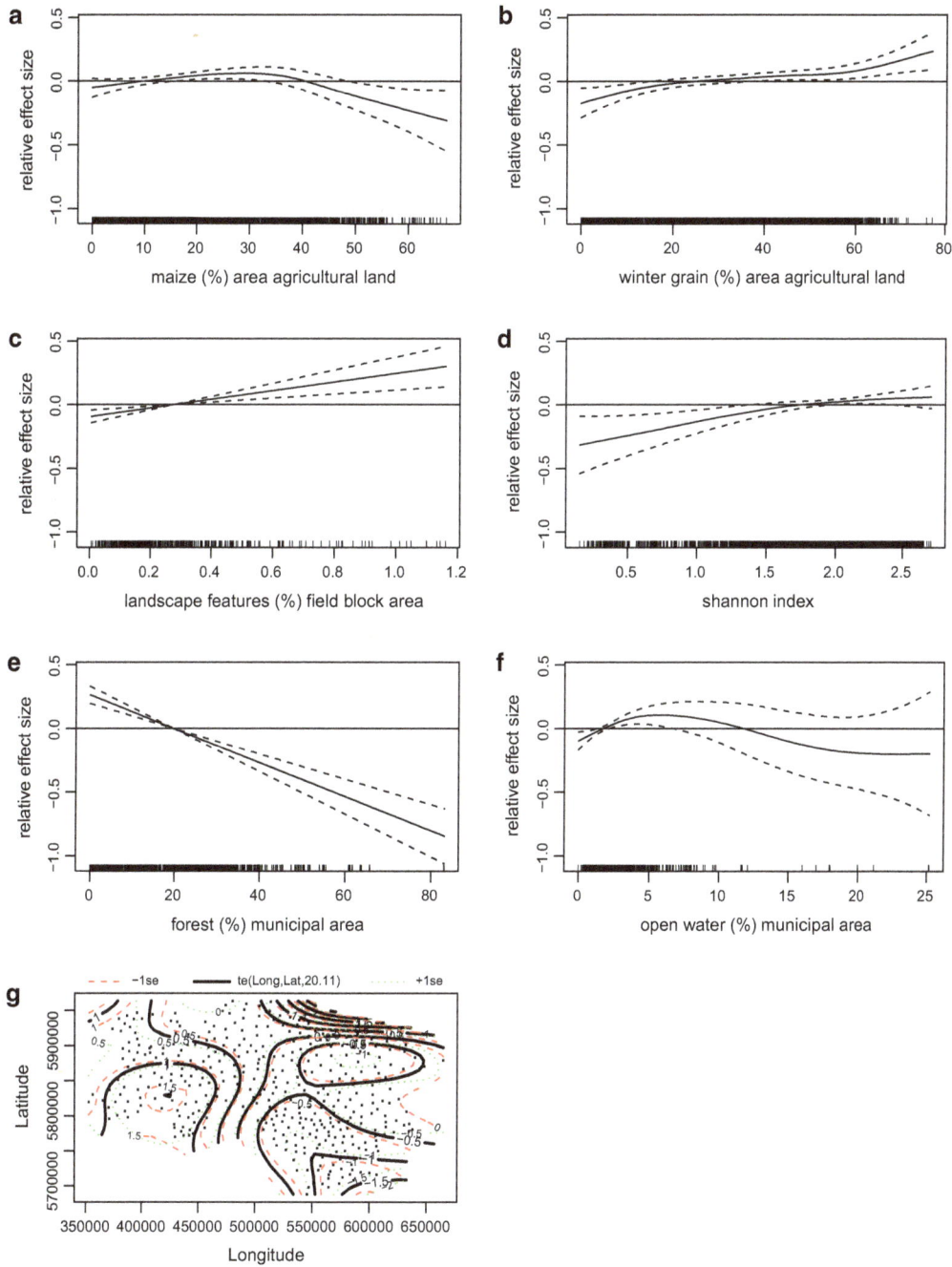

Fig. 5 Minimum adequate habitat model of the common pheasant hens. Figure displays results of GAMM showing significant smoothers: **a** % maize/agricultural area, **b** % winter grains /agricultural area, **c** % landscape features/field block area, **d** Shannon Index, **e** % forest/municipal area, **f** % open water/municipal area, **g** longitude × latitude. R^2 adjusted = 0.82

features, field block size, Shannon index, water expanse, winter grain and maize are all of similar importance and are the more conclusive results. In many respects, the grey partridge's habitat preferences are comparable to the pheasant. They also avoid municipalities with a higher percentage of woodland. However they prefer municipalities with <3 % open water bodies. Pheasants show some preference to municipalities with higher proportions of open water, which is a plausible result. If available pheasants are often seen in reed (*Phragmites australis*) edges of water courses and lakes [38]. These findings are well-known differences in habitat selection [39].

176

Biodiversity and Ecology

Winter grains were beneficial to both species and may provide shelter and food over the winter which may increase winter survival. Grey partridges need high energy food in winter, especially if winters are harsh [40]. However, attempts with food supplementation over the winter did not show any improvement of reproductive success in the following reproductive season in northern France, [41] and autumn and winter diet analyses support no fodder scarceness for grey partridges during these seasons in Poland [42].

Maize was found to be favorable for the pheasant up to a percentage of roughly 20–30 % of arable land. For higher percentages (>50 %) it has negative effects, however the sample size was somewhat low at the highest percentages and the standard error increased. This is not as convincing of an effect as previously found for the farmland birds skylark (*Alauda arvensis*), yellow wagtail (*Motacilla flava*), corn bunting (*Miliaria calandra*) and northern lapwing (*Vanellus vanellus*) [4]. Overall, maize offers landscape and functional diversity in contrast to areas only dominated by winter grains or grassland. Landscape diversity was found more important than

low intense farming techniques for vertebrates including farmland birds [43]. Functional diversity in habitat types also increases functional diversity of insects [44] and increases overall densities and biomass of insects [45]. The grey partridge, however, does not show any preference or avoidance of maize. At different spatial or temporal scales, the effects might be different and we do not rule out adverse effects of increasing maize cultivation. Municipalities with high diversity of crops were of advantage for both species.

Landscape features such as hedges and tree rows provide shelter for the pheasants and a positive effect is plausible [46]. The overall picture of the landscape feature data seems adequate in that the areas of Lower Saxony that have higher proportions of hedges also apply for more subsidies, but it is difficult to evaluate whether or not it is proportionate to the actually present features. The grey partridge does not show any relation to landscape features in the models. Other authors describe a relation of hedgerows or permanent cover and grey partridges [47, 48]. These differences may be due to scale, differences in structures between England, Poland and Germany or the known deficits of the used landscape feature data. The grey partridge, however, shows preferences to relatively large field blocks. This may be an effect of adaptation to typical arable land with generally larger fields. Historically and geographically, it was precluded that smaller fields are of advantage for farmland birds [3]. However, northern France supported our findings with on average larger fields encompassing higher population densities of the grey partridge [49]. Whether the grey partridge is an exception, or the results are confounded with underlying effects is difficult to decide and may be of limited importance, considering the non-significant effect. Yet, it may be potentially interpreted as beneficial as grey partridges use the middle of fields during the night as a predator avoiding behavior [50]. Thus, with larger field blocks they may escape predators more easily as most nocturnal predators predominantly search for prey at field margins.

Trend analysis

Increasing winter grains were found to have a positive effect on pheasant population growth, but no effect on grey partridge trends. Areas with the most fertile soils dominated by winter grains i.e. most notably the Börde and the southernmost areas supporting the lowest population densities (Fig. 2a), showed overall inconsistent trends but no convincing population decline (Fig. 2b).

The Shannon index is the most important smoother in the grey partridge model and also significant in the pheasant model. But the two dimensional smoothers explain most of the variance in the pheasant model and improves model fit in the grey partridge model.

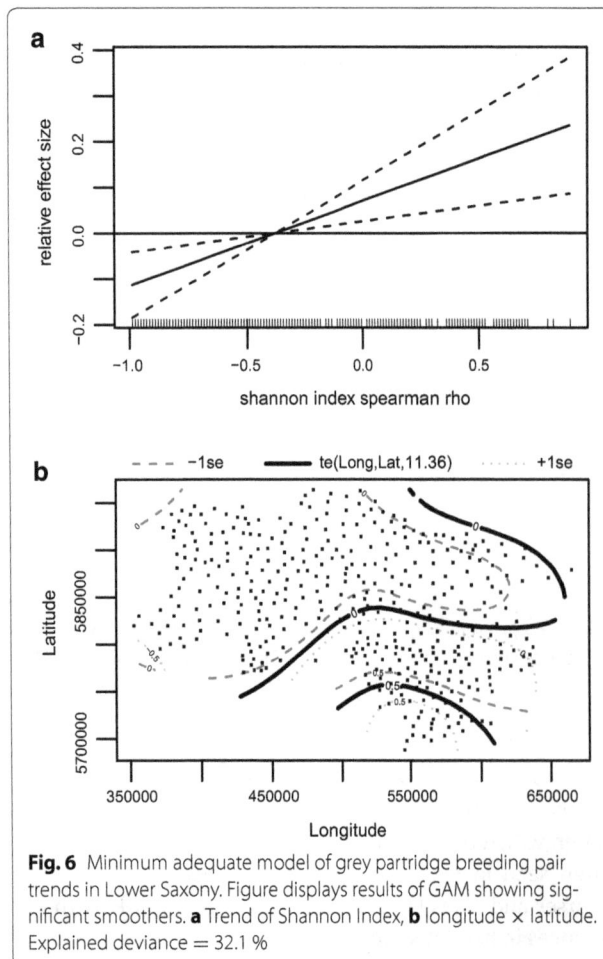

Fig. 6 Minimum adequate model of grey partridge breeding pair trends in Lower Saxony. Figure displays results of GAM showing significant smoothers. **a** Trend of Shannon Index, **b** longitude × latitude. Explained deviance = 32.1 %

Table 2 Summary of GAMM showing habitat preferences of common pheasant hens as modelled by % share of arable crop groups and other important landscape features per municipality

	Estimate	SE	t value	Pr(>|t|)	
Parametric coefficients					
(Intercept)	1.38362	0.01975	70.069	<0.001	***
Year = 2008	0.06254	0.01075	5.82	<0.001	***
Year = 2009	0.02158	0.01388	1.555	0.120	
Year = 2010	0.03784	0.01668	2.269	0.023	*
Year = 2011	−0.04684	0.01871	−2.504	0.012	*
Year = 2012	−0.092	0.01968	−4.676	<0.001	***
Year = 2013	−0.15274	0.01955	−7.815	<0.001	***
Year = 2014	−0.23899	0.02024	−11.81	<0.001	***
	edf	**Ref.df**	**F**	**p value**	
Approximate significance of smooth terms					
s(Maize)	4.331	4.331	6.007	<0.001	***
s(Winter grain)	4.126	4.126	3.870	0.005	**
s(Lanscape features)	1.000	1.000	14.074	<0.001	***
s(Shannon index)	2.391	2.391	3.994	<0.013	*
s(Forest)	1.000	1.000	61.792	<0.001	***
s(Open water)	3.304	3.304	4.347	0.004	**
te(Longitude, latitude)	20.112	20.112	57.676	<0.001	***

R^2 adjusted = 0.87

Table 3 Summary of grey partridge trend model showing parametric coefficients and summary of smooth terms

	Estimate	SE	t value	Pr(> |t|)	
Parametric coefficients					
(Intercept)	−0.755	0.024	−30.980	<0.001	***
	edf	**Ref.df**	**F**	**p value**	
Approximate significance of smooth terms					
s(rho Shannon index)	1.000	1.000	10.036	0.002	**
te(longitude, latitude)	11.370	14.100	8.053	<0.001	***

Explained deviance 32.1 %

The CAP (Common Agricultural Policy) reform after 2013, which installed mandatory greening measures, should increase crop diversity and thus alleviate negative population trends. At this stage it is too early to confirm any benefits and unfortunately most experts expect small effects on sustainability and biodiversity gain [51, 52], therefore we encourage a more combined effort of rural actors to increase a diversity of ecological niches in agricultural lands.

Conclusion

The most conclusive result is the overall importance of diverse crops, which supports our initial expectation and the findings of several studies [e.g. 3, 53]. Other indications of a general intensification of agricultural land use did not show as strong effects. Mean field size increased marginally (from 0.39 to 0.4 ha, including non-typical crops) over the 10 years, but did not explain grey partridge or pheasant habitat or trend models. Data on landscape features are not complete; nevertheless pheasants show some preference to areas with a higher abundance of hedges and tree rows, but we have no data on changes of feature abundance.

Generally within the last 10 years, most signs of agricultural intensification including crop yield, proportion of cultivated to uncultivated land, pesticide and nitrogen application and livestock density have stabilized in most of western Europe or even decreased [54], however this may not rule out higher effectivity of applied compounds. Both, insecticides as well as herbicides have

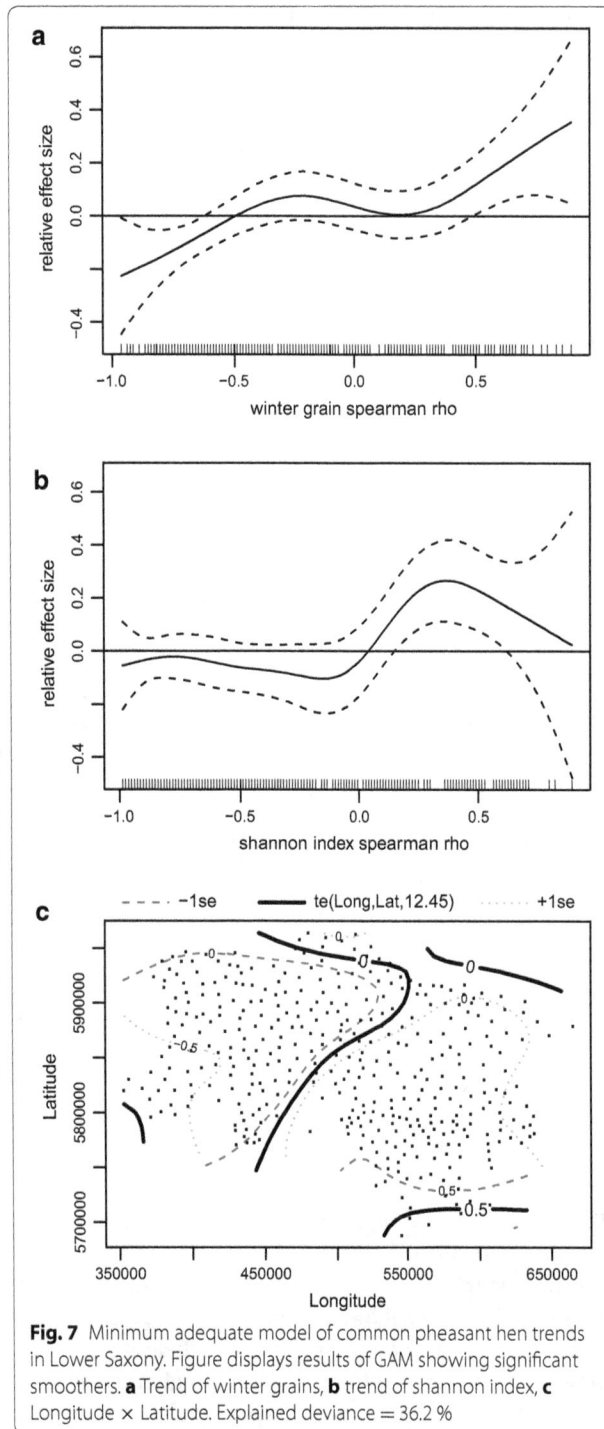

Fig. 7 Minimum adequate model of common pheasant hen trends in Lower Saxony. Figure displays results of GAM showing significant smoothers. **a** Trend of winter grains, **b** trend of shannon index, **c** Longitude × Latitude. Explained deviance = 36.2 %

limitation. Fodder limitation in autumn and winter was found to be unproblematic for grey partridges in Poland [42]. But a density dependent reproductive success was found for the grey partridge at larger population densities in northern France, which was attributed to a lack of favorable habitats [22]. Chick survival rate was found to be crucial for the population decline in several European countries [24, 58, 59] and decreased significantly after the introduction of pesticides [60]. Pheasants at very high abundances may also alter insect abundance [61], however comparable densities are not found in the study area.

Specific crops, especially maize, are held responsible for the latest decline of small game species in Germany. The pheasant habitat model found some weak indication of negative impacts of very high dominance of maize, but the trend model did not confirm a negative effect of increasing maize cultivation. Maize cultivation had already increased before 2005, consequently some effects may have escaped our observation and population declines may have a delayed response to the actual change. Moreover, it is likely that a combination of adverse effects may lead to a dramatic decline rather than a single cause [62]. However, we believe that the widely accepted detrimental effect does not apply to these two species. At different spatial scales unfavorable attributes of maize may be more evident though, as scale significantly affects pheasant habitat models [63]. Radio tracked grey partridges use maize fields, but prefer wild flower strips and sunflower fields in summer and hedges in spring and winter [64]. Despite the lack of significant effects of maize, the analyses showed that monocultures are negative for population trends and municipalities with a 50–80 % of agricultural area with maize cultivation are certainly undesirable for wildlife as shown by the significant effects of crop diversity.

Set-aside fields did not show any significant relation to grey partridge or pheasant trends. Since 2008, no subsidies were given to most categories of set-aside fields, thus most categories of set-aside fields no longer turn up in the statistics. Most farmers converted their fields to arable land; however, some might have left it as it was; which we cannot assess correctly. As an overall consequence, our data might overestimate the negative trend and variations between municipalities may not show up adequately.

Winter grains were found to be beneficial, whereas summer grains showed no effect. Summer grains are relatively strongly correlated to Shannon index (r = 0.4 for the habitat variables and r = 0.56 for the rho values). Thus, it may be difficult in parts to differentiate between the higher diversity and the effect of summer grains. Generally cereals provide nutrient rich fodder, which applies to both groups.

adverse effects on farmland birds impeding the manifold direct and indirect benefits of weeds and invertebrates to wildlife [45]. Food availability models [55] improve habitat models and fodder scarceness for chicks [56] or nest site limitation [57] are possible density dependent causes that can be summarized under the term of resource

Table 4 Summary of trend model for common pheasant hens showing parametric coefficients and summary of smooth terms

	Estimate	Std. Error	t value	Pr(> |t|)	
Parametric coefficients					
(Intercept)	−0.500	0.024	−20.960	<0.001	***
	edf	**Ref.df**	**F**	**p value**	
Approximate significance of smooth terms					
s(rho winter grains)	3.665	4.550	3.460	0.006	**
s(rho Shannon Index)	5.220	6.353	2.366	0.030	*
s(longitude, latitude)	12.448	15.134	6.174	<0.001	***

Explained deviance 36.2 %

Other causes of decline were not tested. Predators were discussed as crucial for population dynamics [65]. The fox however, as main predator of pheasants in Lower Saxony (Voigt unpublished data) were observed to be relatively stable over the time period [17]. A potential effect of climate change should not be locally concentrated, but may explain a general regression of species distribution ranges [66, 67]. For the population dynamic of the two species, climate change was relatively improbable as an effect as Lower Saxony is not at the edge of their climatic niche.

A density dependent decline may also be due to an epidemic. Municipalities with the strongest negative trend are also the municipalities with the highest density in poultry farms within Germany. Only the administrative district of Vechta inhabits over 4 million laying hens in an area of 2018 km^2. Mutual infections between wild galliformes and laying hens are here one among many possibilities. In pheasants ongoing investigations found a high amount of antibodies against infections also typical of poultry farms. Whether these are attributable to the same strains and whether they are at all pathogenic to pheasants is currently being investigated. Chicks were especially found to suffer from diverse infections and parasites [17].

Additional files

Additional file 1. Mean proportions of main habitat types (% forest, % open water, % pastures and meadows, shannon index, field block size, % landscape elements) per natural region (Meynen et al. 1962, modified after Strauss).

Additional file 2. List of crops eligible to payments schemes between 2005 and 2014.

Additional file 3. Process of model selection for the Habitat (GAMMs) and Trend (GAMs).

Abbreviations

IACS: Integrated Administration and Control System; WTE: Wildlife survey Lower Saxony (Wildtiererfassung Niedersachsen); EEG: renewable energy directive (Erneuerbare-Energien-Gesetz); LEA-Portal: internet portal for rural development and agricultural subsidies (Landentwicklung und Agrarförderung); SLA: service center for rural development and agricultural subsidies (Servicezentrum Landentwicklung und Agrarförderung); PCA: principal component analysis; CCA: canonical correlation analysis; NMDS: non-metric multidimensional scaling; GAMM: generalized additive mixed model; GAM: generalized additive model; CAP: common agricultural policy.

Authors' contributions

KR handled data acquisition, performed statistical analyses and drafted the manuscript. ES developed the WTE (Wildtiererfassung Niedersachsen) questionnaires in 1991 and coordinates and maintains the database since. Moreover, he conceived of the study. US critically revised the manuscript and participated in its design and coordination. All authors read and approved the final manuscript.

Author details

[1] Institute for Terrestrial and Aquatic Wildlife Research, University of Veterinary Medicine Hannover, Foundation , Bischofsholer Damm 15, 30173 Hannover, Germany. [2] Hunting Association of Lower Saxony, Schopenhauerstraße 21, 30625 Hannover, Germany.

Acknowledgements

We thank the Ministry for Nutrition, Agriculture and customer protection of Lower Saxony for permission to use IACS data and providing funding through the "Jagdabgabemittel des Landes Niedersachsen". Ralf Hahn from the SLA compiled the relevant IACS data for us. Moreover we would like to thank all participating hunters for their support and Angelika Niebuhr for database management. Special thanks also to Michael Bren and Taren Heintz for improving our English.

Competing interests

The authors declare that they have no competing interests.

Funding

Funding was provided by the Ministry for Nutrition, Agriculture and customer protection of Lower Saxony, through the means of hunting fees (Jagdabgabemittel des Niedersächsischen Ministeriums für Ernährung, Landwirtschaft und Verbraucherschutz (Project number: 406-04032/1-1487).

References

1. Stoate C, Baldi A, Beja P, Boatman ND, Herzon I, van Doorn A, de Snoo GR, Rakosy L, Ramwell C. Ecological impacts of early 21st century agricultural change in Europe—a review. J Environ Manage. 2009;91(1):22–46.

2. Robinson RA, Sutherland WJ. Post-war changes in arable farming and biodiversity in Great Britain. J Appl Ecol. 2002;39(1):157–76.

3. Guerrero I, Morales MB, Onate JJ, Aavik T, Bengtsson J, Berendse F, Clement LW, Dennis C, Eggers S, Emmerson M, et al. Taxonomic and functional diversity of farmland bird communities across Europe: effects of biogeography and agricultural intensification. Biodivers Conserv. 2011;20(14):3663–81.

4. Everaars J, Frank K, Huth A. Species ecology and the impacts of bioenergy crops: an assessment approach with four example farmland bird species. GCB Bioenergy. 2014;6(3):252–64.

5. Brandt K, Glemnitz M. Assessing the regional impacts of increased energy maize cultivation on farmland birds. Environ Monit Assess. 2014;186(2):679–97.

6. Henderson IG, Cooper J, Fuller RJ, Vickery J. The relative abundance of birds on set-aside and neighbouring fields in summer. J Appl Ecol. 2000;37(2):335–47.

7. Henderson IG, Holland JM, Storkey J, Lutman P, Orson J, Simper J. Effects of the proportion and spatial arrangement of un-cropped land on breeding bird abundance in arable rotations. J Appl Ecol. 2012;49(4):883–91.

8. Aebischer NJ, Ewald JA. Grey Partridge *Perdix perdix* in the UK: recovery status, set-aside and shooting. Ibis. 2010;152(3):530–42.

9. Meyer S, Wesche K, Krause B, Leuschner C. Dramatic losses of specialist arable plants in Central Germany since the 1950s/60s-a cross-regional analysis. Divers Distrib. 2013;19(9):1175–87.

10. Hald AB. The impact of changing the season in which cereals are sown on the diversity of the weed flora in rotational fields in Denmark. J Appl Ecol. 1999;36(1):24–32.

11. Aceves-Bueno E, Adeleye AS, Bradley D, Brandt WT, Callery P, Feraud M, Garner KL, Gentry R, Huang YX, McCullough I, et al. Citizen science as an approach for overcoming insufficient monitoring and inadequate stakeholder buy-in in adaptive management: criteria and evidence. Ecosystems. 2015;18(3):493–506.

12. Nov O, Arazy O, Anderson D. Scientists@Home: what drives the quantity and quality of online citizen science participation? PLoS ONE. 2014;9(4):90375.

13. Strauß E. Unterschiedliche Wildtiererfassungenssysteme in Deutschland—eine kritische Betrachtung. Zeitschrift für Jagdwissenschaft. 2000;46:193–7.

14. Strauß E. Pohlmeyer: populationsdichte des Feldhasen (Lepus europaeus PALLAS, 1778) und die Bejagungsaktivität in Niedersachsen. Zeitschrift für Jagdwissenschaft. 2001;47:43–62.

15. Tillmann JE, Beyerbach M, Strauss E. Do hunters tell the truth? Evaluation of hunters' spring pair density estimates of the grey partridge *Perdix perdix*. Wildl Biol. 2012;18:113–20.

16. Krüger T, Oltmanns B. Rote Liste der in Niedersachsen und Bremen gefährdeten Brutvögel. Niedersachsen: Niedersächsischer Landesbetrieb für Wasserwirtschaft, Küsten- und Naturschutz; 2007. p. 52.

17. Strauß E, Gräber R, Klages I, Curland N. Niederwild. In: Gräber R, Strauß E, Johanshon S, editors. Wild und Jagd Landesjagdbericht 2013/2014. Hannover: Niedersächsisches Ministerium für Ernährung, Landwirtschaft und Verbraucherschutz; 2014. p. 46–87. ISSN: 2197-9839.

18. Strauß E, Gräber R, Klages I, Curland N. Niederwild. In: Gräber R, Strauß E, Johanshon S, editors. Wild und Jagd Landesjagdbericht 2014/2015. Hannover: Niedersächsisches Ministerium für Ernährung, Landwirtschaft und Verbraucherschutz; 2015. p. 48-51. ISSN: 2197-9839.

19. Hagen R, Heurich M, Kröschel M, Herdtfelder M. Synchrony in hunting bags: reaction on climatic and human induced changes? Sci Total Environ. 2014;468–469:140–6.

20. Holá M, Zíka T, Šálek M, Hanzal V, Kušta T, Ježek M, Hart V. Effect of habitat and game management practices on ring-necked pheasant harvest in the Czech Republic. Eur J Wildl Res. 2015;61(1):73–80.

21. Rödel HG, Dekker JJA. Influence of weather factors on population dynamics of two lagomorph species based on hunting bag records. Eur J Wildl Res. 2012;58(6):923–32.

22. Bro E, Deldalle B, Massot M, Reitz FO, Selmi S. Density dependence of reproductive success in grey partridge *Perdix perdix* populations in France: management implications. Wildl Biol. 2003;9(2):93–102.

23. De Leo GA, Focardi S, Gatto M, Cattadori IM. The decline of the grey partridge in Europe: comparing demographics in traditional and modern agricultural landscapes. Ecol Model. 2004;177(3–4):313–35.

24. Panek M. Density-dependent brood production in the grey partridge *Perdix perdix* in relation to habitat quality. Bird Study. 1997;44:235–8.

25. Ewald JA, Aebischer NJ, Richardson SM, Grice PV, Cooke AI. The effect of agri-environment schemes on grey partridges at the farm level in England. Agricult Ecosyst Environ. 2010;138(1–2):55–63.

26. Benton TG, Bryant DM, Cole L, Crick HQP. Linking agricultural practice to insect and bird populations: a historical study over three decades. J Appl Ecol. 2002;39(4):673–87.

27. Sotherton NW, Aebischer NJ, Ewald JA. Research into action: grey partridge conservation as a case study. J Appl Ecol. 2014;51(1):1–5.

28. Geiger F, Bengtsson J, Berendse F, Weisser WW, Emmerson M, Morales MB, Ceryngier P, Liira J, Tscharntke T, Winqvist C, et al. Persistent negative effects of pesticides on biodiversity and biological control potential on European farmland. Basic Appl Ecol. 2010;11(2):97–105.

29. Hallmann CA, Foppen RPB, van Turnhout CA, de Kroon H, Jongejans E. Declines in insectivorous birds are associated with high neonicotinoid concentrations. Nature. 2014;511(7509):341.

30. Whittaker RH. Evolution and measurement of species diversity. Taxon. 1972;21(2/3):213–51.

31. Oksanen J, Blanchet FG, Kindt R, Legendre P, Minchin PR, O'Hara RB, Simpson GL, Solymos P, Stevens MHH, Wagner H. Vegan: community ecology package. R package version 2.3-0. 2015: http://CRAN.R-project.org/package=vegan.

32. R CoreTeam. R: A language and environment for statistical computing. Vienna: R Foundation for Statistical Computing; 2014.

33. Wood SN. Fast stable restricted maximum likelihood and marginal likelihood estimation of semiparametric generalized linear models. J R Stat Soc. 2011;73(1):3–36.

34. Wood SN. Stable and efficient multiple smoothing parameter estimation for generalized additive models. J Am Stat Assoc. 2004;99:673–86.

35. Dormann CF, Elith J, Bacher S, Buchmann C, Carl G, Carre G, Marquez JRG, Gruber B, Lafourcade B, Leitao PJ, et al. Collinearity: a review of methods to deal with it and a simulation study evaluating their performance. Ecography. 2013;36(1):27–46.

36. Zuur AF, Ieno EN, Walker NJ, Saveliev AA, Smith GM. Mixed effects models and extensions in ecology with R. New York: Springer Science +Business Media; 2009.

37. Barry S, Elith J. Error and uncertainty in habitat models. J Appl Ecol. 2006;43(3):413–23.

38. von Blotzheim UNG, Bauer KM, Bezzel E. Galliformes und Gruiformes, vol. 5. Wiesbaden: Aula-Verlag Wiesbaden; 1994.

39. Krüger T, Ludwig J, Pfützke S, Zang H. Atlas der Brutvögel ind Niedersachsen und Bremen 2005–2008, vol. 48. Hannover: Niedersächsischer Landschaftsbetrieb für Wasserwirtschaft-, Küsten- und Naturschutz (NLWKN)—Fachbehörde für Naturschutz; 2014.

40. Westerskov K. Winter food and feeding habits of the partridge (Perdix perdix) in the Canadian prairie. Can J Zool. 1966;44(2):303–22.

41. Bro E, Mayot P, Reitz F. Effectiveness of habitat management for improving grey partridge populations: a BACI experimental assessment. Anim Biodiv Conserv. 2012;35(2):405–13.

42. Orłowski G, Czarnecka J, Panek M. Autumn–winter diet of Grey Partridges Perdix perdix in winter crops, stubble fields and fallows. Bird Study. 2011;58(4):473–86.

43. Gonthier DJ, Ennis KK, Farinas S, Hsieh HY, Iverson AL, Batary P, Rudolphi J, Tscharntke T, Cardinale BJ, Perfecto I. Biodiversity conservation in agriculture requires a multi-scale approach. Proc R Soc B Biol Sci. 2014;281:1791.

44. Holland JM, Storkey J, Lutman PJW, Birkett TC, Simper J, Aebischer NJ. Utilisation of agri-environment scheme habitats to enhance invertebrate ecosystem service providers. Agr Ecosyst Environ. 2014;183:103–9.

45. Marshall EJP, Brown VK, Boatman ND, Lutman PJW, Squire GR, Ward LK. The role of weeds in supporting biological diversity within crop fields. Weed Res. 2003;43(2):77–89.

46. Figala J, Prchalova J, Tester JR. GIS assessment of the decline of gray partridge (*Perdix perdix*) nesting habitat in the Elbe River Lowlands, the Czech Republic, 1949–1996. Ekol Bratislava. 2001;20(2):209–18.

47. Rands MRW. Effect of hedgerow characteristics on partridge breeding densities. J Appl Ecol. 1986;23(2):479–87.

48. Panek M. The effect of agricultural landscape structure on food resources and survival of grey partridge *Perdix perdix* chicks in Poland. J Appl Ecol. 1997;34(3):787–92.

49. Joannon A, Bro E, Thenail C, Baudry J. Crop patterns and habitat preferences of the grey partridge farmland bird. Agron Sustain Dev. 2008;28(3):379–87.

50. Tillmann JE. Fear of the dark: night-time roosting and anti-predation behaviour in the grey partridge (*Perdix perdix* L.). Behaviour. 2009;146:999–1023.

51. Hauck J, Schleyer C, Winkler KJ, Maes J. Shades of greening: reviewing the impact of the new EU Agricultural Policy on Ecosystem Services. Change Adapt Soc Ecol Syst. 2014;1:51–62.

52. Hodge I, Hauck J, Bonn A. The alignment of agricultural and nature conservation policies in the European Union. Conserv Biol. 2015;29(4):996–1005.

53. Fischer C, Flohre A, Clement LW, Batary P, Weisser WW, Tscharntke T, Thies C. Mixed effects of landscape structure and farming practice on bird diversity. Agr Ecosyst Environ. 2011;141(1–2):119–25.

54. van der Sluis T, Pedroli B, Kristensen SBP, Lavinia Cosor G, Pavlis E. Changing land use intensity in Europe: recent processes in selected case studies. Land Use Policy. 2015;57:777–85.

55. Ponce C, Bravo C, Alonso JC. Effects of agri-environmental schemes on farmland birds: do food availability measurements improve patterns obtained from simple habitat models? Ecol Evol. 2014;4(14):2834–47.

56. Tillmann JE, Ronnenberg K. Assessment of habitat-specific food availability using human imprinted Grey Partridge chicks *Perdix perdix*. Ornis Fennica. 2015;92(2):87.

57. Bro E, Reitz F, Clobert J, Mayot P. Nesting success of grey partridges (*Perdix perdix*) on agricultural land in North-Central France: relation to nesting cover and predator abundance. Game Wildl Sci. 2000;17:199–218.

58. Panek M. The effect of environmental-factors on survival of gray partridge (*Perdix perdix*) chicks in Poland during 1987–89. J Appl Ecol. 1992;29(3):745–50.

59. Potts GR. Partridges: countryside barometer. London: Harper Collins; 2012.

60. Potts GR, Aebischer NJ. Population-dynamics of the grey partridge *Perdix perdix* 1793–1993—monitoring, modeling management. Ibis. 1995;137:S29–37.

61. Neumann JL, Holloway GJ, Sage RB, Hoodless AN. Releasing of pheasants for shooting in the UK alters woodland invertebrate communities. Biol Conserv. 2015;191:50–9.

62. Bradshaw CJ, Brook BW, Delean S, Fordham DA, Herrando-Pérez S, Cassey P, Early R, Sekercioglu CH, Araújo MB. Predictors of contraction and expansion of area of occupancy for British birds. Proc R Soc Lond B Biol Sci. 2014;281:1786.

63. Jorgensen CF, Powell LA, Lusk JJ, Bishop AA, Fontaine JJ. Assessing landscape constraints on species abundance: does the neighborhood limit species response to local habitat conservation programs. PLoS ONE. 2014;9(6):99339.

64. Buner F, Jenny M, Zbinden N, Naef-Daenzer B. Ecologically enhanced areas—a key habitat structure for re-introduced grey partridges *Perdix perdix*. Biol Conserv. 2005;124(3):373–81.

65. Tillmann JE. An ethological perspective on defecation as an integral part of anti-predatory behaviour in the grey partridge (*Perdix perdix* L.) at night. J Ethol. 2009;27(1):117–24.

66. Lemoine N, Bauer HG, Peintinger M, Bohning-Gaese K. Effects of climate and land-use change on species abundance in a central European bird community. Conserv Biol. 2007;21(2):495–503.

67. Lindstrom A, Green M, Paulson G, Smith HG, Devictor V. Rapid changes in bird community composition at multiple temporal and spatial scales in response to recent climate change. Ecography. 2013;36(3):313–22.

68. Meynen E, Schnmidthüsen J, Gellert J, Neff E, Müller-Miny H, Schultze JH. Handbuch der Naturräumlichen Gliederung Deutschlands. Bad Godesberg. 1953–1962.

15

Moose–tree interactions: rebrowsing is common across tree species

Karen Marie Mathisen[1]*[iD], Jos M. Milner[2] and Christina Skarpe[1]

Abstract

Background: Plant strategies to resist herbivory include tolerance and avoidance. Tolerance strategies, such as rapid regrowth which increases the palatability of new shoots, can lead to positive feedback loops between plants and herbivores. An example of such a positive feedback occurs when moose (*Alces alces*) browse trees in boreal forests. We described the degree of change in tree morphology that accumulated over time in response to repeated browsing by moose, using an index of accumulated browsing. We evaluated whether accumulated browsing could predict the probability and extent of current browsing across woody species in a Norwegian boreal forest, and how our accumulated browsing index related to changes in tree height, shoot availability and shoot size.

Results: The probability and extent of current browsing increased with the degree of accumulated browsing in all tree species. Plants highly modified by previous browsing were the most attractive, with no indication of decreased preference with repeated browsing over time. The preference for previously browsed trees is most likely driven by increased relative availability of shoots within browsing height and maybe increased palatability. This response to previous browsing was general for both preferred and avoided forage species, in both conifers and deciduous trees.

Conclusions: Our results suggest that the adaptation for rapid regrowth after browsing does not reduce herbivory on trees. Rather, our results indicate that plant responses to browsing increase the probability of future herbivory. This response could potentially lead to higher plant mortality where cervid populations are maintained at stable high densities and has implications for plant population dynamics and forestry practices.

Keywords: Tolerance, Compensatory growth, Palatability, Regrowth, Brooming, Accumulated browsing, Height, Alces alces

Background

Plant–herbivore interactions are important drivers of population and ecosystem dynamics, and affect ecosystem processes such as nutrient cycling and succession [1]. Furthermore, an understanding of small scale plant–herbivore interactions is important for understanding larger scale dynamics [2]. Herbivore selection within and between individual plants can affect large scale processes by, for example, changing the rate or direction of succession depending on the successional stage of selected species [3].

Plants have evolved a diverse set of strategies to avoid or tolerate predation from herbivores [4]. Plant responses to herbivory are context dependent, varying with plant species [5], competition [6], season [7–9], time since previous browsing [10] and frequency and intensity of browsing [5, 11], as well as the plant part browsed [8] and habitat productivity [5, 10]. Avoidance strategies include having thorns and small leaves, and responding to herbivory by increasing these traits to reduce intake rate and bite size, thus deterring herbivory on the same plant [12]. Similarly, constitutive or induced chemical defenses deter herbivory by affecting taste, reducing digestibility or by being toxic to the herbivore [13]. These chemical or structural defense responses reduce the probability of herbivory, creating a negative feedback loop between the plant and the herbivore. By contrast, plant tolerance strategies involve responses to herbivory

*Correspondence: karen.mathisen@inn.no
[1] Department of Forestry and Wildlife Management, Faculty of Applied Ecology and Agricultural Sciences, Inland Norway University of Applied Sciences, Pb 400, 2418 Elverum, Norway
Full list of author information is available at the end of the article

such as increased growth rate, increased shoot size and increased resource allocation from root to shoot, allowing plants to compensate for herbivory without deterring herbivores [14, 15]. Tolerance responses may increase the risk of future herbivory if plants produce larger or more vigorous shoots that have a higher nutrient concentration or lower concentration of defense compounds [16]. As many herbivores feed preferentially on such plants or plant modules [17], this can create a positive feedback loop between plants and herbivores [16].

Our study focuses on plant tolerance responses and the positive feedback driven by browsing and re-browsing. How plants respond to previous browsing may in turn affect future browse selection and biomass removal, with implications for plant species composition and dynamics. An example of a tolerance response that increases the probability of future browsing occurs when browsing on leading shoots reduces the apical dominance of leading meristems, an adaptation to plant competition [18, 19]. Removal of dominant meristems reduces nutrient competition with apical shoots, and in turn benefits the browser by increasing shoot production at lower, readily available, heights [8, 20]. In addition, rapid regrowth reduces the synthesis of secondary metabolites leading to more palatable shoots for browsers [16].

A positive feedback between plant and herbivore has been observed in several studies of re-browsing by moose (*Alces alces*) in the boreal forest ecosystem [21–24]. Both the probability of a tree being browsed and browsing pressure may increase with previous browsing [21, 22, 25, 26]. In addition, bite size may increase as a response to increased shoot size and palatability [9, 26]. The pattern of moose responses to previous browsing may also differ between trees with different growth patterns. Biomass production of browsed birch (*Betula pubescence* and *B. pendula*) may increase with moderate moose browsing, whilst biomass production of Scots pine (*Pinus sylvestris*) decreases [5]. Differences in responses to browsing may be linked to determinate versus indeterminate growth patterns, and to different sites of nutrient storage in deciduous and evergreen trees [27–30].

Repeated browsing generally reduces tree height growth in both coniferous and deciduous species [22]. The number of shoots available per tree has been observed to decrease with browsing in birch and pine [31]. However, browsing often increases the production of branched shoots in birch [32], as well as the number of shoots available to moose in rowan [33], hence the overall availability of shoots in deciduous trees may either increase or decrease with previous browsing. Shoot morphology and chemistry may also change in response to browsing. Annual shoot size has been observed to either increase [6, 20, 34] or decrease [6, 33] in response to

moose browsing depending on plant species and time scale, which may affect the size of available bites on previously browsed trees [9, 26]. Browsing may affect the concentration of nutrients and secondary compounds in shoots. Increases in structural carbohydrates may be required to support the growth of large compensatory shoots but reduces their digestibility [7]. In contrast, regrowth from browsed shoots in willow (*Salix phylicifolia*) was less toxic and more digestible than growth from unbrowsed shoots [34].

As current browsing is related to previous browsing through positive feedback loops between plants and herbivores, an index of previous browsing is expected to be a strong predictor of current browsing [35, 36]. In this study, we have used an index of accumulated browsing [37] which describes the degree of change in tree structure that accumulates over time in response to repeated browsing by moose. We quantified the degree of accumulated browsing occurring across tree species in young boreal forest managed for timber production in south-eastern Norway and investigated three specific questions: (i) whether accumulated browsing could predict the probability and extent of current browsing; (ii) how current browsing differed in response to accumulated browsing between trees with different growth pattern and (iii) how the accumulated browsing index reflected changes in shoot availability, tree height and bite sizes.

Within species, we predicted that moose would respond to accumulated browsing by increasing their selection of trees with higher levels of previous browsing (i.1). Consequently, we expected an increase in the number of recently browsed shoots (i.2) and bite diameter (i.3) as the level of previous browsing increased. Given the higher capability of deciduous trees for compensatory growth and the production of more palatable biomass after browsing [5], we expected moose preference for birch (indeterminate growth) over pine (determinate growth) to increase as accumulated browsing increased (ii). Hence, we expected that browse selection (ii.1) and intensity (ii.2 and ii.3) would be higher for birch than pine at high levels of accumulated browsing. Based on previous work [31, 32], we predicted that the number of shoots available per tree would increase with accumulated browsing for birch, but decrease for pine (iii.4). We also predicted that tree height would decrease (iii.5), and shoot size would increase (iii.6: diameter; iii.7: length) with increasing accumulated browsing.

Methods

The aim for this study was to investigate how accumulated browsing in the past can affect current moose browsing on young trees. We quantified the degree of accumulated browsing occurring across tree species in young boreal forest managed for timber production in south-eastern Norway.

Study areas

This study was carried out in the counties of Oppland and Hedmark in south-eastern Norway (~61°N, 11°E, Fig. 1).

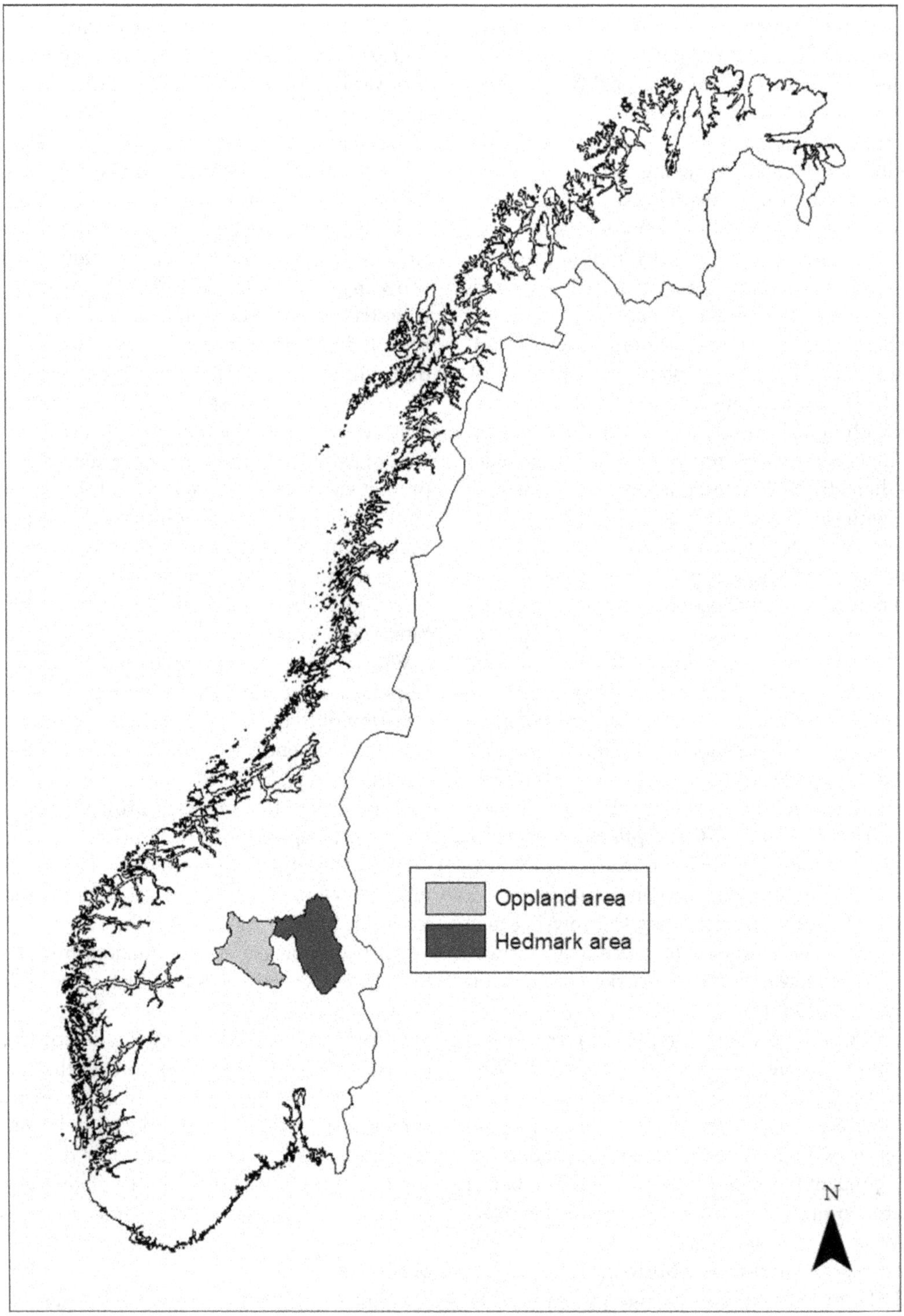

Fig. 1 Map of Norway with study area indicated. Young Scots pine (*Pinus sylvestris*) stands were surveyed for moose browsing in the indicated areas in Hedmark (2010) and Oppland (2011)

Within these study areas, forest stands were located in Stor-Elvdal, Åmot and Rendalen municipalities in Hedmark, and Gausdal, Sør-Fron, Nord-Fron, Sel and Vågå municipalities in Oppland. The vegetation was primarily boreal forest [38] below the commercial timberline, managed for Scots pine and Norway spruce (*Picea abies*) timber and pulp production. Pine stands regenerate naturally, so the young pine stands in this study contained commercial and non-commercial tree species, both of which provide forage for moose. The site productivity index for pine in both areas was low to medium [39]. Stands were dominated by Scots pine, Norway spruce, and downy birch interspersed with silver birch, grey alder (*Alnus incana*), rowan (*Sorbus aucuparia*), aspen (*Populus tremula*), willows (*Salix* spp.) and juniper (*Juniperus communis*). The field layer vegetation was dominated by dwarf shrubs such as *Vaccinium* spp. The Hedmark study area was situated between 250 and 1100 m above sea level with 30-year mean summer (May–September) and winter (October–April) temperatures of 10.6 °C and −5.8 °C, respectively, in the valley bottom. The 30-year mean annual precipitation was 628 mm and the mean snow depth (October–April) was 39 cm [40]. The Oppland study area had a slightly higher elevation (515–920 m a s l), with a mean annual precipitation of 564 mm, winter temperature of −5.0 °C, summer temperature of 7.0 °C (30-year mean) and snow depth of 67 cm (average for the last 10 years). The study area was characterized by valleys and mountains and in both cases; moose tend to migrate down to the less snowy valley bottoms during winter. In the Hedmark area, winter density was approximately 1.3 moose per km² [41], for Oppland there were no records on moose density.

Field survey

We selected young forest stands based on age and tree species composition. In Hedmark, young forest stands of pure and mixed Scots pine were identified from satellite maps of forest stands from the Norwegian Forest and Landscape Institute [42]. Spruce dominated stands were excluded, as spruce is rarely eaten by moose [25, 43]. As supplementary feeding of moose is common in this area, only stands >1 km from supplementary feeding stations were included, to avoid confounding effects on browsing. Previous studies have shown that supplementary feeding sites affect moose browsing intensity at a local scale (<1 km from feeding sites) but not at the landscape scale [44, 45]. In Oppland, young stands were identified from forestry maps from Statskog (Norwegian state-owned forest company), the main landowner in the study area. All stands were visited to confirm that they were dominated by Scots pine and had trees of the desired height (0.5–5 m), with live branches within moose browsing

height ≤3 m [26]. The resulting sample consisted of 69 stands in Hedmark and 42 stands in Oppland.

Forest stands were surveyed in June–July 2010 in Hedmark and May–June 2011 in Oppland using 50 m² circular plots. In Hedmark, four plots were surveyed within each stand, distributed 20 m from the centre point in each cardinal direction (N, S, E, and W). In Oppland, six plots were surveyed within each stand, laid out systematically in a grid using ArcGIS software. Each plot was at least 20 m from the edge, and at least 20 m from each other with the distance between plots increasing with the size of the stand.

Within all plots, moose pellet groups from the previous winter were counted to provide an index of moose density [46]. We classified plot vegetation type according to Moen et al. [38], based on the dominant field-layer vegetation species, and used this as an index of forest productivity (Additional file 1) ranging from low to medium to high [38, 39, 47].

Within each plot, we counted all trees taller than 0.5 m and assessed them for moose browsing (in total 12,565 trees were measured, see Table 1 for sample sizes per species). Trees below 0.5 m height were assumed to be covered by snow during winter, when most browsing occurs. For each measured tree, we recorded the total number of shoots (defined as twigs >1 cm long) from the last growing season, within moose browsing height (0.5–3 m). We classified shoots as either moose browsed or unbrowsed.

Table 1 Number of measured trees in the accumulated browsing index (ABI) categories for all tree species

Species	ABI 0	ABI 1	ABI 2	ABI 3	Total sum
Scots pine *Pinus sylvestris*	1797	1195	1623	708	5323
Downy birch *Betula pubescens*	1104	877	939	304	3224
Norway spruce *Picea abies*	1444	60	39	4	1547
Silver birch *Betula pendula*	284	201	368	43	896
Willows *Salix* sp.	118	41	255	100	514
Juniper *Juniperus communis*	330	45	75	14	464
Rowan *Sorbus aucuparia*	15	33	143	220	411
Aspen *Populus tremula*	*9*	*1*	42	70	122
Grey alder *Alnus incana*	24	29	11	*0*	64
Total sum	5125	2482	3495	1463	12,565

ABI 0 no previous browsing by moose. *ABI 1* previously browsed, but structure of the tree has not changed, *ABI 2* previous browsing has caused a change in tree structure, *ABI 3* strongly modified structure due to previous browsing. Combinations with low sample size (<10 trees) are indicated by italic

Tree height was measured to the nearest 10 cm for trees ≤5 m. On each tree, we measured the diameters of 3–5 browsed shoots (if present) at the point of browsing (hereafter called bite diameter). We assigned each tree a qualitative accumulated browsing index (hereafter abbreviated to ABI, [37, 44], to describe the cumulative effect of previous browsing on tree structure (i.e. excluding browsing during the most recent winter). The scores were as follows: ABI 0 = no previous browsing, ABI 1 = previous browsing visible but the tree structure was mainly unchanged, ABI 2 = previous browsing had visibly modified the structure of the tree (such as crooked stem, increased branching), ABI 3 = previous browsing had strongly modified the structure of the tree (i.e. multiple leader stems, hedged state, brooming). Trees that had modified structure due to other causes and showed no old bite marks, were classified as ABI 0. Old bite marks were usually visible on leading shoots (ABI 2 and 3), broom-shaped shoots (ABI 3), or on side shoots (ABI 1). The ABI incorporated a time effect as trees in class 3 showed signs of repeated browsing over multiple years, whilst trees in class 1–2 may have only been browsed in 1 year.

To evaluate the relationship between ABI and shoot diameter and length (predictions iii.6–7), we sampled shoots of all tree species, except spruce and alder (the least browsed species), in Hedmark. We sampled 1087 shoots from 554 randomly selected trees, by stopping every 500 m along a forest road, and walking 50 m into the forest stand alternating between left and right side of the road, and selecting the closest 3 trees of each species found. We measured diameter and length on 3 randomly selected shoots per tree, by choosing the closest shoot in each height class above ground, if available (0.5–1.0 m, 1.1–1.5 m, 1.6–2.0 m). The diameter was measured at the base of the shoot to the nearest 0.1 mm and the length

was measured from the base of the shoot to the base of the terminal bud to the nearest 0.1 cm. Only 12 of the recorded shoots were branched so these were subsequently excluded from analyses.

Statistical analysis

The effects of ABI on current moose browsing and tree morphology were analyzed in R 3.1.0 [48], using mixed models within the nlme [49] and lme4 [50] packages. The models and explanatory variables required to test each prediction are shown in Table 2. To analyze whether browse selection of individual trees increased with ABI (prediction i.1 and ii.1), we used a generalized linear mixed model, with the occurrence of browsing fitted as a binomial response variable (0/1) and a logit-link function. Predictions i.2–iii.7 were analyzed using linear mixed models, verifying assumptions of normality with residual plots. Numbers of shoots browsed (prediction i.2) and shoots available (prediction iii.4) were log_n-transformed, other response variables were normally distributed. ABI and site productivity indices were fitted as categorical variables. We also controlled for variation in the variables pellet group density and forest productivity, by fitting them as additional fixed effects. We used plot identity within stand identity and study area (Oppland/Hedmark) as nested random intercept terms to account for unbalanced sample sizes between different plots and stands, and to control for non-independence within plots and stands.

Sample sizes among tree species and ABI categories varied greatly (0-1960) and were unbalanced (Table 1). Consequently models that included the interaction species*ABI would not converge. Therefore, predictions i.1–3 were analyzed individually for all tree species. Then we investigated the interaction between ABI and species separately for Scots pine and downy birch for which we

Table 2 Overview of variables included in linear and generalised linear mixed models to test each prediction

Prediction	Response variable	Predictor variables	Random intercept
i.1	Probability of browsing (0/1)	ABI, moose, prod	Area/stand/plot
i.2	Ln(browsed.shoots)	ABI, moose, prod, ln(av.shoots + 1)	Area/stand/plot
i.3	Bite diameter (mm)	ABI, moose, prod	Area/stand/plot
ii.1	Probability of browsing (0/1)	ABI*sp, moose, prod	Area/stand/plot
ii.2	Ln(browsed.shoots)	ABI*sp, moose, prod, ln(av.shoots + 1)	Area/stand/plot
ii.3	Bite diameter (mm)	ABI*sp, moose, prod	Area/stand/plot
iii.4	Ln(av.shoots +1)	ABI*ln(tree height), prod	Area/stand/plot
iii.5	Tree height (standardized)	ABI*stand height, prod	Area/stand/plot
iii.6	Shoot diameter	ABI*sp, height above ground	Plot/tree ID
iii.7	Shoot length	ABI*sp, height above ground	Plot/tree ID

Predictions i.1–ii.3 investigate the moose response (current browsing) to accumulated browsing (ABI), while predictions iii.4–7 investigate the tree's morphological response to previous browsing. Prediction i.1–3 were analysed separately for each individual tree species. Prediction ii.1–3 and iii.4–5 were analysed for birch and pine only, because they provided sufficient data. For prediction iii.6–7 all tree species were grouped together, excluding spruce and alder due to insufficient data

Sp species, *moose* moose pellet groups, *prod* productivity index from vegetation type, *av. shoots* available shoots in browsing height (0.5–3 m)

had sufficient data across ABI categories. We used Scots pine and downy birch as examples of different growth forms (evergreen with determinate growth versus deciduous with indeterminate growth) to test prediction ii, the effect of interaction between ABI and growth form on current moose browsing (predictions ii.1–3). We evaluated the effect of fixed effects on response variables using a comparison of likelihoods between nested models in a backward step selection procedure [51]. We only present the effects of accumulated browsing on current browsing and morphology as this was the focus of our study.

A small number of trees above 5 m in height were present in the stands (e.g. seed trees, trees left after logging), but these were excluded from all analysis as we wanted to focus on trees with live branches within browsing height for moose (0.5–3 m). For analyses of the occurrence of browsing (i.1 and ii.1), shoots available (iii.4) and tree height (iii.5), all trees ≤5 m were included. For the analyses of shoots browsed (i.2 and ii.2) and bite diameter (i.3 and ii.3), only trees browsed by moose the current winter were included. The relationship between ABI and tree morphology [number of shoots available (iii.4), tree height (iii.5)] was investigated further for Scots pine and downy birch. For the analysis of effects of ABI on tree height, the height of each tree was subtracted from the average stand height to account for differences among stands in height development, and the analysis was performed on standardized tree height. The effect of the interaction between ABI and stand height on relative tree height was included to see how ABI was related to height development among trees. Number of shoots browsed was positively related to number of shoots available within browsing height, so the interaction between ABI and shoots available was included in the analysis of shoots browsed (i.2 and ii.2), to investigate if the slope between shoots available and browsed changed with ABI. Tree height was also strongly positively correlated with number of shoots available at browsing height, so the interaction between tree height and ABI was included in this analysis (iii.5), to investigate if the slope between tree height and shoot production changed with ABI.

Results

The degree of accumulated browsing differed markedly between tree species (Fig. 2). In highly preferred tree species such as rowan and aspen, 80–90% of the trees were categorized as structurally modified or heavily modified by previous browsing (ABI 2 and 3, Fig. 2). In contrast, <5% of spruce trees had previously been browsed. The two birch species and pine were intermediate, with around 40–45% of trees in classes ABI 2 and 3. Across all species, 40% of trees were previously unbrowsed by moose.

Effects of accumulated browsing on current browsing of all tree species

As predicted (i.1 and i.2), ABI was a positive predictor of current moose browsing. Both the probability of

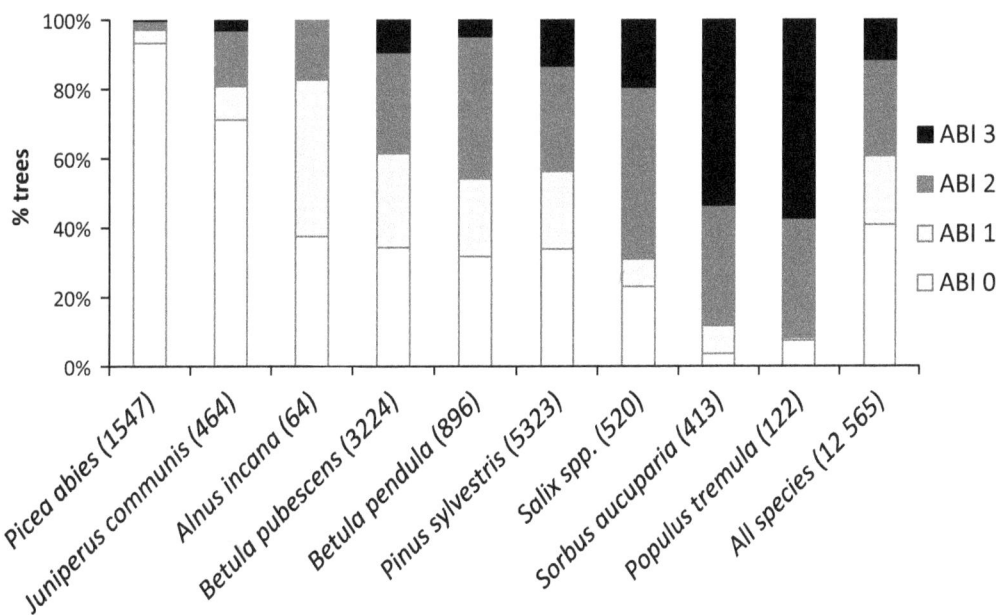

Fig. 2 Percent (%) of trees per species in each accumulated browsing index (ABI) category, sorted from low to high values. ABI 0 = no previous browsing by moose. ABI 1 = previously browsed, but structure of the tree has not changed, ABI 2 = previous browsing has caused a change in tree structure, ABI 3 = strongly modified structure due to previous browsing. Sample size per species added in brackets. See Table 1 for common names

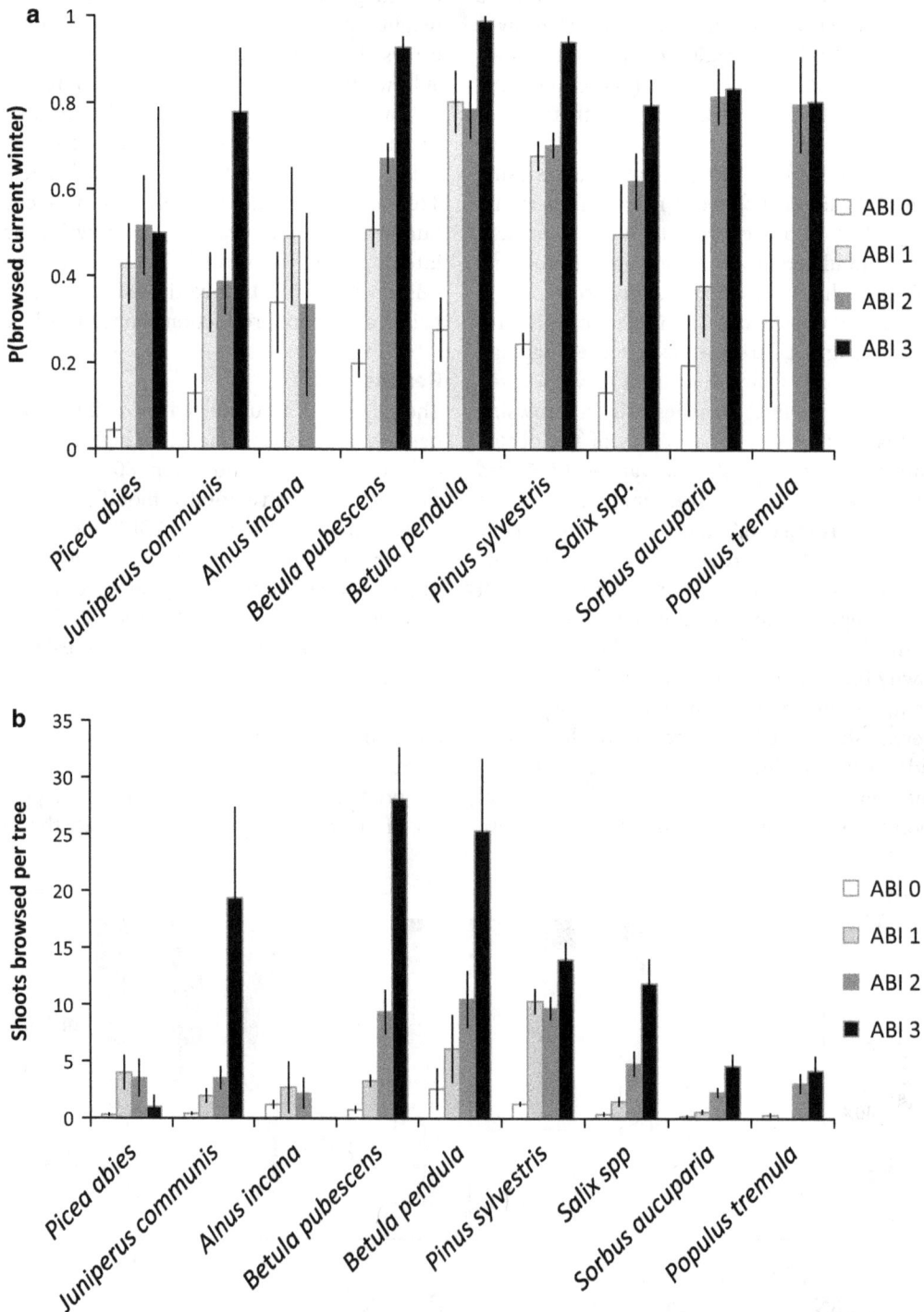

Fig. 3 Moose browsing in the current winter in relation to accumulated browsing (see Fig. 2 for definition) in the past for woody plant species in young boreal forest. **a** Proportion of trees browsed by moose in the current winter (mean ± SE). **b** Number of browsed shoots per tree (mean ± SE) on browsed trees. See Table 1 for sample size and common names

current browsing (i.1, Fig. 3a) and the number of recently browsed shoots per tree (i. 2, Fig. 3b) increased significantly with increasing ABI across all species (Table 3).

However, the moose response to the degree of accumulated browsing differed between tree species. For juniper, birches, pine and willow, trees strongly modified

(ABI 3) by previous browsing had the highest probability of being re-browsed, whilst for rowan and aspen, modified (ABI 2) and strongly modified (ABI 3) trees had an equal probability of rebrowsing (Fig. 3a). For spruce, all previously browsed trees had an equal probability of being re-browsed, but only 4 trees were classified as ABI 3 (Fig. 3a). For alder there were no trees in ABI 3, and in general there was little data to evaluate this species. The number of browsed shoots per tree showed a strong increase in highly modified trees (ABI 3) relative to other classes in juniper, birches and willow (Fig. 3b). In general, and contrary to prediction i.3, bite diameter showed no relationship with ABI (model 3, Table 3). Rowan and aspen were exceptions but small samples sizes within the unbrowsed classes meant these results should be interpreted with caution (see Tables 1, 3).

Differences in current browsing responses between Scots pine and downy birch

The relationship between the probability of moose browsing in the current winter and ABI differed between Scots pine and downy birch trees (interaction—species*ABI: $\chi^2 = 43.86$, df = 3, p < 0.001). Although the probability of current browsing increased with the degree of ABI for both species, the observed pattern was not as predicted in ii.1. Current browsing of the lightly browsed class was much higher for pine than birch, while at high levels of accumulated browsing there was little difference in current browsing probability between species (Fig. 4a).

The number of shoots browsed per tree also differed between birch and pine in relation to ABI (interaction—species*ABI: $\chi^2 = 150.18$, df = 3, p < 0.001). As predicted (ii.2), in highly modified trees (ABI 3) the number of

Table 3 Results from linear and generalized mixed models (Table 2) analyzing the effects of the fixed effects; accumulated browsing index (ABI), moose density (pellet groups) and site productivity (Additional file 1) on moose browsing in the current winter for all tree species in young forest stands ≤5 m high

Species	Response variable	Acc. browsing Ind. ABI (dF = 3)	Moose density (dF = 1)	Productivity (dF = 2)
Pine	Probability of browsing (0/1)	$\chi^2 = 936.01, p < 0.001$	$\chi^2 = 5.18, p = 0.023$	$\chi^2 = 0.99, p = 0.609$
	Ln(browsed.shoots)	$L = 190.46, p < 0.001$	$L = 17.85, p < 0.001$	$L = 1.62, p = 0.445$
	Bite diameter (mm)	$L = 6.29, p = 0.098$	$L = 0.61, p = 0.435$	$L = 4.49, p = 0.106$
Downy birch	Probability of browsing (0/1)	$\chi^2 = 431.44, p < 0.001$	$\chi^2 = 3.99, p = 0.046$	$\chi^2 = 0.28, p = 0.868$
	Ln(browsed.shoots)	$L = 283.45, p < 0.001$	$L = 12.59, p < 0.001$	$L = 5.679, p = 0.058$
	Bite diameter (mm)	$L = 8.12, p = 0.506$	$L = 1.63, p = 0.202$	$L = 8.17, p = 0.017$
Silver birch	Probability of browsing (0/1)	$\chi^2 = 130.58, p < 0.001$	$\chi^2 = 0.74, p = 0.391$	$\chi^2 = 0.35, p = 0.552$
	Ln(browsed.shoots)	$L = 83.97, p < 0.001$	$L = 15.48, p < 0.001$	$L = 1.63, p = 0.201$
	Bite diameter (mm)	$L = 1.60, p = 0.660$	$L = 1.08, p = 0.299$	$L = 0.98, p = 0.321$
Rowan	Probability of browsing (0/1)	$\chi^2 = 46.00, p < 0.001$	$\chi^2 = 1.39, p = 0.239$	$\chi^2 = 5.54, p = 0.019$
	Ln(browsed.shoots)	$L = 22.93, p < 0.001$	$L = 0.79, p = 0.375$	$L = 0.18, p = 0.675$
	Bite diameter (mm)	$L = 11.35, p = 0.010$	$L = 2.01, p = 0.156$	$L = 0.22, p = 0.634$
Willows	Probability of browsing (0/1)	$\chi^2 = 97.22, p < 0.001$	$\chi^2 = 3.69, p = 0.055$	$\chi^2 = 0.03, p = 0.859$
	Ln(browsed.shoots)	$L = 26.47, p < 0.001$	$L = 7.20, p = 0.007$	$L = 0.01, p = 0.942$
	Bite diameter (mm)	$L = 4.61, p = 0.203$	$L = 0.47, p = 0.492$	$L = 0.73, p = 0.392$
Aspen	Probability of browsing (0/1)	$\chi^2 = 14.48, p < 0.001$	$\chi^2 = 2.34, p = 0.126$	$\chi^2 = 1.44, p = 0.486$
	Ln(browsed.shoots)	$L = 6.82, p = 0.033$	$L = 0.54, p = 0.461$	$L = 3.93, p = 0.140$
	Bite diameter (mm)	$L = 15.67, p < 0.001$	$L = 4.54, p = 0.033$	$L = 1.04, p = 0.560$
Juniper	Probability of browsing (0/1)	$\chi^2 = 46.36, p < 0.001$	$\chi^2 = 0.45, p = 0.504$	$\chi^2 = 0.77, p = 0.379$
	Ln(browsed.shoots)	$L = 20.98, p < 0.001$	$L = 3.65, p = 0.056$	$L = 0.03, p = 0.866$
	Bite diameter (mm)	$L = 4.69, p = 0.196$	$L = 2.11, p = 0.147$	$L = 2.96, p = 0.086$
Spruce	Probability of browsing (0/1)	$\chi^2 = 92.74, p < 0.001$	$\chi^2 = 1.24, p = 0.266$	$\chi^2 = 9.74, p = 0.008$
	Ln(browsed.shoots)	$L = 6.51, p = 0.089$	$L = 0.21, p = 0.645$	$L = 0.00, p = 0.953$
	Bite diameter (mm)	$L = 1.93, p = 0.587$	$L = 1.61, p = 0.205$	$L = 2.13, p = 0.145$
Grey alder[a]	Probability of browsing (0/1)	$\chi^2 = 4.83, p = 0.089$	$\chi^2 = 0.46, p = 0.497$	$\chi^2 = 4.50, p = 0.034$

For each fixed effect, nested models including/excluding the variable were compared in a likelihood ratio test, and the Likelihood ratio (L), dF and p value (<0,05 in italic) is presented for linear models, and a similar Chi square (χ^2) test for the binomial model for browsing probability. For sample sizes and scientific names, see Table 1. ABI 0 = no previous browsing by moose. ABI 1 = previously browsed, but structure of the tree has not changed, ABI 2 = previous browsing has caused a change in tree structure, ABI 3 = strongly modified structure due to previous browsing

[a] For grey alder, the data on browsed trees were to scarce to analyze shoots browsed and bite diameter

recently browsed shoots was higher for birch than pine, but, contrary to expectation, the opposite was true of lightly modified trees (ABI 1, Fig. 4b). Hence, for pine, the main effect was a difference in the number of browsed shoots between previously browsed and unbrowsed trees, but for birch the number of shoots browsed increased gradually with increasing ABI. As the number of recently browsed shoots was positively related to the number of shoots available, we investigated the interaction between ABI and available shoots on the number of moose browsed shoots for pine and downy birch separately. The interaction was significant for both species (pine: $L = 77.20$, df = 3, $p < 0.001$, birch: $L = 119.57$, df = 3, $p < 0.001$) such that the number of browsed shoots increased more steeply in relation to available shoots for higher levels of ABI (Fig. 5a, b). However, the relationship increased gradually between ABI classes in pine whereas in birch it was steeper for strongly modified trees (ABI 3) than other classes (Fig. 5a, b).

Contrary to prediction ii.3, there was no effect of the interaction between ABI and species on bite diameter (model 3 interaction—tree type*ABI: $\chi^2 = 1.07$, df = 3, $p = 0.785$). However, bite diameter was larger in pine (3.98 ± 0.15 mm) than birch (2.16 ± 0.15 mm; $\chi^2 = 1346.28$, df = 3, $p < 0.001$).

Morphological plant responses to previous browsing

The number of shoots available within browsing height increased with tree height for both pine and downy birch. In pine, there was also a significant interaction between ABI and tree height affecting the number of shoots available ($L = 245.08$, df = 3, $p < 0.001$). Previously unbrowsed trees had more shoots available per height increment than previously browsed trees (Fig. 5d), indicating that accumulated browsing led to reduced shoot production in pine. For downy birch, the effect of the interaction between tree height and ABI on number of shoots available within browsing height was also significant but less strong and in the opposite direction ($L = 7.96$, df = 3, $p = 0.047$). Therefore, as predicted (iii.4), the number of shoots per height increment increased with ABI in birch but decreased for pine (Fig. 5c, d).

There was also an effect of the interaction between average tree height per stand and ABI on relative tree height for both pine ($L = 286.94$, df = 3, $p < 0.001$) and downy birch ($L = 155.59$, df = 3, $p < 0.001$; Fig. 5e, f). The direction of the effect was similar in both species. Trees that had not been previously browsed by moose tended to be shorter than the average tree in the stand, while trees that had been lightly browsed but showed no change in structure (ABI 1) tended to be taller than the average tree. Trees where browsing had modified (ABI 2), or strongly modified (ABI 3) tree structure, tended to be

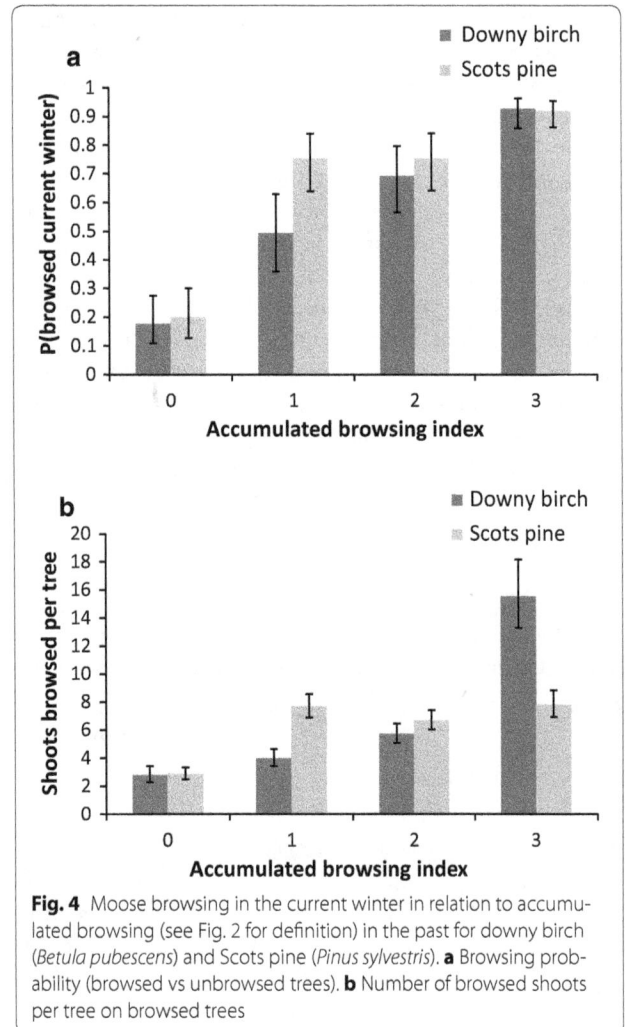

Fig. 4 Moose browsing in the current winter in relation to accumulated browsing (see Fig. 2 for definition) in the past for downy birch (*Betula pubescens*) and Scots pine (*Pinus sylvestris*). **a** Browsing probability (browsed vs unbrowsed trees). **b** Number of browsed shoots per tree on browsed trees

shorter than the average tree in agreement with prediction iii.5 (Fig. 5e, f). The difference in height among ABI classes became apparent above 1 m height, and increased with increasing stand height.

The diameter of annual shoots was not affected by the interaction between species and ABI ($L = 25.18$, df = 18, $p = 0.120$) or by ABI class alone ($L = 2.34$, df = 3, $p = 0.504$). Similarly shoot length did not vary significantly with the interaction between species and ABI ($L = 20.78$, df = 18, $p = 0.290$) or with ABI alone ($L = 6.04$, df = 3, $p = 0.110$). Predictions that shoot diameter (iii.6) and shoot length (iii.7) would increase with ABI were therefore not supported.

Discussion

Moose and tree responses to accumulated browsing

Moose browsing, in terms of browsing probability and number of shoots browsed, increased with the accumulation of past browsing in all tree species studied in

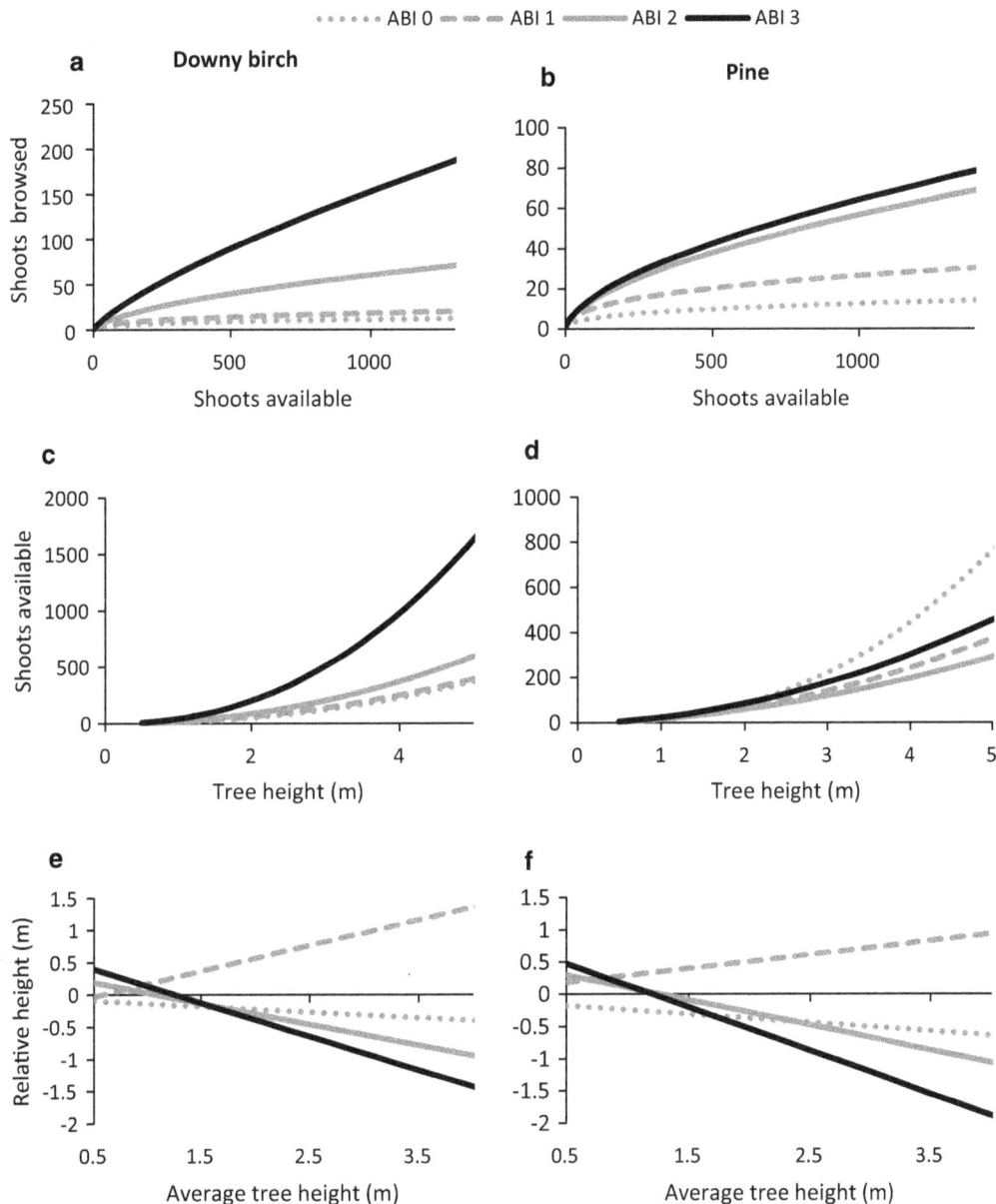

Fig. 5 Responses to varying degrees of accumulated browsing (ABI—definition in Fig. 2) in downy birch (*Betula pubsecencs*, *left*) and pine (*Pinus sylvestris*, *right*). **a, b** The relationship between number of shoots browsed in the current winter by moose per tree and the number of shoots available within browsing height. **c, d** The relationship between tree height and shoots available in browsing height (0.5–3.0 m) for trees of height 0.5–5 m with different degree of ABI in the past. **e, f** Effect of the interaction between average tree height in the stand and ABI on the relative tree height (height-average tree height)

our boreal forest system. Our study is the first to show that this relationship occurred in both highly preferred (rowan, aspen, willow) and less preferred (spruce and alder) browse species [25, 52], and in both deciduous and coniferous trees, suggesting it may be a general pattern. Highly modified trees (ABI 3) were the most preferred, with no indication of decreased preference with repeated browsing over time. However, we found no support for

our prediction that bite size would increase with accumulated browsing. Previous studies have shown a tendency for rebrowsing on the same tree [21–26, 34, 35, 53], which could have been caused by selection for larger or more palatable shoots, or higher browse availability at foraging height [20, 54]. Our lack of a bite size effect indicates that height, availability of shoots and maybe palatability may be more important than shoot size.

Moose reinforce height variation among trees

Both birch and pine trees with a high level of accumulated browsing (2 and 3) were of below average height and relatively shorter than trees with previous light browsing. In addition, moose preferred highly modified trees to unbrowsed trees of a similar height, indicating that height was not the only factor behind greater moose preference for modified trees. Apical shoots were frequently browsed, with top shoot browsing reported on 59% of pines and 66% of birches in Hedmark [44]. This breaking of apical dominance would reduce vertical growth. Apical dominance is an adaption to ensure rapid growth in height when competing for light in a dense forest [18]. Removal of the apical meristem, and release of lateral meristems from hormonal control, may lead to compensatory growth responses that make browsed trees more palatable [8, 18, 20, 54, 55]. For example, mobilization of stored carbon for regrowth means less carbon is available for production of secondary defense compounds [16]. In addition, rapid regrowth is considered to be a response mechanism enabling plants to grow out of browsing height [14, 28, 56]. However, if regrowth is insufficient to escape browsing height, it makes the plant more attractive to herbivores the following year by creating a high availability of forage within browsing height [8, 29].

Hence adaptations to plant competition may in turn reduce plant competitiveness under certain conditions. Other studies have also found that repeated browsing reduces height growth in boreal tree species [22]. Energy constraints may limit regrowth over multiple years and repeated pruning of a woody plant may reduce its carbon reserves, preventing both compensatory growth and production of carbon–based defense compounds [28]. This is particularly likely in boreal forest with medium–low productivity [5, 57], as was typical of our study area.

By contrast, lightly browsed trees were taller than average, whilst unbrowsed trees were shorter. There are several possible explanations for this. Unbrowsed trees may have become established in the stand later and had their subsequent growth suppressed by competition from taller trees. According to the plant vigor hypothesis [17], moose prefer fast-growing trees with larger and more nutritious shoots [58] so suppressed trees may be less attractive. Lightly browsed trees may have become established in the stand early on and developed rapidly in height due to low competition, or having been browsed once, responded with sufficient compensatory growth to escape above browsing height if browsing pressure was low [6, 9]. As differences in height between trees with different degrees of accumulated browsing increased with the average stand height, lightly browsed trees were likely to form the future stand canopy.

Our results show that accumulated browsing by moose leads to or reinforces height variation within young stands [22], and could reduce plant competition for lightly browsed trees. As these grow above browsing height, browsing patches of trees with a high degree of accumulated browsing will eventually be overtopped and outcompeted for light by other trees, leaving patches of strongly modified dead trees under the canopy. By creating habitat for insects and fungi in managed forests that otherwise lack this type of habitat, herbivore selection for previously browsed trees may increase habitat heterogeneity [2], with positive effects for conservation of biodiversity.

Shoot availability increased in birch, but decreased in pine

We predicted that deciduous trees would have a higher capability for compensatory regrowth than conifers [27]. This was supported by an increase in shoot availability with increasing accumulated browsing in birch, but a reduction in pine. In birch, the increase in shoot availability may be due to increased shoot sprouting at lower meristems on previously browsed trees [32, 58] and/or increased branching. Production of branched shoots could also lead to a larger number of bites being available for moose within browsing height. The higher availability of birch shoots likely explained the greater number of shoots browsed by moose on highly modified birch than pine trees (ABI 3). The reduction in shoot availability with increased accumulated browsing in pine has been shown previously [59] and can be explained by the deterministic growth pattern of pine. Each year a whorl of shoots is added to the main axis and all branch axes, and new shoots form by elongation of terminal buds, formed in the previous season [8]. Therefore in pine, the ability to produce new shoots after browsing is restricted to a few meristems. In birch the growth form is more flexible, allowing dormant and short shoots along the stem and branches to convert into long-shoots and new shoots to form after browsing [58]. In addition, as nutrients are stored in the needles in pine during winter but in the roots and stems in birch [27], pine suffers proportionally higher losses due to winter browsing, and has fewer resources available for compensatory growth.

Shoot and bite size did not increase with accumulated browsing

We found no evidence of larger shoots on previously browsed trees. Although some other studies have shown a decrease in shoots size in response to winter browsing [7, 33], most previous studies have shown increases in shoot size [9, 10, 23, 26, 32, 34], or needle size in pine

[24], which we did not measure. The discrepancy may have arisen because we studied multiple species in natural forest stands with low productivity and recurring browsing, whilst most other studies focused on only one species in one season, and, in some cases, were simulated browsing experiments. Plants that have been repeatedly browsed over several years may have depleted resources, and be less able to compensate for browsing by producing large shoots, particularly in low productivity sites. Edenius et al. [10] found an initial increase in pine shoot size in the first year of simulated moose browsing, but a decrease in shoot size in the second and third years. Low nutrient availability in our study area may have limited compensatory growth of large shoots [5]. Furthermore, small differences in which plant part is removed can create different responses [8], showing that response patterns are complex.

Moose have previously been reported to browse on larger shoots of moderately than lightly browsed trees [26]. The greater preference for previously browsed trees was therefore partly explained by the selection of large shoots [60, 61] in order to maximize net energy gain [62]. We predicted that moose bite diameters would increase with accumulated browsing, but did not find this relationship in any tree species. In our study, this was most likely explained by the fact that we found no increase in shoot diameter with accumulated browsing. Therefore, our results do not support the idea that moose selected previously browsed trees in order to gain larger bites.

Plant chemical responses

We have not investigated plant chemical composition in this study. However, the observed preference for trees with high accumulated browsing could be caused by increased nutrient concentration or reduced concentration of plant secondary compounds. Previous studies have found that the carbon demands of regrowth reduces production of secondary defense compounds in previously browsed trees [7, 16, 28, 34], while increased nutrient concentration can occur because of the reduced number of meristems [18, 19]. Moose selection of winter browse is known to be negatively related to the concentration of specific phenolics in *Salix phylicifolia*, and concentrations were lower in previously browsed shoots [63]. Nitrogen concentration in needles has also been found to be slightly higher in browsed than unbrowsed pine trees [24], but most studies of effects on birch show a neutral or negative impact of winter browsing on nutrient concentration of shoots [9]. Plant chemical responses to rebrowsing require further research across a range of species and environmental conditions.

Implications of rebrowsing

Both preferred and avoided tree species, and tree species with different growth form responded to rebrowsing in similar ways, supporting the hypothesis that some tolerance traits are a general adaptation against disturbance (drought, fire, herbivory), and not specifically an adaptation to resist herbivory [64]. In the past, evolutionary pressure from competition has likely been stronger than evolutionary pressure from herbivory. However, over recent decades densities of large herbivores have increased dramatically [65, 66] with the consequence that increased attractiveness to herbivores due to compensatory growth may become maladaptive. Indeed, most studies conclude that rebrowsing reduces flowering, seed production and long-term plant survival, although in some cases it may increase biomass at smaller spatial or temporal scales [26, 67, 68]. In other cases, producing attractive shoots for herbivores may be adaptive at the individual plant level. For modular organisms, it has been suggested that producing shoots of differing palatability to herbivores can be a two-level strategy. By offering some attractive shoots to herbivores, reproduction and growth are concentrated on other highly defended shoots, which then escape herbivory [34]. Further monitoring of the long-term survival and fitness of plants with different degrees of rebrowsing, is needed to answer these questions.

Rebrowsing may be beneficial for moose, as they can return to the same place every year, and browse on previously browsed trees that now have an increased availability of palatable shoots within browsing height. It has even been suggested that rebrowsing by moose may be an example of resource regulation, with the food quality being improved for the accompanying next generation [64]. Rebrowsing by one herbivore may also facilitate other herbivores in the community, as browsing at the foraging height of a tall herbivore may increase shoot production lower in the canopy [36]. As birch shows better regrowth than pine, and birch leaves are important summer forage for moose, rebrowsing may increase the relative availability of summer forage for moose in more heavily browsed stands. However, although patch quality and relative forage availability may improve, total forage biomass generally decreases with browsing over the longer term [5, 33], potentially increasing searching time. In addition, rebrowsing might increase the speed of vegetation succession [69], leading to shading of forage patches and reduced forage availability.

Management implications

At a stand level, rebrowsing leads to the non-random distribution of browsing and increases the variation in height growth among trees [22]. This may have both positive and negative impacts on timber production, as moose carry out the thinning operations for the forester but create an uneven distribution of trees and may not achieve the desired stand density. We recommend forest managers keep preferred rebrowsed trees within the stand, rather than removing them in pre-commercial thinning, in order to reduce browsing on unbrowsed timber crop trees.

In natural systems, large unmanaged herbivore populations tend to naturally fluctuate between periods of high and low density [70], allowing trees to regenerate in pulses [35]. However, often game management aims to maintain large stable populations which might keep preferred browse species in a hedged state, with knock-on effects for biodiversity. Furthermore, the positive feedback loop between plants and herbivores may lead to a more rapid depletion of forage resources than expected, as patches of heavily browsed trees become overtopped by less browsed trees. In order to co-manage forage resources and cervid populations sustainably, it would be beneficial to include the effect of rebrowsing in models of forest development to predict forage availability for cervids.

Conclusions

We found that preferences for all tree species increased with previous browsing. Similar feedback loops between woody plants and browsers have been observed not only in low productive boreal forest, but also in temperate forest and semi-arid savanna [11, 37, 71]. Herbivore selection among woody plants seems to be strongly related to their response to previous browsing, and may be a general feature of tree-browser interactions in forest communities. In our study, all tree species showed a similar direction of compensatory response to browsing, leading to increased herbivore use with increasing accumulated browsing. This indicates the absence of induced qualitative defenses in response to browsing, and occurs regardless of preference among tree species. Our results therefore support the view that plant responses to browsing attract rather than deter future browsing. This type of tolerance response may potentially be maladaptive for the plant, at least in areas with low productivity and high herbivore browsing pressure. If cervid populations are managed at stable high densities, consequences for plant population dynamics should be expected, together with possible evolutionary effects on plant defenses.

Abbreviations

ABI: accumulated browsing index; ABI 0: no previous browsing; ABI 1: previous browsing visible but the tree structure was mainly unchanged; ABI 2: previous browsing had visibly modified the structure of the tree (such as crooked stem, increased branching); ABI 3: previous browsing had strongly modified the structure of the tree (i.e. multiple leader stems, hedged state, brooming).

Authors' contributions

CS and JM conceived and designed the study. CS applied for and received the research grant for the project. KMM carried out the fieldwork and analysis, and the main writing of the manuscript. CS and JM contributed to discussions of the analysis, interpretation and writing of the manuscript. All authors read and approved the final manuscript.

Author details

[1] Department of Forestry and Wildlife Management, Faculty of Applied Ecology and Agricultural Sciences, Inland Norway University of Applied Sciences, Pb 400, 2418 Elverum, Norway. [2] School of Biological Sciences, University of Aberdeen, Tillydrone Avenue, Aberdeen, AB24 2TZ, UK.

Acknowledgements

We are greatly thankful to all the students from Inland Norway University of Applied Sciences who helped us carry out the fieldwork. We are also grateful to the local municipalities, Statskog, Stor-Elvdal land-owner association and Langmorkje Almenning for providing assistance, information and maps. Thanks also go to Zea Walton and too anonymous reviewers in Peerage of Science for reading and commenting the manuscript.

Competing interests

The authors declare that they have no competing interests.

Funding

The data collection for this study was funded by the regional research fund for Oppland and Hedmark counties (Grant No. 212525). Design of the study, data collection, analysis, interpretation and writing has been funded by Inland Norway University of Applied Sciences.

References

1. Hobbs NT. Modification of ecosystems by ungulates. J Wildl Manag. 1996;60(4):695–713.
2. De Jager NR, Pastor J, Hodgson AL. Scaling the effects of moose browsing on forage distribution, from the geometry of plant canopies to landscapes. Ecol Monogr. 2009;79(2):281–97.
3. Davidson DW. The effects of herbivory and granivory on terrestrial plant succession. Oikos. 1993;68(1):23–35.
4. Skarpe C, Hester A. Plant traits, browsing and grazing herbivores, and vegetaiton dynamics. In: Gordon IJ, Prins HHT, editors. The ecology of browsing and grazing. Berlin: Springer; 2008. p. 217–61.
5. Persson IL, Bergström R, Danell K. Browse biomass production and regrowth capacity after biomass loss in deciduous and coniferous trees: responses to moose browsing along a productivity gradient. Oikos. 2007;116(10):1639–50.
6. Edenius L, Danell K, Bergström R. Impact of herbivory and competition on compensatory growth in woody-plants—winter browsing by moose on Scots pine. Oikos. 1993;66(2):286–92.

7. Rea RV, Gillingham MP. Initial effects of brush cutting and shoot removal on willow browse quality. Rangel Ecol Manag. 2007;60(6):566–73.

8. Honkanen T, Haukioja E, Suomela J. Effects of simulated defoliation and debudding on needle and shoot growth in Scots pine (*Pinus sylvestris*)—implications of plant source–sink relationships for plant–herbivore studies. Funct Ecol. 1994;8(5):631–9.

9. Danell K, Bergström R, Edenius L. Effects of large mammalian browsers on architecture, biomass, and nutrients of woody plants. J Mammal. 1994;75(4):833–44.

10. Edenius L. Browsing by moose on Scots pine in relation to plant resource availability. Ecology. 1993;74(8):2261–9.

11. Makhabu SW, Skarpe C. Rebrowsing by elephants three years after simulated browsing on five woody plant species in northern Botswana. S Afr J Wildl Res. 2006;36(1):99–102.

12. Cooper SM, Owen-Smith N. Effects of plant spinescence on large mammalian herbivores. Oecologia. 1986;68(3):446–55.

13. Stamp N. Out of the quagmire of plant defense hypotheses. Q Rev Biol. 2003;78(1):23–55.

14. Strauss SY, Agrawal AA. The ecology and evolution of plant tolerance to herbivory. Trends Ecol Evol. 1999;14(5):179–85.

15. Jaremo J, Nilsson P, Tuomi J. Plant compensatory growth: herbivory or competition? Oikos. 1996;77(2):238–47.

16. du Toit JT, Bryant JP, Frisby K. Regrowth and palatability of Acacia shoots following pruning by African savanna browsers. Ecology. 1990;71(1):149–54.

17. Price PW. The plant vigor hypothesis and herbivore attack. Oikos. 1991;62(2):244–51.

18. Aarssen LW. Hypotheses for the evolution of apical dominance in plants: impliations for the interpretation of overcompensation. Oikos. 1995;74(1):149–56.

19. Senn J, Haukioja E. Reactions of the mountain birch to bud removal—effects of severity and timing, and implications for herbivores. Funct Ecol. 1994;8(4):494–501.

20. Danell K, Haukioja E, Huss-Danell K. Morphological and chemical responses of mountain birch leaves and shoots to winter browsing along a gradient of plant productivity. Ecoscience. 1997;4(3):296–303.

21. Bergström R. Rebrowsing on birch (*Betula pendula* and *B. pubescens*) stems by moose. Alces. 1984;19:3–13.

22. Bergqvist G, Bergström R, Edenius L. Effects of moose (*Alces alces*) rebrowsing on damage development in young stands of Scots pine (*Pinus sylvestris*). For Ecol Manag. 2003;176(1–3):397–403.

23. Bowyer JW, Bowyer RT. Effects of previous browsing on the selection of willow stems by Alaskan moose. Alces. 1997;33:11–8.

24. Löyttyniemi K. On repeated browsing of Scots pine saplings by moose (*Alces alces*). Silva Fenn. 1985;19(4):387–91.

25. Shipley LA, Blomquist S, Danell K. Diet choices made by free-ranging moose in northern Sweden in relation to plant distribution, chemistry, and morphology. Can J Zool. 1998;76(9):1722–33.

26. Danell K, Huss-Danell K, Bergström R. Interactions between browsing moose and 2 species of birch in Sweden. Ecology. 1985;66(6):1867–78.

27. Millard P, Hester A, Wendler R, Baillie G. Interspecific defoliation responses of trees depend on sites of winter nitrogen storage. Funct Ecol. 2001;15(4):535–43.

28. Bryant JP, Chapin FS, Klein DR. Carbon nutrient balance of boreal plants in relation to vertebrate herbivory. Oikos. 1983;40(3):357–68.

29. Danell K, Bergström R, Edenius L, Ericsson G. Ungulates as drivers of tree population dynamics at module and genet levels. For Ecol Manag. 2003;181(1–2):67–76.

30. Pallardy SG, Kozlowski TT. Physiology of woody plants. 3rd ed. Burlington: Elsevier Science; 2010.

31. De Jager NR, Pastor J. Effects of moose Alces alces population density and site productivity on the canopy geometries of birch *Betula pubescens* and *B. pendula* and Scots pine *Pinus sylvestris*. Wildl Biol. 2008;14(2):251–62.

32. Bergström R, Danell K. Effects of simulated winter browsing by moose on morphology and biomass of 2 birch species. J Ecol. 1987;75(2):533–44.

33. De Jager NR, Pastor J. Effects of simulated moose Alces alces browsing on the morphology of rowan Sorbus aucuparia. Wildl Biol. 2010;16(3):301–7.

34. Stolter C. Intra-individual plant response to moose browsing: feedback loops and impacts on multiple consumers. Ecol Monogr. 2008;78(2):167–83.

35. Brandner TA, Peterson RO, Risenhoover KL. Balsam fir on Isle Royale—effects of moose herbivory and population density. Ecology. 1990;71(1):155–64.

36. Makhabu SW, Skarpe C, Hytteborn H. Elephant impact on shoot distribution on trees and on rebrowsing by smaller browsers. Acta Oecol. 2006;30(2):136–46.

37. Skarpe C, Bergström R, Braten AL, Danell K. Browsing in a heterogeneous savanna. Ecography. 2000;23(5):632–40.

38. Moen A, Lillethun A, Odland A. Vegetation. Hønefoss: Norwegian mapping authority; 1999.

39. Larsson JY, Hylen G. Statistikk over skogforhold og skogressurser I Norge registrert I perioden 2000–2004. Ås: Norsk institutt for skog og landskap; 2007.

40. Metrological Institute N: eKlima. http://eklima.met.no. Accessed 16 Sept 2014.

41. Milner JM, Storaas T, van Beest FM, Lien G. Final report of the project Improving moose forage with benefits for the hunting, forestry and farming sectors (In Norwegian with English summary). In: Commissioned report, vol. 1. Elverum: Hedmark University College; 2012.

42. Gjertsen AK. Accuracy of forest mapping based on Landsat TM data and a kNN-based method. Remote Sens Environ. 2007;110(4):420–30.

43. Bergström R, Hjeljord O. Moose and vegetation interactions in northwestern Europe and Poland. Swed Wildl Res Suppl. 1987;1:213–27.

44. Mathisen KM, Milner JM, Van Beest FM, Skarpe C. Long-term effects of supplementary feeding of moose on browsing impact at a landscape scale. For Ecol Manag. 2014;314:104–11.

45. van Beest FM, Gundersen H, Mathisen KM, Milner JM, Skarpe C. Long-term browsing impact around diversionary feeding stations for moose in Southern Norway. For Ecol Manag. 2010;259:1900–11.

46. Månsson J, Andren H, Pehrson A, Bergström R. Moose browsing and forage availability: a scale-dependent relationship? Can J Zool. 2007;85(3):372–80.

47. Fremstad E. Vegetasjonstyper i Norge, vol. 12. Trondheim: Norsk institutt for naturforskning; 1997.

48. R Development Core Team. R: a language and environment for statistical computing. In., 3.1.0. edn. Vienna: R Foundation for Statistical Computing; 2016.

49. Pinheiro J, Bates D, DebRoy S, Sarkar D. Team RC: nlme: linear and nonlinear mixed effects models. In: vol. R package version 3.1-131; 2017.

50. Bates D, Maechler M, Bolker B, Walker S. Fitting linear mixed-effects models using lme4. J Stat Softw. 2015;67(1):1–48.

51. Zuur AF, Ieno EN, Walker NJ, Saveliev AA, Smith GM. Mixed effects models and extensions in ecology with R. New York: Springer; 2009.

52. Månsson J, Kalen C, Kjellander P, Andren H, Smith H. Quantitative estimates of tree species selectivity by moose (*Alces alces*) in a forest landscape. Scand J For Res. 2007;22(5):407–14.

53. Heikkila R. Moose browsing in a Scots pine plantation mixed with deciduous tree species. Acta For Fenn. 1991;224:1–13.

54. Honkanen T, Haukioja E, Kitunen V. Responses of *Pinus sylvestris* branches to simulated herbivory are modified by tree sink/source dynamics and by external resources. Funct Ecol. 1999;13(1):126–40.

55. Hjältén J, Danell K, Ericson L. Effects of simulated herbivory and intraspecific competition on the compensatory ability of birches. Ecology. 1993;74(4):1136–42.

56. McNaughton SJ. Compensatory plant-growth as a response to herbivory. Oikos. 1983;40(3):329–36.

57. Tamm CO. Nitrogen cycling in undisturbed and manipulated boreal forest. Philos Trans R Soc Lond B Biol Sci. 1082;1982(296):419–25.

58. Danell K, Huss-Danell K. Feeding by insects and hares on birches earlier affected by moose browsing. Oikos. 1985;44(1):75–81.

59. Persson IL, Danell K, Bergström R. Different moose densities and accompanied changes in tree morphology and browse production. Ecol Appl. 2005;15(4):1296–305.

60. Belovsky GE. Food plant-selection by a generalist herbivore—the moose. Ecology. 1981;62(4):1020–30.

61. Shipley LA, Gross JE, Spalinger DE, Hobbs NT, Wunder BA. The scaling of intake rate in mammalian herbivores. Am Nat. 1994;143(6):1055–82.

62. Pyke GH, Pulliam HR, Charnov EL. Optimal foraging—selective review of theory and tests. Q Rev Biol. 1977;52(2):137–54.

63. Stolter C, Ball JP, Julkunen-Tiitto R. Seasonal differences in the relative importance of specific phenolics and twig morphology result in contrasting patterns of foraging by a generalist herbivore. Can J Zool. 2013;91(5):338–47.

64. Craig TP. The resource regulation hypothesis and positive feedback loops in plant–herbivore interactions. Popul Ecol. 2010;52(4):461–73.

65. Côté SD, Rooney TP, Tremblay JP, Dussault C, Waller DM. Ecological impacts of deer overabundance. Annu Rev Ecol Evol Syst.
2004;35:113–47.

66. McShea WJ, Underwood HB, Rappole JH. The science of overabundance:
deer ecology and population management. Washington, D.C.: Smithso-
nian Institution Press; 1997.

67. Pastor J, Danell K. Moose–vegetation–soil interactions: a dynamic system.
Alces. 2003;39:177–92.

68. Fornara DA, du Toit JT. Browsing lawns? Responses of *Acacia nigrescens* to
ungulate browsing in an African savanna. Ecology. 2007;88(1):200–9.

69. Pastor J, Naiman RJ. Selective foraging and ecosystem processes in boreal
forests. Am Nat. 1992;139(4):690–705.

70. Illius AW. Linking functional responses and foraging behaviour to population dynamics. In: Danell K, Bergstrom R, Duncan P, Pastor J, editors. Large
herbivore ecology, ecosystem dynamics and conservation. Cambridge:
Cambridge University Press; 2006. p. 71–96.

71. Cromsigt JPGM, Kuijper DPJ. Revisiting the browsing lawn concept:
evolutionary interactions or pruning herbivores? Perspect Plant Ecol Evol
Syst. 2011;13(3):207–15.

16

Application of fundamental equations to species—area theory

Xubin Pan

Abstract

Background: Species—area relationship (SAR), endemics-area relationship (EAR) and overlap-area relationship (OAR) are three important concepts in biodiversity study. The application of fundamental equations linking the SAR, EAR and OAR, can enrich the axiomatic framework of the species—area theory and deepen our understanding of the mechanisms of community assembly.

Results: Two fundamental equations are derived and extended to power law model and random replacement model of species—area distribution. Several important parameters, including the overlap index and extinction rate, are defined and expressed to enrich the species—area theory. For power law model, both EAR and OAR have three parameters, with one more parameter of the total area than SAR does. The EAR equation is a monotonically increasing function for parameter c and z, and a monotonically decreasing function for parameter A. The extinction rate, with two parameters, is a monotonically increasing function for parameter z, and a monotonically decreasing function for parameter A. The overlap index is a monotonically increasing function for parameter A, and a monotonically decreasing function for parameter z, independent of parameter c.

Conclusions: The general formats of SAR, EAR, OAR, overlap index, overlap rate, sampling rate and extinction rate, are derived and extended to power law model and random replacement model as the axiomatic framework of species—area theory. In addition, if the total area is underestimated, the extinction rate will be overestimated.

Keywords: Endemics-area relationship, Overlap-area relationship, Power law, Random replacement, Real total area, Sampling rate, Overlap rate, Extinction rate, Overlap index

Background

Species—area relationship (SAR) is a core concept in biodiversity and species distribution [1], and endemics-area relationship (EAR) is a useful tool in biodiversity conservation and habitat preservation [2–5]. Besides SAR and EAR, overlap-area relationship (OAR), which refers to the number of overlap species in two areas, is also a relevant and important concept [6–8]. To link SAR and EAR and develop a complete species—area theory, two fundamental equations are established to describe species distribution and interrelation between two compensatory areas [7]. Now the species—area theory has been reconstructed by the set theory, integrating SAR, EAR, OAR, alpha diversity, beta diversity, and gamma

diversity [8]. Although OAR curves for two areas of the same size are described and zeta diversity as the average number of species shared by multi-assemblages is proposed, the expanding concept that compares two or more areas of different sizes has not been fully discussed yet [8, 9]. Furthermore, to investigate the spatial characteristics of species richness, it is necessary to integrate the two fundamental equations into the species—area model with distribution information or assumption. Then more parameters can be defined and expressed with empirical data, which can enrich the axiomatic framework of the species—area theory and deepen our understanding of the mechanisms or processes of community assembly.

In addition, debate still exists over the estimation of extinction rate based on the SAR, which is higher than observed extinction rate [10–13]. One explanation for the overestimation is that some species are "committed to

*Correspondence: xubin.hu.pan@gmail.com
Institute of Plant Quarantine, Chinese Academy of Inspection and Quarantine, Beijing 100029, China

extinction" instead of going extinct due to habitat clear-ing [14–16]. However, another reason has been ignored in this debate [8]. According to the power law model, the SAR is a two-parameter equation, whereas the EAR is a three-parameter equation. It does not consider total area in SAR, while total area and its corresponding total species number are crucial factors to determine species disappearing and extinction rate. However, the impact of total area on the extinction rate is still unknown without the specific species—area model and sensitivity analysis.

In this paper, power law and random replacement func-tions, both of which are widely used species—area mod-els, were selected for the application of two fundamental equations [17–19]. Then several important parameters were defined and expressed to enrich the species—area theory. For power law model, sensitivity analysis of parameters was conducted for EAR, extinction rate and overlap index, and the extinction rate based on different total areas were assessed for overestimate comparison.

Methods

The relationships among SAR, EAR and OAR have been shown in Fig. 1, where S_a is the number of species in area a, E_a is the number of species that will disappear when habitat area a is cleared, $O_{a, A-a}$ is the number of overlap species in two areas a and $A - a$,

$$S_A (= E_A) \tag{1}$$

is the number of total species in the total area A. These relationships can be connected by two fundamental equations [7]:

$$S_a + E_{A-a} = S_A, \tag{2}$$

and

$$O_{a,A-a} = S_a - E_a. \tag{3}$$

These fundamental Eqs. (2 and 3) for species—area theory were applied to power law model and random replacement model of SAR. To enrich the species—area theory, several parameters were proposed, including overlap index, overlap rate, sampling rate and extinction rate, which were defined by equations in the general for-mat, power law model and random replacement model.

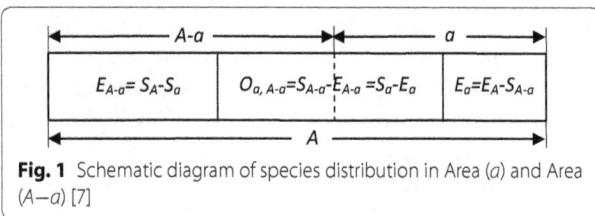

Fig. 1 Schematic diagram of species distribution in Area (a) and Area ($A-a$) [7]

For power law model, sensitivity analysis of param-eters was conducted for EAR, extinction rate and over-lap index, and the extinction rate based on different total areas was assessed for overestimate comparison. The data can be downloaded from the Supplementary of Data.

Results
Application of two fundamental equations to power law model
Power law model has been widely used for species—area relationship:

$$S_a = ca^z, \tag{4}$$

where S_a is the number of species in area a, and c ($a = 1$, $c = S_1$) and z ($0 \leq z \leq 1$) are fitted constants [11, 17]. However, the corresponding function of endemics-area relationship (EAR) has not received much attention, such as in He and Hubbell' paper [12],

$$S_{loss} = S_A - S_{A-a} = cA^z - c(A - a)^z, \tag{5}$$

where S_{loss} is the number of species that disappear when habitat area a is cleared, A is the total area, and S_A is the total number of the species in area A. In fact,

$$E_a = S_{loss} = cA^z - c(A - a)^z, \tag{6}$$

where E_a is the number of species that exist only in area a, but not in area $A - a$. Because

$$a + (A - a) = A, \quad S_a + E_{A-a} = ca^z + cA^z$$
$$- c(A - (A - a))^z = cA^z = S_A = E_A \tag{7}$$

(E_A is the total number of specific species in area A), another method to get EAR is to derive it based on the relationship between SAR and EAR. If area $A - a$ is cleared while area a remains, the number of species that will disappear is $cA^z - ca^z$. Thus, we can get the endem-ics-area curves,

$$E_{A-a} = cA^z - ca^z, \quad E_a = cA^z - c(A - a)^z, \tag{8}$$

the same as those that are derived with the former method. SAR and EAR are rotationally symmetrical, with the center at $(A/2, S/2)$ [7]. The OAR ($O_{a, A-a}$) between the area a and $A - a$ can be calculated as

$$O_{a,A-a} = S_a + S_{A-a} - S_A$$
$$= ca^z + c(A - a)^z - cA^z \tag{9}$$
$$= S_a - E_a.$$

When it comes to $a = A - a$, $O_{a, A-a}$ attains its peak maxi-mum value of $2c\left(\frac{A}{2}\right)^z - cA^z$. When $A = 1280$, $c = 25$, and $z = 0.25$, the SAR, EAR and OAR curves are shown in Fig. 2.

For power law format of SAR, both EAR and OAR equations have three parameters: c, z and A. The EAR equation derived from power law function is different from the previous power law format of "endemics-area relationship":

$$E_a = c\prime a^{z\prime}, \tag{10}$$

where

$$c\prime = cA^{(z-z\prime)} \tag{11}$$

$$1/2^{z\prime} = 1 - 1/2^z \tag{12}$$

which is based on the community-level fractal model [4, 12, 20]. A comparison of these two EAR curves is shown in Fig. 3, where three intersecting points are: (0, 0), $(A/2, cA^z - cA^z/2^z)$, and (A, cA^z). In the interval (0, $A/2$), power law format of "endemics-area relationship" based on fractal model underestimates the number of endemic species; in the interval ($A/2$, A), however, the power law format overestimates the number of endemic species. Both formats are three-parameter equations, but the latter equation is derived based on community-level fractal and power-law assumptions, which has decreased the accuracy of the model. Additionally, the former equation is simple and easy for parameter fitting. In the EAR curve, the species number decreases slowly at the beginning of habitat loss. Due to accumulation effect of habitat loss, the extinction rate of endemic species will speed up until all species disappear. Although the species extinction rate seems small at the beginning of land clearing, it will be too late to conserve biodiversity when most of the habitat disappears.

Overlap index h is proposed here,

$$h = \frac{O_{a,A-a}}{S_a} = \frac{S_a - E_a}{S_a} = 1 - \frac{E_a}{S_a}. \tag{13}$$

When $a = A/2$, h reaches $2 - 2^z$, the same value as obtained with bisection scheme [6]. We can get the overlap index for power law model as

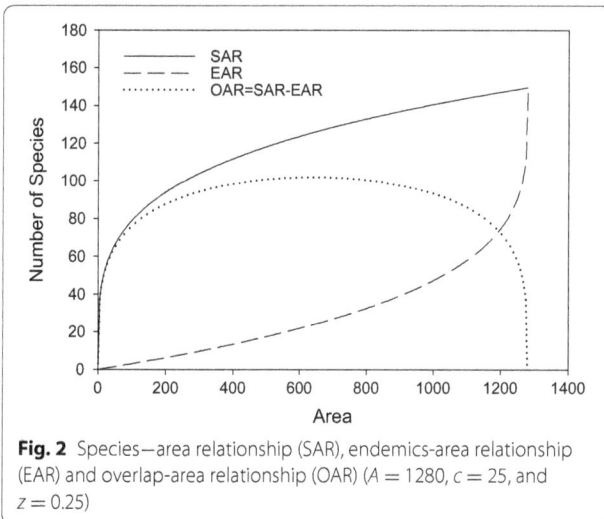

Fig. 2 Species–area relationship (SAR), endemics-area relationship (EAR) and overlap-area relationship (OAR) ($A = 1280$, $c = 25$, and $z = 0.25$)

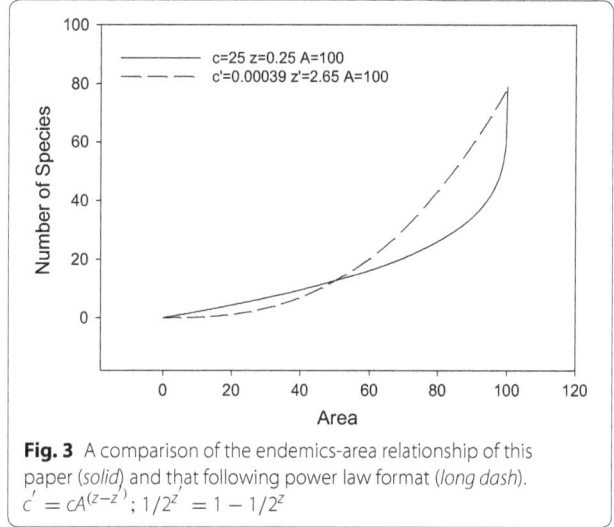

Fig. 3 A comparison of the endemics-area relationship of this paper (*solid*) and that following power law format (*long dash*). $c\prime = cA^{(z-z\prime)}$; $1/2^{z\prime} = 1 - 1/2^z$

$$
\begin{aligned}
h &= \frac{O_{a,A-a}}{S_a} \\
&= \frac{ca^z + c(A-a)^z - cA^z}{ca^z} \\
&= \frac{a^z + (A-a)^z - A^z}{a^z} \\
&= 1 + \left(\frac{A}{a} - 1\right)^z - \left(\frac{A}{a}\right)^z \\
&= 1 + \left(\frac{1}{x} - 1\right)^z - \left(\frac{1}{x}\right)^z,
\end{aligned} \tag{14}
$$

where $x = \frac{a}{A}$ is the ratio of area a over total area A. The overlap index h is the function of z and x, independent of c. Then parameter z reflects the overlapping or self-similarity properties of species number in power law function of SAR [6]. The ratio of overlapping species number over the total species number, overlap rate, is

$$h\prime = \frac{O_{a,A-a}}{S_A} = \frac{S_a - E_a}{S_A}. \tag{15}$$

Since S_A is constant, $h\prime$ has a similar shape with $O_{a,A-a}$, and will reach its peak value at $2^{1-z} - 1$ when $a = A/2$. The ratio of S_a over S_A, sampling rate, is

$$\eta = \frac{S_a}{S_A} = c\left(\frac{a}{A}\right)^z = cx^z. \tag{16}$$

The extinction rate is

$$\lambda = \frac{E_a}{S_A} = \frac{cA^z - c(A-a)^z}{cA^z} = 1 - (1-x)^z, \tag{17}$$

the format of which coincides with the previous extinction rate in the species–area curves [3].

Application of two fundamental equations to random replacement model

Random replacement models for SAR and EAR are

$$S_a = S_a^1 = S_A - \sum_{i=1}^{S_A} \left(1 - \frac{a}{A}\right)^{N_i}, \tag{18}$$

$$E_a = S_a^N = \sum_{i=1}^{S_A} \left(\frac{a}{A}\right)^{N_i}, \tag{19}$$

respectively, where N_i is the number of individuals of species i, and S_a^1 and S_a^N are species–area curve and endemics-area curve across all species in A, respectively [12, 17].

$$S_a + E_{A-a} = S_A - \sum_{i=1}^{S_A} \left(1 - \frac{a}{A}\right)^{N_i} + \sum_{i=1}^{S_A} \left(\frac{A-a}{A}\right)^{N_i} \tag{20}$$

$$= S_A = E_A,$$

corresponding with the fundamental equation.

$$O_{a,A-a} = S_a - E_a$$

$$= S_A - \sum_{i=1}^{S_A} \left(1 - \frac{a}{A}\right)^{N_i} - \sum_{i=1}^{S_A} \left(\frac{a}{A}\right)^{N_i} \tag{21}$$

$$= S_A - \sum_{i=1}^{S_A} \left[\left(1 - \frac{a}{A}\right)^{N_i} + \left(\frac{a}{A}\right)^{N_i}\right]$$

$$O_{a,A-a} = S_a + S_{A-a} - S_A = S_A - \sum_{i=1}^{S_A} \left(1 - \frac{a}{A}\right)^{N_i}$$

$$+ S_A - \sum_{i=1}^{S_A} \left(\frac{a}{A}\right)^{N_i} - S_A \tag{22}$$

$$= S_A - \sum_{i=1}^{S_A} \left[\left(1 - \frac{a}{A}\right)^{N_i} + \left(\frac{a}{A}\right)^{N_i}\right]$$

$$h = \frac{O_{a,A-a}}{S_a} = \frac{S_a - E_a}{S_a}$$

$$= 1 - \frac{E_a}{S_a} \tag{23}$$

$$= 1 - \frac{\sum_{i=1}^{S_A} \left(\frac{a}{A}\right)^{N_i}}{S_A - \sum_{i=1}^{S_A} \left(1 - \frac{a}{A}\right)^{N_i}}$$

$$h' = \frac{O_{a,A-a}}{S_A} = \frac{S_a + S_{A-a} - S_A}{S_A}$$

$$= \frac{2S_A - \sum_{i=1}^{S_A} \left[\left(1 - \frac{a}{A}\right)^{N_i} + \left(\frac{a}{A}\right)^{N_i}\right]}{S_A} - 1 \tag{24}$$

When $a = A/2$, $O_{a, A-a}$ and h' reaches its maximum value

$$O_{a,A-a} = S_A - 2\sum_{i=1}^{S_A} \left(\frac{1}{2}\right)^{N_i} \tag{25}$$

$$h = \frac{2S_{A/2} - S_A}{S_{A/2}}$$

$$= 2 - \frac{S_A}{S_{A/2}}$$

$$= 2 - \frac{S_A}{S_A - \sum_{i=1}^{S_A} \left(\frac{1}{2}\right)^{N_i}} \tag{26}$$

$$= 2 - \frac{1}{1 - \frac{\sum_{i=1}^{S_A} \left(\frac{1}{2}\right)^{N_i}}{S_A}}$$

$$h' = \frac{2\left[S_A - \sum_{i=1}^{S_A} \left(\frac{1}{2}\right)^{N_i}\right]}{S_A} - 1 \tag{27}$$

$$\eta = \frac{S_a}{S_A} = 1 - \frac{E_{A-a}}{S_A} = 1 - \frac{\sum_{i=1}^{S_A} \left(1 - \frac{a}{A}\right)^{N_i}}{S_A} \tag{28}$$

$$\lambda = \frac{E_a}{S_A} = \frac{\sum_{i=1}^{S_A} \left(\frac{a}{A}\right)^{N_i}}{S_A} \tag{29}$$

The general format, power law model and random replacement model for SAR, EAR and OAR are shown in Table 1.

Sensitivity analysis for power law model

The EAR equation is a monotonically increasing function for parameter c. In Fig. 4a, when c increases from 10 to 50, the number of extinct species increases from 32 to 158. The EAR equation is a monotonically increasing function for parameter z. In Fig. 4b, when z increases from 0.1 to 0.5, the number of extinct species increases from 40 to 250. The EAR equation is a monotonically decreasing function for parameter A. In Fig. 4c, when A increases from 9 to 100, the number of extinct species in the same area decreases, providing an important theoretical support for large habitat preservation.

Extinction rate is a monotonically increasing function for parameter z in the interval (0, 1). In Fig. 5a, when the fraction of habitat loss x increases from 0 to 1, the extinction rate also increases from 0 to 1. The extinction rate derived from EAR equation displays a similar pattern for parameter A. In Fig. 5b, when A increases from 9 to 100, the percentage of extinct species in the same area decreases.

Table 1 General format, power law and random placement models for SAR, EAR and OAR

Parameters	General format	Power law model	Random placement model
SAR	$S_a + E_{A-a} = S_A = E_A$	$S_a = ca^z$	$S_a = S_a^1 = S_A - \sum_{i=1}^{S_A}\left(1-\frac{a}{A}\right)^{N_i}$
EAR	$S_{A-a} + E_a = S_A = E_A$	$E_a = cA^z - c(A-a)^z$	
			$E_a = S_a^N = \sum_{i=1}^{S_A}\left(\frac{a}{A}\right)^{N_i}$
OAR	$O_{a,A-a} = S_a - E_a$ $O_{a,A-a} = S_a + S_{A-a} - S_A$ $O_{a,A-a} = S_A - E_a - E_{A-a}$ $O_{a,A-a} = O_{A-a,a}$	$O_{a,A-a} = ca^z + c(A-a)^z - cA^z$	$O_{a,A-a} = S_A - \sum_{i=1}^{S_A}\left[\left(1-\frac{a}{A}\right)^{N_i} + \left(\frac{a}{A}\right)^{N_i}\right]$
h	$h = \frac{O_{a,A-a}}{S_a} = \frac{S_a - E_a}{S_a} = 1 - \frac{E_a}{S_a}$	$h = 1 - \frac{A^z-(A-a)^z}{a^z}$	$h = 1 - \frac{\sum_{i=1}^{S_A}\left(\frac{a}{A}\right)^{N_i}}{S_A - \sum_{i=1}^{S_A}\left(1-\frac{a}{A}\right)^{N_i}}$
$h, a = A/2$	$h = \frac{2S_{A/2}-S_A}{S_{A/2}} = 2 - \frac{S_A}{S_{A/2}}$	$h = 2 - 2^z$	$h = 2 - \frac{1}{1-\frac{\sum_{i=1}^{S_A}\left(\frac{1}{2}\right)^{N_i}}{S_A}}$
h'	$h' = \frac{O_{a,A-a}}{S_A} = \frac{S_a + S_{A-a}}{S_A} - 1$	$h' = \frac{ca^z + c(A-a)^z}{cA^z} - 1$	$h' = \frac{2S_A - \sum_{i=1}^{S_A}\left[\left(1-\frac{a}{A}\right)^{N_i}+\left(\frac{a}{A}\right)^{N_i}\right]}{S_A} - 1$
$h', a = A/2$	$h' = \frac{O_{a,A-a}}{S_A} = \frac{2S_{A/2}}{S_A} - 1$	$h' = 2^{1-z} - 1$	
η	$\eta = \frac{S_a}{S_A} = 1 - \frac{E_{A-a}}{S_A}$	$\eta = c\left(\frac{a}{A}\right)^z$	$\eta = 1 - \frac{\sum_{i=1}^{S_A}\left(1-\frac{a}{A}\right)^{N_i}}{S_A}$
λ	$\lambda = \frac{E_a}{S_A} = 1 - \frac{S_{A-a}}{S_A}$ $\lambda = (1-h)\eta$	$\lambda = 1 - c\left(1-\frac{a}{A}\right)^z$	$\lambda = \frac{\sum_{i=1}^{S_A}\left(\frac{a}{A}\right)^{N_i}}{S_A}$

SAR species–area relationship; *EAR* endemic-area relationship; *OAR* overlap-area relationship; S_a is the number of species in area a; E_a is the number of species only in the area a, but not in the area $A-a$; $O_{a,A-a}$ is the number of species both in the area a and $A-a$; c ($a = 1, c = S_1$) and z ($0 \leq z \leq 1$) are constants; A is the total area, S_A is total number of the species in the area A, and E_A is the total number of specific species in area A; h is overlap index; h' is the ratio of overlapping species number over the total species number, overlap rate; η is ratio of S_a over S_A, sampling rate; λ is extinction rate; N_i is the number of individuals of the specific species i; S_a^1 and S_a^N are species–area curve and endemics-area curve across all species in A [3, 6, 7, 10, 12, 16, 17]

A sensitivity analysis of two parameters (z and A) is conducted for the overlap index h. The h equation is a monotonically increasing function for parameter A in Fig. 6a, when A increases from 320 to 1280. The h equation is a monotonically decreasing function for parameter z in Fig. 6b, when z increases from 0.15 to 0.5.

Discussion

Table 2 shows the extinction rate estimate and overestimate comparison. If the real total area is 100, overestimate would occur when the total area is set to, say, 9 or 49. If the total area A is underestimated for 51 %, the overestimate of extinction rates will be 1.04 and 1.36 for 0.52 and 25 % of real total habitat loss, respectively [13, 21, 22]. However, the power law format of EAR will underestimate the extinction rate for the 0.52 and 25 % of real total habitat loss. If the total area A is underestimated for 91 %, the overestimate of extinction rates will be 10.34 and 40.91 for 0.52 and 9 % of real total habitat loss, respectively. Thus for field estimate of extinction rate, species–area relationship is an important tool, plus boundary identification of real total area, which can be assisted by remote sensing and geographic information system. If a smaller total area is adopted compared with the actual area, the species extinction rate estimate will be overstated.

Based on EAR, however, small total area for habitat preservation does lead to potential high species extinction rate. Thus large total area should be adopted for the Natural Protected Areas (NPAs). UNESCO-MAB World Network of Biosphere Reserves, suggests to apply a zonation system to NPAs, which consists of a core zone, a buffer zone and a transition zone. Normally, both the buffer zone and transition zone do not have any different or concerned species that are not in the core zone, thus the total number of species will not increase when the protected area is expanded from core zone to include the buffer zone and transition zone. But both the buffer zone and transition zone can relieve the impact of anthropic activities on the core zone, and this result can be derived from the species–area theory.

Since EAR and OAR involve species in two complementary areas, one more parameter, the total area, has been added in their expressions compared with SAR. If the concepts of EAR and OAR are expanded to arbitrary two areas (they can be treated as complementary in the point of mathematics), then the h' will be

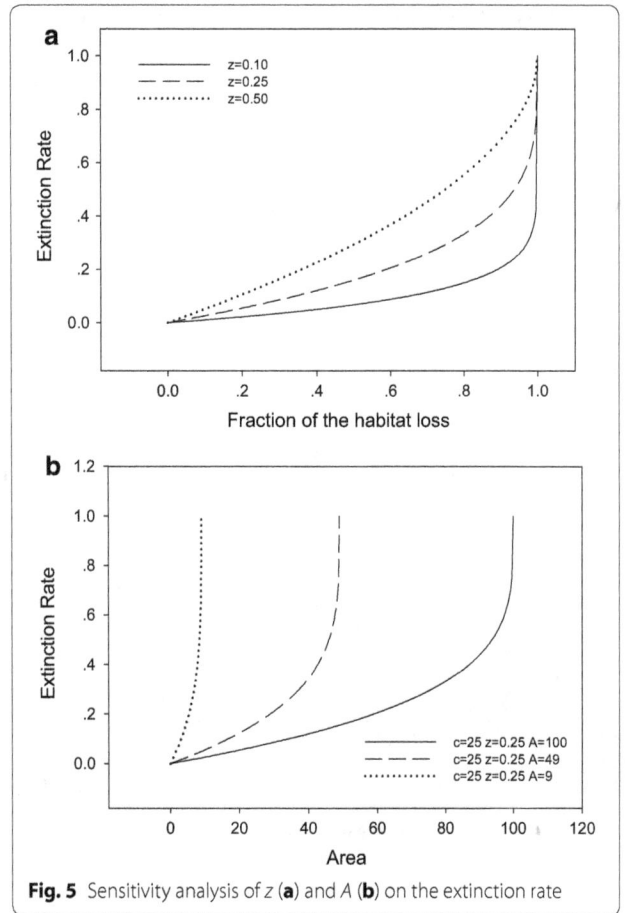

Fig. 4 Sensitivity analysis of c (**a**), z (**b**), and A (**c**) on the power law format of endemics-area equation

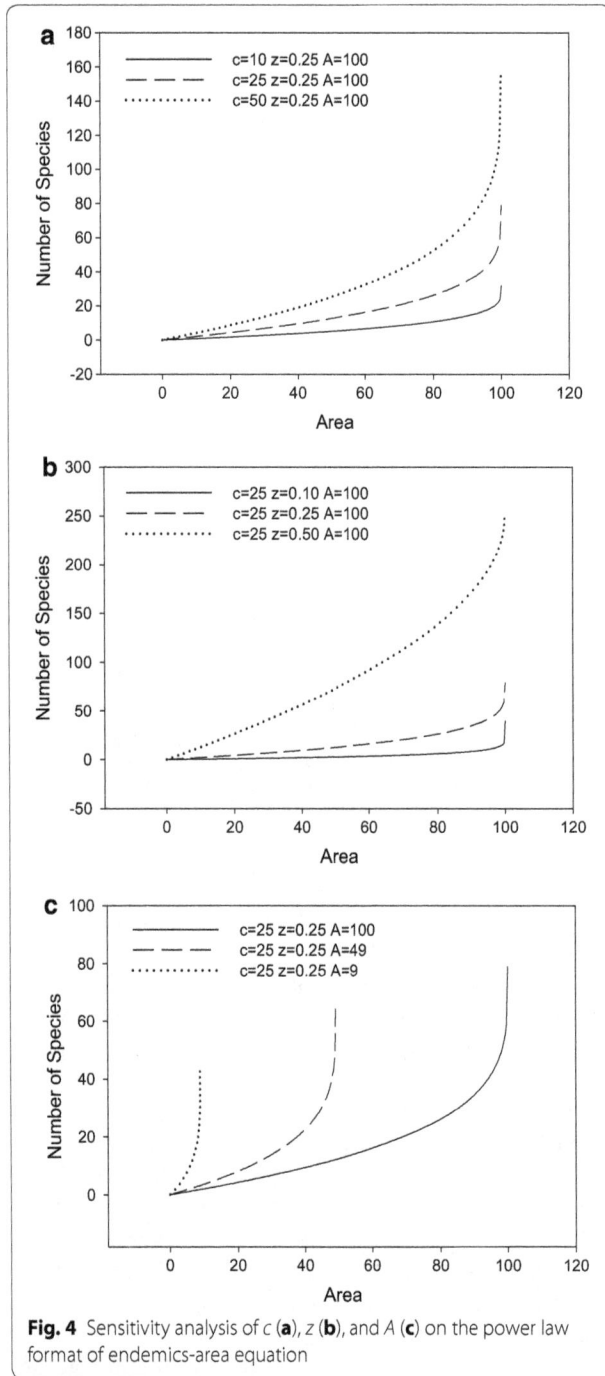

Fig. 5 Sensitivity analysis of z (**a**) and A (**b**) on the extinction rate

transferred to the Jaccard index, and Sørensen index can also be expressed by $\frac{2O_{a,A-a}}{S_A + 2O_{a,A-a}}$ [23–25]. If the concepts of EAR and OAR are expanded to more areas, then zeta diversity and new beta diversity can handle this circumstance [8, 9].

Conclusions

Fundamental equations for species–area theory are applied to power law model and random replacement model of SAR. To enrich the species–area theory, several parameters are proposed, including overlap index, overlap rate, sampling rate and extinction rate, which are defined by equations in the general format, power law model and random replacement model. For power law model, both EAR and OAR have three parameters, with one more parameter of the total area than SAR does. If the total area is underestimated, the extinction rate will be overestimated. The EAR equation is a monotonically increasing function for parameter c and z, and a monotonically decreasing function for parameter A. Extinction rate, which has two parameters, is a monotonically increasing function for parameter z, and a monotonically decreasing function for parameter A. The overlap index is a monotonically increasing function for parameter A,

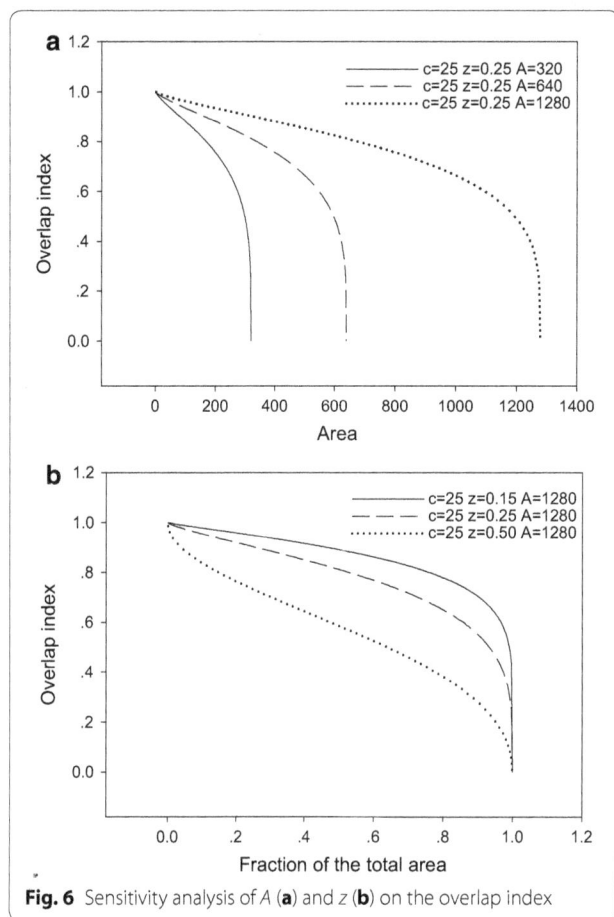

Fig. 6 Sensitivity analysis of A (**a**) and z (**b**) on the overlap index

Table 2 Extinction rate estimate and overestimate comparison

Area of habitat loss	$c = 25$ $z = 0.25$ $A = 9$		$c = 25$ $z = 0.25$ $A = 49$		$c = 25$ $z = 0.25$ $A = 100$
	λ (%)	Overestimate[a]	λ (%)	Overestimate[a]	λ_{100} (%)
0.52	1.48	10.34	0.27	1.04	0.13
1.00	2.90	10.56	0.51	1.05	0.25
9.00	100.00	41.91	4.95	1.12	2.33
25.00			16.34	1.36	6.94

[a] Overestimate $= (\lambda - \lambda_{100})/\lambda_{100}$

and a monotonically decreasing function for parameter z, independent of parameter c.

Abbreviations
SAR: species–area relationship; EAR: endemics-area relationship; OAR: overlap-area relationship.

Acknowledgements
Thanks go to Mr. Fengqiao Liu for language polishing, and the editor and two anonymous reviewers for their insightful comments on the manuscript.

Competing interests
The author declares that he has no competing interests.

Funding
This work is supported by Beijing Nova Programme (Z1511000003150107). The author also wants to thank the financial support for his study from Chinese Government Award for Outstanding Self-Financed Students Abroad, Texas A&M University- Kingsville, Tsinghua University and Chinese Academy of Inspection and Quarantine.

References
1. Tjørve E, Turner WR. The importance of samples and isolates for species–area relationships. Ecography. 2009;32:391–400.
2. WCMC. Global biodiversity: status of the earth's living resources. London: Chapman & Hall; 1992.
3. Kinzig AP, Harte J. Implications of endemics-area relationships for estimates of species extinctions. Ecology. 2000;81:3305–11.
4. Green JL, Ostling A. Endemics-area relationships: the influence of species dominance and spatial aggregation. Ecology. 2003;84:3090–7.
5. Millennium Ecosystem Assessment (2005) Ecosystems and human well-being: biodiversity synthesis. World Resources Institute.
6. Tjørve E, Tjørve KMC. The species–area relationship, self-similarity, and the true meaning of the z-value. Ecology. 2008;89:3528–33.
7. Pan XB. Fundamental equations for species–area theory. Sci Rep. 2013;3:1334.
8. Pan XB. Reconstruct species–area theory using set theory. Natl Acad Sci Lett. 2015;38:173–7.
9. Hui C, McGeoch MA. Zeta diversity as a concept and metric that unifies incidence-based biodiversity patterns. Am Nat. 2014;184:684–94.
10. Pimm SL, Askins RA. Forest losses predict bird extinctions in eastern North America. Proc Natl Acad Sci USA. 1995;92:9343–7.
11. Rosenzweig ML. Species diversity in space and time. Cambridge: Cambridge University Press; 1995.
12. Harte J, Kinzig AP. On the implications of the species–area relationships for endemism, spatial turnover, and food web patterns. Okios. 1997;80:417–27.
13. He F, Hubbell SP. Species–area relationships always overestimate extinction rates from habitat loss. Nature. 2011;473:368–71.
14. Heywood VH, Mace GM, May RM, Stuart SN. Uncertainties in extinction rates. Nature. 1994;368:105.
15. Tilman D, May RM, Lehman CL, Nowak MA. Habitat destruction and the extinction debt. Nature. 1994;371:365–6.
16. Mace GM, Gittleman JL, Purvis A. Preserving the tree of life. Science. 2003;300:1707–9.
17. Arrhenius O. Species and area. J Ecol. 1921;9:95–9.
18. Coleman B. On random placement and species–area relationships. Math Biosci. 1981;54:191–215.
19. Tjørve E, Kunin WE, Polce C, Tjørve KMC. The species–area relationship: separating the effects of species-abundance and spatial distribution. J Ecol. 2008;96:1141–51.
20. Harte J, Kinzig AP, Green J. Self-similarity in the distribution and abundance of species. Science. 1999;284:334–6.
21. Achard F, et al. Determination of deforestation rates of the world's humid tropical forests. Science. 2002;297:999–1002.
22. World Resources Institute. Millennium ecosystem assessment. Ecosystems and human well-being: biodiversity synthesis; 2005.
23. Jaccard P. The distribution of the flora in the alpine zone. New Phytol. 1912;11:37–50.
24. Tuomisto H. A diversity of beta diversities: straightening up a concept gone awry. Part 1. Defining beta diversity as a function of alpha and gamma diversity. Ecography. 2010;33:2–22.
25. Sørensen TA. A method of establishing groups of equal amplitude in plant sociology based on similarity of species content, and its application to analyses of the vegetation on Danish commons. Kongelige Danske Videnskabernes Selskabs Biol Skr. 1948;5:1–34.

PERMISSIONS

LIST OF CONTRIBUTORS

Jean-Jacques Godon, Jean-Philippe Steyer and Jérôme Hamelin
UR0050, Laboratoire de Biotechnologie de l'Environnement, INRA, 102avenue des étangs, 11100 Narbonne, France

Pugazhendi Arulazhagan
UR0050, Laboratoire de Biotechnologie de l'Environnement, INRA, 102avenue des étangs, 11100 Narbonne, France
Centreof Excellence in Environmental Studies, King Abdulaziz University, Jeddah, Saudi Arabia

Canan Karakoç
Department of Environmental Microbiology, Helmholtz Centre for Environmental Research-UFZ, Permoserstraße 15, 04318 Leipzig, Germany

Hauke Harms and Antonis Chatzinotas
Department of Environmental Microbiology, Helmholtz Centre for Environmental Research-UFZ, Permoserstraße 15, 04318 Leipzig, Germany
Centre for Integrative Biodiversity Research (iDiv) Halle-Jena-Leipzig, Deutscher Platz 5e, 04103 Leipzig, Germany

Karin Johst
Department of Ecological Modelling, Helmholtz Centre for Environmental Research- UFZ, Permoserstraße 15, 04318 Leipzig, Germany

Alexander Singer
Department of Ecological Modelling, Helmholtz Centre for Environmental Research- UFZ, Permoserstraße 15, 04318 Leipzig, Germany
Swedish Species Information Centre, Swedish University of Agricultural Sciences, 75007 Uppsala, Sweden

Kristine J. Teichman
Department of Geography, University of Victoria, STN CSC, Victoria, BC V8W 3R4, Canada.
Biology Department, University of British Columbia, 3333 University Way, Kelowna, BC V1V 1V7, Canada

Bogdan Cristescu
Department of Biological Sciences, University of Cape Town, Private Bag X3, Rondebosch 7701, South Africa

Chris T. Darimont
Department of Geography, University of Victoria, STN CSC, Victoria, BC V8W 3R4, Canada.
Raincoast Conservation Foundation, Bella Bella, BC V0T 1B0, Canada
Hakai Institute, Heriot Bay, BC V0P 1H0, Canada

Weihong Ji
Human Wildlife Interaction Research Group, Institute of Natural and Mathematical Sciences, Massey University, Private Bag 102 904, Albany, Auckland 0745, New Zealand

Raj Kumar Koirala
Human Wildlife Interaction Research Group, Institute of Natural and Mathematical Sciences, Massey University, Private Bag 102 904, Albany, Auckland 0745, New Zealand
Institute of Forestry, Tribhuvan University, Pokhara, Nepal

David Raubenheimer
The Charles Perkins Centre and School of Life and Environmental Sciences, University of Sydney, Sydney, NSW, Australia

Achyut Aryal
The Charles Perkins Centre and School of Life and Environmental Sciences, University of Sydney, Sydney, NSW, Australia
Department of Forestry & Resource Management, Toi Ohomai Institute of Technology , Rotorua 3046, New Zealand
Waste Management NZ Ltd, Auckland, New Zealand

Mitra Lal Pathak
Department of Plant Resources, National Herbarium and Plant Laboratories, Godawari, Nepal

Vesa Selonen
Department of Biology, Section of Ecology, University of Turku, 20014 Turku, Finland

Ralf Wistbacka
Department of Biology, University of Oulu, 90014 Oulu, Finland

Gabriel Ruiz Ayma, Alina Olalla Kerstupp, Antonio Guzmán Velasco and José I. González Rojas
Facultad de Ciencias Biologicas, Universidad Autonoma de Nuevo Leon, Ave. Universidad s/n. Cd. Universitaria, 66455 San Nicolas de los Garza, Nuevo Leon, Mexico

Alberto Macías Duarte
Ley Federal del Trabajo S/N, Universidad Estatal de Sonora, Col. Apolo, 83100 Hermosillo, Sonora, Mexico

Melanie Dammhahn
Animal Ecology, Institute for Biochemistry and Biology, Faculty of Natural Sciences, University of Potsdam, Maulbeerallee 1, 14469 Potsdam, Germany

Toky M. Randriamoria
Association Vahatra, BP 3972, 101 Antananarivo, Madagascar
Mention Zoologie et Biodiversité Animale, Université d'Antananarivo, BP 906, 101 Anta-nanarivo, Madagascar

Steven M. Goodman
Association Vahatra, BP 3972, 101 Antananarivo, Madagascar
Field Museum of Natural History, 1400 South Lake Shore Drive, Chicago, IL 60605, USA

Thomas M. Cullen and David C. Evans
Department of Ecology and Evolutionary Biology, University of Toronto, Toronto, ON, Canada
Department of Natural History, Royal Ontario Museum, 100 Queen's Park, Toronto, ON M5S 2C6, Canada

John L. Berini
Department of Fisheries, Wildlife and Conservation Biology, University of Minnesota, 135 B Skok Hall, 2003 Upper Buford Circle, St. Paul, MN 55108-1052, USA

Catherine Badgley
Department of Ecology and Evolutionary Biology, University of Michigan, 1109 Geddes Avenue, Ann Arbor, MI 48109-1079, USA

S. Veron, P. Clergeau and S. Pavoine
Centre d'Ecologie et des Sciences de la Conservation (CESCO UMR7204), Sorbonne Universités, MNHN, CNRS, UPMC, CP51, 43-61 rue Buffon, 75005 Paris, France

Madlen Ziege, Svenja Bischoff, Anna-Lena Brandt, Mareike Brix and Sandra Wenninger
Department of Ecology and Evolution, Goethe University Frankfurt, Max-von-Laue-Str. 13, 60439 Frankfurt am Main, Germany

David Bierbach
Department of Biology and Ecology of Fishes, Leibniz-Institute of Freshwater Ecology and Inland Fisheries, Müggelseedamm 310, 12587 Berlin, Germany

Bastian Greshake
Department for Applied Bioinformatics, Goethe University Frankfurt, Max-von-Laue-Str. 13, 60439 Frankfurt am Main, Germany

Stefan Merker
Department of Zoology, State Museum of Natural History Stuttgart, Rosenstein 1, 70191 Stuttgart, Germany

Torsten Wronski
Bristol Zoological Society, Conservation Science, Clifton, Bristol BS8 3HA, UK

Martin Plath
College of Animal Science and Technology, Northwest A&F University, Yangling 712100, Shaanxi, China

Dorothee Ehrich and Maite Cerezo
Department of Arctic and Marine Biology, University of Tromsø-The Arctic University of Norway, 9037 Tromsø, Norway

Anna Y. Rodnikova
Faculty of Biology, Lomonosov Moscow State University, GSP-1, Leninskie Gory, Moscow 119991, Russia

Natalya A. Sokolova and Aleksandr A. Sokolov
Arctic Research Station of Institute of Plant and Animal Ecology, Ural Branch, Russian Academy of Sciences, 629400, Zelenaya Gorka Str., 21, Labytnangi, Russia
Arctic Research Center of Yamal-Nenets Autonomous District, Salekhard, Russia

Eva Fuglei
Norwegian Polar Institute, Fram Centre, Langnes, 9296 Tromsø, Norway

Victor G. Shtro
Arctic Research Station of Institute of Plant and Animal Ecology, Ural Branch, Russian Academy of Sciences, 629400, Zelenaya Gorka Str., 21, Labytnangi, Russia

Alex R. Hardisty, Jonathan Giddy, Andrew Jones, Abraham Nieva de la Hidalga and Francisco Quevedo Fernandez
School of Computer Science and Informatics, Cardiff University, Queens Buildings, 5 The Parade, Cardiff CF24 3AA, UK

Finn Bacall, Niall Beard, Donal Fellows, Norman Morrison, Carole Goble, Aleksandra Nenadic and Alan R. Williams
School of Computer Science, University of Manchester, Kilburn Building, Oxford Road, Manchester M13 9PL, UK

Maria-Paula Balcázar-Vargas and Gerard Oostermeijer
Institute for Biodiversity and Ecosystem Dynamics (IBED), University of Amsterdam, 1090 Amsterdam, The Netherlands

Bachir Balech, Francesca De Leo, Bruno Fosso and Monica Santamaria
Institute of Biomembranes and Bioenergetics (IBBE), National Research Council (CNR), via Amendola 165/A, 70126 Bari, Italy

Zoltán Barcza and Laura Dobor
Department of Meteorology, Eötvös Loránd University, Pázmány sétány 1/A, Budapest 1117, Hungary

Sarah J. Bourlat and Matthias Obst
Department of Marine Sciences, University of Gothenburg, 405 30 Gothenburg, Sweden

Renato De Giovanni
Centro de Referência em Informação Ambiental, Avenida Dr. Romeu Tórtima, 388, Campinas, SP 13084-791, Brazil

Yde de Jong
Institute for Biodiversity and Ecosystem Dynamics (IBED), University of Amsterdam, 1090 Amsterdam, The Netherlands
SIB Labs, Joensuu Science Park, University of Eastern Finland, 80101 Joensuu, Finland

Giacinto Donvito
Institute of Nuclear Physics (INFN), Via E. Orabona 4, 70125 Bari, Italy

Antonio Fernandez Guerra
Max Planck Institute for Marine Microbiology, Celsiusstrasse 1, 28359 Bremen, Germany
Jacobs University Bremen GmbH, Campus Ring 1, 28359 Bremen, Germany

Nuno Ferreira and Salvatore Pinto
Stichting EGI (EGI.eu), Science Park 140, 1098 Amsterdam, The Netherlands

Yuliya Fetyukova
SIB Labs, Joensuu Science Park, University of Eastern Finland, 80101 Joensuu, Finland

Anton Güntsch and Cherian Mathew
Botanic Garden and Botanical Museum Berlin, Freie Universität Berlin, Königin-Luise-Strasse 6-8, 14195 Berlin, Germany

Robert Haines
IT Services, University of Manchester, Kilburn Building, Oxford Road, Manchester M13 9PL, UK

Vera Hernández Ernst
Fraunhofer Institute for Intelligent Analysis and Information Systems (IAIS), Schloss Birlinghoven, 53757 Sankt Augustin, Germany

Hannes Hettling
Naturalis Biodiversity Center, Postbus 9517, 2300 Leiden, The Netherlands

Dóra Hidy
MTA-SZIE Plant Ecology Research Group, Szent István University, Páter K. u.1., Gödöllő 2103, Hungary

Ferenc Horváth, Dóra Ittzés and Péter Ittzés
Institute of Ecology and Botany, Centre for Ecological Research, Hungarian Academy of Sciences, Alkotmány u. 2-4, Vácrátót 2163, Hungary

Renzo Kottmann and Pelin Yilmaz
Max Planck Institute for Marine Microbiology, Celsiusstrasse 1, 28359 Bremen, Germany

Robert Kulawik, Axel Poigné and Karl-Heinz Sylla
Fraunhofer Institute for Intelligent Analysis and Information Systems (IAIS), Schloss Birlinghoven, 53757 Sankt Augustin, Germany

Sonja Leidenberger
Swedish Species Information Centre/
ArtDatabanken, Swedish University of Agricultural
Sciences, Bäcklösavägen 10, 750 07 Uppsala, Sweden

Päivi Lyytikäinen-Saarenmaa
Department of Forest Sciences, University of
Helsinki, Helsinki, Finland

Elisabeth Paymal
Fondation pour la Recherche sur la Biodiversité
(FRB), 195, rue Saint-Jacques, 75005 Paris, France

Graziano Pesole
Institute of Biomembranes and Bioenergetics (IBBE),
National Research Council (CNR), via Amendola
165/A, 70126 Bari, Italy
Department of Biosciences, Biotechnology and
Biopharmaceutics, University of Bari "A. Moro",
via Orabona, 1514, 70126 Bari, Italy

Hannu Saarenmaa
SIB Labs, Joensuu Science Park, University of
Eastern Finland, Joensuu, Finland

Gergely Sipos
Stichting EGI (EGI.eu), Science Park 140, 1098
Amsterdam, The Netherlands

Marko Tähtinen
Finnish Museum of Natural History, University of
Helsinki, Helsinki, Finland

Saverio Vicario
Institute of Biomedical Technology (ITB), National
Research Council (CNR), via Amendola 122/D,
70126 Bari, Italy

Rutger Aldo Vos
Institute for Biodiversity and Ecosystem Dynamics
(IBED), University of Amsterdam, 1090 Amsterdam,
The Netherlands
Naturalis Biodiversity Center, Postbus 9517, 2300
Leiden, The Netherlands

Katrin Ronnenberg and Ursula Siebert
Institute for Terrestrial and Aquatic Wildlife
Research, University of Veterinary
Medicine Hannover, Foundation , Bischofsholer
Damm 15, 30173 Hannover, Germany

Egbert Strauß
Institute for Terrestrial and Aquatic Wildlife
Research, University of Veterinary Medicine
Hannover, Foundation , Bischofsholer Damm 15,
30173 Hannover, Germany
Hunting Association of Lower Saxony,
Schopenhauerstraße 21, 30625 Hannover, Germany

Karen Marie Mathisen and Christina Skarpe
Department of Forestry and Wildlife Management,
Faculty of Applied Ecology
and Agricultural Sciences, Inland Norway University
of Applied Sciences, Pb 400, 2418 Elverum, Norway

Jos M. Milner
School of Biological Sciences, University of
Aberdeen, Tillydrone Avenue, Aberdeen, AB24
2TZ, UK

Xubin Pan
Institute of Plant Quarantine, Chinese Academy
of Inspection and Quarantine, Beijing 100029,
China

Index

www.ingramcontent.com/pod-product-compliance
Lightning Source LLC
Chambersburg PA
CBHW080244230326
41458CB00097B/3333